Applying and Extending Oracle Spatial

A practitioner's guide on how to extend, apply, and combine Oracle's Spatial offerings with other Oracle and open source technologies to solve everyday problems

Simon Greener

Siva Ravada

[PACKT] enterprise 88
PUBLISHING   professional expertise distilled

BIRMINGHAM - MUMBAI

Applying and Extending Oracle Spatial

First published: September 2013

Production Reference: 1210913

Published by Packt Publishing Ltd.
Livery Place
35 Livery Street
Birmingham B3 2PB, UK.

ISBN 978-1-84968-636-5

www.packtpub.com

Cover Image by Aashish Variava (aashishvariava@hotmail.com)

Credits

Authors
Simon Greener
Siva Ravada

Reviewers
Paul Thomas Dziemiela
Jan Espenlaub
John O'Toole
Brendan Soustal

Acquisition Editors
Pramila Balan
Rukhsana Khambatta

Lead Technical Editor
Dayan Hyames

Technical Editors
Shashank Desai
Dylan Fernandes
Krishnaveni Haridas
Ankita Thakur

Project Coordinators
Anurag Banerjee
Shiksha Chaturvedi

Proofreaders
Maria Gould
Ameesha Green
Paul Hindle

Indexers
Mariammal Chettiyar
Rekha Nair
Tejal R. Soni

Graphics
Sheetal Aute
Valentina D'silva
Disha Haria
Yuvraj Mannari

Production Coordinator
Conidon Miranda

Cover Work
Conidon Miranda

About the Authors

Simon Greener has university qualifications in geomatics, computing science, database technologies, and project management.

He started his working career with mining and surveying experience. He then found his calling as a computer scientist with his first job as a database programmer on IBM mainframes for Telstra. He switched to GIS three years later through working in a multi-disciplinary GIS research team at Telecom's Research Laboratories (TRL) in Clayton, Victoria. While at TRL, he worked on projects whose outcome saw the creation of what is now Telstra's Sensis group.

After leaving TRL, he worked as a lecturer and consultant for CenSIS (University of Tasmania) under Professor Peter Zwart, writing student and technical training courses. While there, he continued to consult to Telstra's Directory Services and Mobile groups. It was here that he came in contact with the Spatial DataBase Engine (SDBE) from Geographic Technologies Incorporated (GTI) and saw its potential for the management of large scale spatial databases within relational database technologies, a merging of his IT and GIS worlds. This led to the foundation of Salamanca Software Pvt Ltd (SalSoft), for which he was a Director until it was purchased by ESRI Australia in 1996.

Some notable achievements while at SalSoft included helping brokers with the sale of SDBE to ESRI Inc (now ArcSDE), winning the first ArcSDE sale to Telstra to power its White pages/Yellow pages mapping portal, co-authoring a geocoding specification for Spatial Decision Systems (now Sensis), consulting for Geographic Technologies Australia on numerous projects based on Universal Press street directory data, and the creation of GeoCASE/Blueprint, the world's first data modeling tool that enabled the modeling of spatial data and relationships.

In 1997, he was appointed GIS Manager for Forestry Tasmania (FT) in Hobart, Tasmania. While at FT, he architected the complete revamp of FT's GIS systems using Oracle Spatial (being one of the earliest adopters of the `Sdo_Geometry` implementation) as the core data management technology. He was concentrating on embedding geospatial data and processing within business systems via a value-oriented, business-centric computing model. He designed and built numerous systems during those years, the best of which was MapComposer, a three-click web-enabled business map production system that, when he left in 2005, had grown (2000-) to over 320 online uses, producing over 50,000 maps a year from a repository of over 100 different map templates (still in operation in 2013). His years at FT concluded with the writing of a GIS Strategy that saw the use of GIS increase yet the cost of the technology to the organization decrease.

He left FT in September 2005 for the precarious world of self-employment. He was a sometime copyist for Directions Magazine. As a subcontractor to a Spatial distributor in Australia, he wrote a Radius Topology training course and provided Radius Topology and Oracle Spatial consulting services for them at numerous customer sites until May 2006. From May to August 2006, he was engaged by Spatial, Cambridge, UK, under a UK Government Department of Trade and Industry's GlobalWatch program to conduct research and development in relation to enhancing the export potential of their latest product, Radius Studio (this resulted in Radius Studio being integrated with Feature Data Objects – FDO technology to extend its data access capabilities).

In his consulting career, he has written a spatial strategy document and conducted a database performance analysis review for a large Tasmanian Government department. He has conducted a number of Oracle spatial database best practice, tender and system, and return on investment reviews at a number of Victorian Government departments. He wrote and delivered a user requirements document for Enterprise GIS at a large Australian corporation. He also provided guidance and implementation services to an ambulance service helping integrate Oracle Spatial into a data warehouse project that used Oracle Portal, Discoverer, and Data Warehouse Builder. He delivered many solutions for a NSW water authority; and finally, he successfully completed many migration, publication, return on investment, process improvement, and database design contracts for a number of Canberra-based Federal Government departments. Simon makes available a collection of PL/SQL and Java-based sample solutions for the Oracle database via his website. He is also principal programmer for the SQL Developer spatial extension, GeoRaptor. Finally, he was awarded, the 2011 Oracle Spatial Excellence Award for Education and Research by Oracle.

His technical areas of expertise include systems design and architecture (spatial and attribute), data management, and modeling in both the OLTP and OLAP spaces, and he is also an evangelist for O-RDMS-based spatial data. He is available for free-lance geospatial solutions architecture work, Java and PL/SQL programming, and he provides Oracle Spatial benchmarking and performance enhancement services.

His non-technical interests are his family, friends, walking, reading, singing, and motorcycle riding.

I would particularly like to acknowledge those who have helped in many ways that are required to coax a person through the difficult gestation of a new book. The first, of course, is my wife, Anna. Her quiet and unshakable support, through what was also a financially difficult period of time, is something that I, as a man, husband, and father, am deeply thankful for and immensely proud of. Ti Amo, Cara. To my co-author, Siva Ravada, for kindly stepping in to help me write this book when I felt writing the whole book was beyond my resources. The colleagues that I would particularly like to thank include John O'Toole (Spatial Ireland), who has been an amazing source of inspiration and a fantastic sounding board for all things Oracle Spatial and its application for many years. Brendan Soustal, for his support and belief provided over all my consulting years. Martin Davis, JTS architect and chief programmer, for helping with the chapter on Java in the database. I would also like to thank Holger Läbe, who started helping me programming GeoRaptor, for without his help, the Java chapter would never have eventuated. I would also like to thank Jody Garnett for helping me with GeoTools over the years, and thank you to all those who read my meager blog posts or have downloaded my PL/SQL and Java solutions from my website. Your actions confirmed for me that this book was worth writing. Finally, to the reviewers, both official and private, thank you for all your efforts and help.

Siva Ravada earned a PhD degree from the University of Minnesota in the field of Spatial Databases before joining Oracle's Spatial development team. He is now a Senior Director of Development at Oracle Corporation. At Oracle, Siva was one of the founding team members of the Spatial development team before taking over the team management responsibilities. Siva now manages the Spatial and MapViewer development teams at Oracle. He has more than 15 years of experience in spatial databases and application development. He has also co-authored more than 30 articles published in journals, and holds more than 30 patents. He has also presented key-note speeches at several conferences on the topics of spatial databases and GIS.

Oracle is the second largest software company and the number one database company in the world.

I sincerely thank my wife, Manjari, and my daughter, Pallavi, for their support during this book project. Several of my team members and colleagues at Oracle also helped me with examples for this book, and I thank all of them for their help and support during the last few months.

About the Reviewers

Paul Thomas Dziemiela is a geographic information systems professional specializing in RDBMS geospatial solutions.

Jan Espenlaub studied Geography at the University of Freiburg, Germany. He started with GIS in 1980. He has worked for several companies and governmental agencies as a GIS consultant. His main focus is on data management, Oracle Spatial, GIS Interoperability, and open source software. He is currently working with regioData GmbH, leading the GIS Application development group.

John O'Toole is an Oracle database developer with a passion for spatial data. He has been using Oracle technology since 2001 and currently works in Dublin, Ireland, as a consultant with 1Spatial.

He works on projects with cadastral, land registration, and national mapping agencies, helping them to build innovative and scalable solutions with a particular emphasis on spatial data quality.

Among other activities, he actively participates on the OTN Oracle Spatial discussion forum and contributes to the GeoRaptor project.

John is an Oracle Database 10g and 11g Certified Professional and Oracle Spatial 11g Certified Implementation Specialist.

Brendan Soustal joined the workforce in 2002 as a Land Information Systems graduate. He has implemented a comprehensive program to improve the quality and quantity of geospatial and asset information at a local government as well as at state and international levels. These programs have seen the development of innovative systems and substantial improvements in workplace efficiency. They have also established new opportunities for data sharing and data standardization within the utilities industry.

Brendan's fresh approach to his information management roles and his enthusiasm for innovation has helped a number of organizations to embrace new technologies and broaden their range and depth of GIS and spatial applications.

www.PacktPub.com

Support files, eBooks, discount offers and more

You might want to visit www.PacktPub.com for support files and downloads related to your book.

Did you know that Packt offers eBook versions of every book published, with PDF and ePub files available? You can upgrade to the eBook version at www.PacktPub.com and as a print book customer, you are entitled to a discount on the eBook copy. Get in touch with us at service@packtpub.com for more details.

At www.PacktPub.com, you can also read a collection of free technical articles, sign up for a range of free newsletters and receive exclusive discounts and offers on Packt books and eBooks.

PACKTLIB®

http://PacktLib.PacktPub.com

Do you need instant solutions to your IT questions? PacktLib is Packt's online digital book library. Here, you can access, read and search across Packt's entire library of books.

Why Subscribe?

- Fully searchable across every book published by Packt
- Copy and paste, print and bookmark content
- On demand and accessible via web browser

Free Access for Packt account holders

If you have an account with Packt at www.PacktPub.com, you can use this to access PacktLib today and view nine entirely free books. Simply use your login credentials for immediate access.

Instant Updates on New Packt Books

Get notified! Find out when new books are published by following @PacktEnterprise on Twitter, or the *Packt Enterprise* Facebook page.

Table of Contents

Preface

The focus of this book is the application of Oracle Spatial to the sorts of business problems that are better solved inside the database using a fully integrated technology approach. This approach takes as its first point of reference the view that spatial data and processing is no different from using timestamps, numbers, text, or dates to describe business entities (assets), and also that solutions to business problems involving spatial data processing should use relevant, or related, IT technologies first, before introducing specialist GIS software. One of the issues facing all proponents and users of spatial databases is the lack of knowledge within the professional GIS community about the underlying database software being used to manage spatial descriptions of business entities.

This book elucidates a holistic approach to spatial data management by highlighting how spatial data management and processing is enhanced and supported by utilizing all the data storage types that a database offers, but particularly spatial data.

The examples in this book have been drawn from many years of working with Oracle Spatial within a business IT environment. In addition, some examples have been drawn from requests readers of various blogs and forums have made over the years; some though are purely speculative based as they are on the application of theory to problems.

What this book covers

Chapter 1, Defining a Data Model for Spatial Data Storage, provides a SQL schema and functions that facilitate the storage, update, and query of collections of spatial features in an Oracle database. Oracle Spatial mainly consists of the following:

- A schema (MDSYS) that defines the storage, syntax, and semantics of supported geometric (both vector and raster) data types

- A spatial indexing mechanism for faster querying and retrieval
- Operators, functions, and procedures for performing spatial analysis and query operations
- A persistent topology data model for working with data about nodes, edges, and faces in a topology
- A network data model for modeling and working with spatial networks
- A GeoRaster data type and associated functions that let you store, index, query, analyze, and deliver raster data

The spatial component of a real-world feature is the geometric representation of its shape in some coordinate space (either in 2D or 3D), and in vector space, this is referred to as its geometry. Oracle Spatial is designed to make spatial data management easier and more natural to users of location-enabled business applications and geographic information system (GIS) applications. Oracle allows the storage of spatial data in a table using the SDO_GEOMETRY data type, which is just like any other data type in the database. Once spatial data is stored in the Oracle database, it can be easily manipulated, retrieved, and related to all other data stored in the database.

A spatial database should be designed just like any other database with a fully specified model. A fully specified model that is application independent should control spatial data storage. A good data model supports and enhances application access without compromising quality. In addition, database features can be used to support applications that have limited functionality when it comes to table and column design. For example, some applications mandate a single spatial column per table or only a single homogeneous geometry type per spatial column. These limitations can be accommodated quite easily using database features such as views and triggers. In addition, there are a number of issues that arise when designing a data model that directly affect data quality, performance, and access.

The goal of this chapter is to give readers an understanding of how to model spatial data as SDO_GEOMETRY columns within tables, how to support spatial constraints for improved data quality, how to use synchronous and asynchronous triggers for implementing topological constraint checking, and how to present methods for coping with multiple representations for faster web service access. All these issues, with solutions, are covered in this chapter.

Chapter 2, Importing and Exporting Spatial Data, explains how once we have a data model defined, the next step is to load data into it, and after data is loaded, it needs to be checked for cleanliness before indexing and using it. There are many methods for loading data of different types and formats. In this chapter, we describe some of the more common formats and how they can be loaded into the Oracle database using free tools, tools already available with Oracle, and tools from third party vendors. In addition, we also discuss other issues relating to import performance and organization of data for efficient access by applications.

The goal of this chapter is to give you a complete overview of all aspects of data loading from tools, through physical loading techniques, data organization, data quality checking, and indexing:

- Extract, Transform, and Load (ETL) tools: GeoKettle, Oracle Spatial shapefile loader, and MapBuilder
- Using SQL, Application Express, and Excel
- Implementing theoretical storage resolution and the minimal resolution used by functions
- Using ordinate resolution and the effect of rounding
- Using tolerance and precision
- Creating spatial autocorrelation via a Morton key
- Geometry validation and methods to clean imported data
- Coordinate system transformation techniques
- Spatial indexing
- Exporting formats, for example, Shapefile, GML, WKT, and GeoJson

Chapter 3, Using Database Features in Spatial Applications, introduces some of the standard features that the Oracle database makes available to solve some common data processing, auditing, version management, and quality issues related to the maintenance of spatial data. These database features allow developers to keep the data processing operations in the database instead of doing them in the application code:

- Row-level and Statement-level triggers
- Avoiding the mutating table problem
- Using materialized views
- Logging changes independently of application

- Flashback queries
- AWR reports and ADDM
- Database replay
- Workspace manager
- SecureFile compression

Chapter 4, Replicating Geometries, shows the process of copying and maintaining data across different databases in a distributed system. The goal of this chapter is to present a few methods for replicating geometry data. Some of the traditional Oracle replication technologies do not directly support replication of tables with SDO_ GEOMETRY data. The examples given here show alternate ways of replicating tables with geometry data. Replication does not always mean replicating the same data in different databases. In some cases, it also means copying the data in one database into a different database in a different form. For example, data in Online Transaction Processing (OLTP) databases can be converted to Online Analytical Processing (OLAP) databases by replicating or combining data from different OLTP tables into a single table in the OLAP system. We show how to do this conversion of data from a transactional OLTP database to a publication or OLAP database. Starting with the 12cR1 release, Logical Standby support for SDO_GEOMETRY data types is introduced, so we will look at how this feature can be used to replicate geometry data. In this chapter, the following topics will be covered:

- Introducing different types of replication
- Materialized-view based replication
- Streams based replication
- Physical and Logical Standby
- OLTP and OLAP databases

Chapter 5, Partitioning of Data Using Spatial Keys, explains how spatial applications tend to generate large volumes of data, especially as the scale of observation of the world's surface extends to large parts of the Earth's surface. With the increasing level of data, database models have to adapt to deal with large volumes of spatial data that is not seen in traditional GIS applications. GIS applications expect all the related data in one feature layer, even if the feature layer contains millions of features. Oracle database supports a feature called partitioning that can break large tables at the physical storage level to smaller units while keeping the table as one object at the logical level. In this chapter, we cover the following five topics that are useful for managing large volumes of spatial data:

- Introduction to partitioning
- Time-based partitioning
- Spatial key based partitioning
- Implementing space curves based partitioning
- High performance loading of spatial data

Chapter 6, Implementing New Functions, shows us to create new functions that use and extend those offered by Oracle Spatial and locator products. The SDO_GEOMETRY object, its attributes, and its structure must be thoroughly understood. Therefore, this chapter will start by building some functions that will help us understand, access, and process an SDO_GEOMETRY object's attributes. In building these functions, those Oracle SDO_GEOMETRY methods and SDO_GEOM and SDO_UTIL package functions that relate to the processing of the SDO_GEOMETRY attributes and structure will be introduced and used. This chapter will present information that will help the reader understand how to:

- Expose or create additional properties for the SDO_GEOMETRY object
- Manipulate SDO_ORDINATE_ARRAY in SQL and PL/SQL
- Create functions to expose SDO_ELEM_INFO_ARRAY properties
- Use SDO_ELEM_INFO_ARRAY to process SDO_ORDINATE_ARRAY correctly
- Organize functions via object types and packaging
- Sort geometries

Later chapters will build on the knowledge gained in this chapter when creating functions that solve specific real-world problems.

Chapter 7, Editing, Transforming, and Constructing Geometries, explains Desktop GIS, CAD, and Extract Transform and Load (ETL) software that provide a rich set of tools that the experienced operator can use to construct, edit, or process geometric objects. But few realize that creating and applying such functionality within the database is also possible, and this can be more effective, efficient, and less complicated.

While the Oracle database SDO_GEOMETRY data type provides an excellent storage, search, and processing engine for spatial data, what users often overlook is its ability to provide geometry modification and processing capabilities that can be used in database objects such as views, materialized views, and triggers for the implementation of specific business functionality, with that functionality being available to any software product that connects to the database.

Chapter 8, Using and Imitating Linear Referencing Functions, shows how to use the functions created in *Chapter 6, Implementing New Functions* and *Chapter 7, Editing, Transforming, and Constructing Geometries* to build new functions that can be used to solve business problems relating to managing linear assets. The main business problems that need this functionality are road, cycle way, or track management, geocoding street addresses, survey, inventory, condition assessment, and water management applications.

Oracle Spatial has a robust linear referencing package (SDO_LRS) that can be applied to all the above problems. The SDO_LRS package can only be licensed for use by purchasing the Spatial package and deploying it within an Oracle Enterprise database. In addition, SDO_LRS cannot be purchased separately from the whole Spatial package. Finally, SDO_LRS cannot be used with locator.

The functions created in this chapter will provide SDO_LRS functionality where licensing of the SDO_LRS package is beyond the user's resources. These functions will support simple linear processing against measured and non-measured geometries (a normal 2D linestring is a non-measured geometry). The following are the uses of linear processing:

- Snapping a point to a line
- Splitting a line using a known point that is on or off the line
- Adding, modifying, and removing measures to and from a linestring
- Finding linear centroids
- Creating a point at a known distance along, and possibly offset from, a line
- Extracting segments of linear geometries
- Linear analysis of point data

The examples presented in this chapter will include real-world situations to demonstrate the power of developing and using these types of functions. The SQL statements that demonstrate the use of the functions developed in this chapter are available in an SDO_LRS equivalent form in the SQL scripts shipped with this book for this chapter; they will not be included in the actual chapter.

Chapter 9, Raster Analysis with GeoRaster, shows how spatial features can be represented in vector or raster format. So far, we have discussed the vector related features of Oracle Spatial, and we introduce the raster related features called GeoRaster in this chapter. Traditional GISs propose to store the raster data as BLOBs in the database. This approach might be sufficient if the raster data is only used as backdrop images in maps.

But if any raster data processing and analysis is required, storing raster data as GeoRaster objects offers many features and advantages over storing this data just as BLOBs. Loading and storing any raster data inside a database simply for the purpose of storage or visualization provides limited utility. Storing raster data for use within a transactional system has engendered a view that one must see all data as a part of a complete model; the data loaded must be seen in relation to all other data under control of that model.

The goal of this chapter is to demonstrate how to use raster data in conjunction with all the data in the database to answer questions that otherwise could not be answered in the database. The following topics are covered in this chapter to show how this goal can be achieved:

- Introduction to GeoRaster
- Loading and storing raster data inside a database
- Raster data for visualization applications
- Raster data for analytical applications
- Mapping between raster and vector space

Chapter 10, Integrating Java Technologies with Oracle Spatial, shows how to embrace and extend the standard functionality available with Oracle Locator and Spatial using PL/SQL. PL/SQL is the programming language that is native to Oracle. Oracle also supports the creation of Java Stored Procedures. This chapter explores the application of Java to spatial processing involving the SDO_GEOMETRY type.

In particular, this chapter will cover the following topics:

- Why Java and Oracle Spatial?
- Available Java spatial technologies
- Matching requirements to source code project
- Strengths and limitations of using Java
- How to download, modify, compile, and install external libraries
- Calling an external method
- Converting an SDO_GEOMETRY to a Java object
- Exposing **Java Topology Suite (JTS)** functionality:
    - One-sided buffers
    - Snapping geometries
    - Building polygons from lines
- Performance of Java-based SQL processing

Chapter 11, SQL/MM – A Basis for Cross-platform, Inter-operable, and Reusable SQL, explains how Oracle's SDO_GEOMETRY type is very widely used even by the open source community, which promotes the virtues of standards conformance and compliance. However, many in the geospatial industry still criticize Oracle's SDO_GEOMETRY for its lack of perceived standards compliance. Whether this criticism is based on ignorance or maleficence, SDO_GEOMETRY is standards compliant in its storage, geometry description, and with some functions; but, its API is not fully compliant.

SDO_GEOMETRY is not the whole story however. Oracle also provides an ST_GEOMETRY object type which is an implementation that is based on the ISO/IEC FCD 13249-3 Spatial (ISO 13249-3, Information technology - Database languages - SQL Multimedia and Application Packages - Part 3: Spatial.) standard (hereafter known as the SQL/MM standard). This chapter aims to show that such criticism of standards compliance of Oracle is limited and ill informed through exposure of the benefits of ST_GEOMETRY to practitioners.

ST_GEOMETRY is of special importance in situations where a business IT environment has a heterogeneous database environment (for example, Oracle, SQL Server, PostgreSQL). It can be a most useful mechanism for implementing cross-platform spatial data processing and developing highly reusable skills. This latter aspect of skills development is important because reusability and training, which increases and improves an individual's skill-set, is an important ingredient in staff training and development.

This chapter will present two aspects of how the use of standards: OGC SFA 1.x OpenGIS® Implementation Specification for Geographic information - Simple Feature Access - Part 1: Common architecture, Version 1.1 and 1.2. Previously known as Simple Features – SQL (SFS). (hereafter known as OGC SFA or OGC SFA 1.1 or 1.2, if function is only available in one of the standards) and SQL/MM, to aid cross-platform interoperability:

- A demonstration of how the spatial data types offered by three databases: Oracle, SQL Server 2012 (Express Edition), and PostgreSQL 9.x with the **PostGIS** 2.x extension. PostgreSQL can be used to create constrained geometry storage. This does not repeat the material presented in *Chapter 1, Defining a Data Model for Spatial Data Storage*. Rather, it presents the material in light of the OGC SFA 1.x and SQL/MM standards and the benefits for database interoperability.

- How the standardized methods common to the three databases can be used to develop SQL statements as well as stored procedures and functions that can be ported to other platforms with relative ease.

Appendix A, Table Comparing Simple Feature Access/SQL and SQL/MM–Spatial provides a comparison of SFA-SQL 1.2 (`http://portal.opengeospatial.org/files/?artifact_id=25354`) and SQL/MM-Spatial (ISO 13249-3, Information technology - Database)

Appendix B, Use of TREAT and IS OF TYPE with ST_GEOMETRY examines the need for `TREAT` in more detail. In the `ST_GEOMETRY` hierarchy, a `POINT` object can be created in the following two ways:

```
MDSYS.ST_GEOMETRY.FROM_WKT('POINT(6012578.005 2116495.361)',2872)
MDSYS.ST_POINT.FROM_WKT('POINT(6012578.005 2116495.361)',2872).
```
The result in both cases is not an `ST_POINT`, rather it is an `ST_GEOMETRY`object.

What you need for this book

If the reader of this chapter wishes to use SQL and the example code that ships with this book, then the following technologies are required:

- Oracle database 11*g* Release 1 or 2
- SQL Developer Version 3.2.x (not Version 4.x)
- For visualizing `SDO_GEOMETRY` data in SQL Developer queries, either use SQL Developer's integrated Spatial extension or the open source extension GeoRaptor (download from SourceForge)
- SQL Developer's embedded Data Modeling extension
- JDeveloper 11.x

Who this book is for

This book is aimed at the experienced practitioner who is already literate with Oracle Locator or Spatial and who has at least heard of or used the standard Oracle database technologies this book uses to solve problems.

The reader should be familiar with using SQL Developer or a similar product to execute the SQL examples, though for visualization, SQL Developer and the free GeoRaptor extension are preferred.

The reader should be at least familiar with physical database modeling, and even if they have not used SQL Developer's Data Modeler tool, should be willing to learn how to use it.

For the programming aspects of this book, it is preferable but not mandatory that the reader has some experience in writing relatively simple PL/SQL and a base level of knowledge of writing Java. In respect of Java, while the actual Java Stored Procedures are relatively simple in structure, the use of external source code and JAR files to construct the complete solution is something that requires some experience of working within a larger Java framework. As such, familiarity with JDeveloper or a similar Integrated Development Environment (IDE) is recommended.

Conventions

In this book, you will find a number of styles of text that distinguish between different kinds of information. Here are some examples of these styles, and an explanation of their meaning.

Code words in text are shown as follows: "Oracle allows the storage of spatial data in a table using the SDO_GEOMETRY data type, which is just like any other data type in the database".

A block of code is set as follows:

```
grant connect,resource to Book identified by <password>;
grant connect, resource to book;
grant create table to book;
grant create view to book;
grant create sequence to book;
grant create synonym to book;
grant create any directory to book;
grant query rewrite to book;
grant unlimited tablespace to book;
```

When we wish to draw your attention to a particular part of a code block, the relevant lines or items are set in bold:

```
Enter value for report_type: html

Type Specified: html

Instances in this Workload Repository schema
```

Any command-line input or output is written as follows:

```
-- this is the setup for UNIX/Linux machines
setenv clpath $ORACLE_HOME/jdbc/lib/ojdbc5.jar:$ORACLE_HOME/md/jlib/
sdoutl.jar:$ORACLE_HOME/md/jlib/sdoapi.jar
```

New terms and **important words** are shown in bold. Words that you see on the screen, in menus or dialog boxes for example, appear in the text like this: "Once the table is created, go to the APEX home page and then navigate to the **SQL Workshop and Utilities** tab".

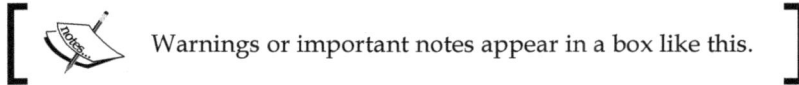

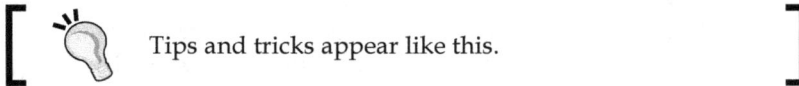

[Warnings or important notes appear in a box like this.]

[Tips and tricks appear like this.]

Reader feedback

Feedback from our readers is always welcome. Let us know what you think about this book—what you liked or may have disliked. Reader feedback is important for us to develop titles that you really get the most out of.

To send us general feedback, simply send an e-mail to feedback@packtpub.com, and mention the book title via the subject of your message.

If there is a topic that you have expertise in and you are interested in either writing or contributing to a book, see our author guide on www.packtpub.com/authors.

Customer support

Now that you are the proud owner of a Packt book, we have a number of things to help you to get the most from your purchase.

Downloading the example code

You can download the example code files for all Packt books you have purchased from your account at http://www.packtpub.com. If you purchased this book elsewhere, you can visit http://www.packtpub.com/support and register to have the files e-mailed directly to you.

Downloading the color images of this book

We also provide you a PDF file that has color images of the screenshots/diagrams used in this book. The color images will help you better understand the changes in the output. You can download this file from: http://www.packtpub.com/sites/default/files/downloads/6365EN_ColoredImages.pdf

Errata

Although we have taken every care to ensure the accuracy of our content, mistakes do happen. If you find a mistake in one of our books—maybe a mistake in the text or the code—we would be grateful if you would report this to us. By doing so, you can save other readers from frustration and help us improve subsequent versions of this book. If you find any errata, please report them by visiting http://www.packtpub.com/submit-errata, selecting your book, clicking on the **errata submission form** link, and entering the details of your errata. Once your errata are verified, your submission will be accepted and the errata will be uploaded on our website, or added to any list of existing errata, under the Errata section of that title. Any existing errata can be viewed by selecting your title from http://www.packtpub.com/support.

Piracy

Piracy of copyright material on the Internet is an ongoing problem across all media. At Packt, we take the protection of our copyright and licenses very seriously. If you come across any illegal copies of our works, in any form, on the Internet, please provide us with the location address or website name immediately so that we can pursue a remedy.

Please contact us at copyright@packtpub.com with a link to the suspected pirated material.

We appreciate your help in protecting our authors, and our ability to bring you valuable content.

Questions

You can contact us at questions@packtpub.com if you are having a problem with any aspect of the book, and we will do our best to address it.

1
Defining a Data Model for Spatial Data Storage

Oracle Spatial and Graph provides a SQL schema and functions that facilitate the storage, update, and query of collections of spatial features in an Oracle database. Oracle Spatial and Graph is the new name for the feature formerly known as Oracle Spatial. In this book, we refer to this feature as Oracle Spatial for the sake of simplicity. We also focus exclusively on spatial feature of Oracle Spatial and Graph in this book. Oracle Spatial mainly consists of the following:

- A schema (MDSYS derived from Multi-Dimensional System) that defines the storage, syntax, and semantics of the supported geometric (both vector and raster) data types
- A spatial indexing mechanism for faster querying and retrieval
- Operators, functions, and procedures for performing spatial analysis and query operations
- A persistent topology data model for working with data about nodes, edges, and faces in a topology
- A network data model for modeling and working with spatial networks
- A GeoRaster data type and associated functions that let you store, index, query, analyze, and deliver raster data

The spatial component of a real-world feature is the geometric representation of its shape in some coordinate space (either in 2D or 3D), and in vector space, this is referred to as its geometry. Oracle Spatial is designed to make spatial data management easier and more natural to users of location-enabled business applications and **geographic information system (GIS)** applications. Oracle allows the storage of spatial data in a table using the SDO_GEOMETRY data type that is just like any other data type in the database. Once the spatial data is stored in the Oracle database, it can be easily manipulated, retrieved, and related to all other data stored in the database.

A spatial database should be designed just like any other database with a fully specified model. A fully specified model that is application independent should control the spatial data storage. A good data model supports and enhances application access without compromising the quality. In addition to these features, database features can be used to support applications that have limited functionality when it comes to table and column design. For example, some applications mandate a single spatial column per table or only a single homogeneous geometry type per spatial column. These limitations can be accommodated quite easily using database features such as views and triggers. In addition, there are a number of issues that arise when designing a data model that directly affects the data quality, performance, and access.

The goal of this chapter is to give readers an understanding of how to model spatial data as SDO_GEOMETRY columns within tables, how to support spatial constraints for improved data quality, how to use synchronous and asynchronous triggers for implementing topological constraint checking, and to present methods for coping with multiple representations for faster web service access. All these issues, with solutions, are covered in this chapter:

- Defining a sample schema
- Using spatial metadata
    - Using Oracle metadata views
    - Using OGC metadata views
- Using different types of geometric representations
    - Implementing tables with homogeneous and heterogeneous columns
    - Implementing multiple representations for a single object
    - Implementing multiple instances of a single column, for example, pre-thinned data for different scales and reprojection for faster web service access
- Restricting data access via views
    - Using views to expose a single geometry type when multiple geometry types are present in the table
    - Using views to expose tables with single geometry columns when multiple geometry columns are present in the table
- Implementing spatial constraints at the database level
    - Restricting geometry types
    - Spatial topological constraints
    - Implementation of synchronous triggers
    - Implementation of asynchronous triggers

Defining a sample schema

We will first define a sample schema that will be used for all the examples in this book. The schema is intended to model typical spatial assets maintained in a city-level GIS. Oracle Spatial provides all the functionality needed to model or describe the spatial properties of an asset (in modeling, it is often called an entity). This spatial description of an asset should not be treated differently from any other descriptive attribute. In addition, a data model should describe all assets/entities within it independently of any application. This should include, to the best of the ability of SQL, all business rules that define or control these assets/entities within the database, and these rules should be implemented using standard database practices.

Defining the data model

We use a schema with 12 tables to represent a spatial database for a city. This schema has tables to represent administrative areas managed at the city level, such as land parcels and neighborhoods, along with tables to manage natural features such as water boundaries.

The LAND_PARCELS table has information about land at the lowest administrative level of the city. Buildings have to be fully contained in these land parcels. A table called BUILDING_FOOTPRINTS has information about all the buildings in the city. This table has the footprint of each building along with other information, such as name, height, and other attributes. Sets of neighborhoods are defined as a collection of land parcels to create more granular administrative areas. These neighborhoods are stored in the PLANNING_NEIGHBORHOODS table. There is a master table, BASE_ADDRESSES, to store information about all the valid street addresses in the city. Every record in the BUILDING_FOOTPRINTS table must have one parent record in this master address table. Note that the master address table does not list all the addresses of the apartments in a building. Rather, it stores one record for each street level address. So, each record in the BUILDING_FOOTPRINTS table has only one corresponding record in the master address table.

There is also a master table, ROADS, that is used to store information about all the roads in the city. ROADS stores one record for each named road in the city so that all common information for the road can be stored together in one table. This is the only table in the schema without any geometry information. Each road in turn maps to a set of road segments that are stored in the ROAD_CLINES table. This table is used to store the geometric representation of center lines of road segments. This table also stores information about address ranges on these road segments. Road segments typically have different address ranges on the left side of the road and on the right side of the road. Each road segment also has a parent ROAD_ID associated with it from the ROADS table.

A city usually manages sidewalks and other assets, such as street lamps, trashcans, and benches that are placed on these sidewalks. The SIDEWALKS table stores the information for all the sidewalks managed by the city. The CITY_FURNITURE table stores all the data corresponding to the assets, such as benches, streetlights, and trashcans.

The ORTHO_PHOTOS table stores the collected information using aerial photography. The raster information stored in this table can be used to look for changes over time for the built-in features of the city.

The water features of the city are stored in two different tables: the WATER_LINE table is used to store the water line features, such as creeks, rivers, and canals. The WATER_AREA table is used to store area features, such as lakes, rivers, and bays. The following figure shows the **entity-relationship (E-R)** diagram for this data model:

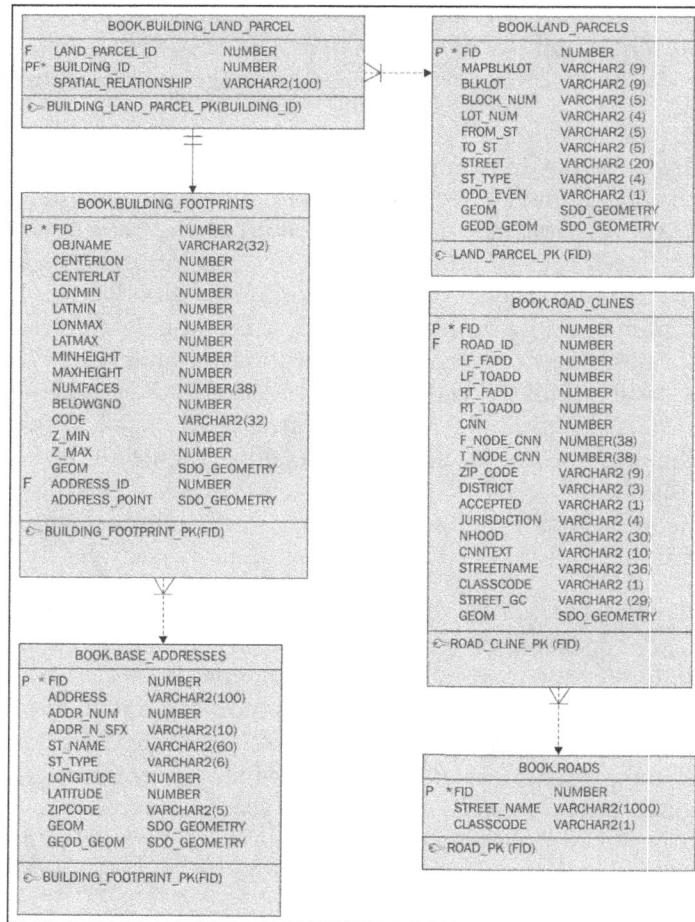

The following figure shows the further E-R diagram for same data model:

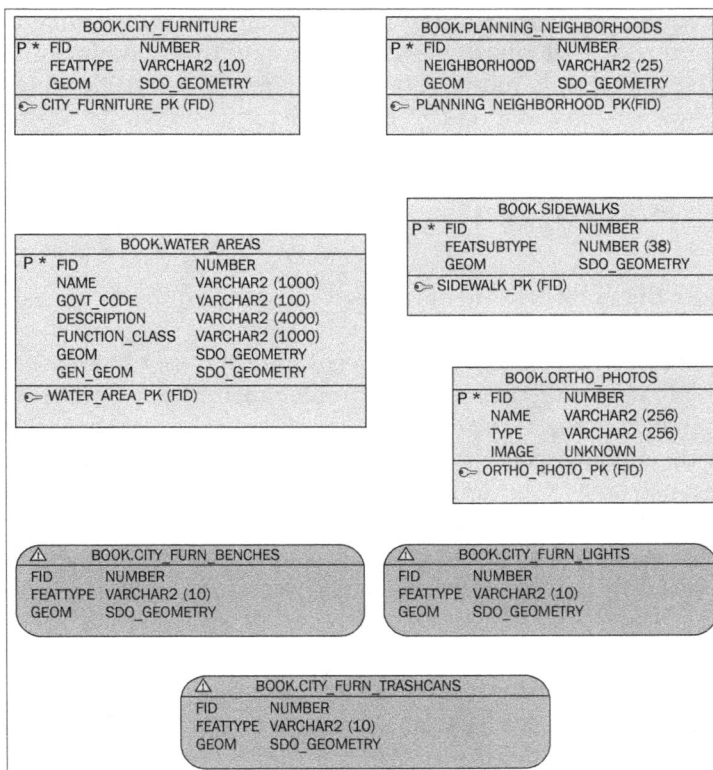

Creating tables in the schema

Create a user called BOOK and assign it a password. Load the script `<schema_load. sql>` and it will create the tables required for running the examples described in this book. It will also create the Oracle Spatial metadata required for these tables. The following privileges are granted to the BOOK user:

```
grant connect,resource to Book identified by <password>;
grant connect, resource to book;
grant create table to book;
grant create view to book;
grant create sequence to book;
grant create synonym to book;
grant create any directory to book;
grant query rewrite to book;
grant unlimited tablespace to book;
```

Understanding spatial metadata

Oracle Spatial requires certain metadata before the spatial data can be meaningfully used by applications. The database views that contain this metadata also act as a catalog for all the spatial data in the database. There are two basic views defined to store this metadata information: USER_SDO_GEOM_METADATA and ALL_SDO_GEOM_METADATA. The USER_ view is used to create a metadata entry for a single SDO_GEOMETRY column within a database table or view. An entry must be created for each SDO_GEOMETRY column within a table; entries for SDO_GEOMETRY columns in views are optional.

If a table has more than one column of type SDO_GEOMETRY, then there is one metadata entry for each column of spatial data in that table. The ALL_ view shows all of the spatial layers that can be accessed by the current user. If a user has the Select grant on another user's table with SDO_GEOMETRY columns, the first user can see the metadata entries for those tables in the ALL_ view. The views are set up so that owners of the spatial tables or views can create the metadata for them. And, the owner of a layer can grant read access to a layer to other users in the system. Granting a Select privilege on the table or view to other users will let them see the metadata for these tables and views. The ALL_ view displays all the spatial tables owned by the user along with other spatial tables for which the current user has read access.

Spatial Reference System

Each SDO_GEOMETRY object has a **Spatial Reference System (SRS)** associated with it, and all the SDO_GEOMETRY objects in a column should have the same SRS. In Oracle Spatial, a **Spatial Reference ID (SRID)** is used to associate an SRS with SDO_GEOMETRY objects. There are cases (for example, engineering drawings) where there is no SRS associated with an SDO_GEOMETRY object. In such cases, a NULL SRID is used to denote that the spatial data has no spatial reference information. An SRS can be geographic or non-geographic. A geographic SRS is used when the spatial data is used to represent features on the surface of the Earth. These types of SRS usually have a reference system that can relate the coordinates of the spatial data to locations on Earth. A unit of measurement is also associated with an SRS so that measurements can be done using a well-defined system. A non-geographic SRS is used when the spatial data is not directly related to locations on Earth. But these systems usually have a unit of measurement associated with them. Building floor plans is a good example of spatial data that is often not directly related to locations on Earth.

A geographic system can be either geodetic or projected. Coordinates in a geodetic system are often described using longitude and latitude. In Oracle Spatial, the convention is to use longitude as the first coordinate and latitude as the second coordinate. A projected system is a Cartesian system that is defined as a planar projection based on the datum and projection parameters.

Before an entry is created for a layer of data, the SRID associated with the data should be identified along with the tolerance to be used for the spatial layer. All the spatial data has an inherent accuracy associated with it. Hence, the tolerance value used for a spatial layer is very important and should be determined based on the accuracy of the data. Once these two values are identified, you are ready to create the metadata for the spatial layer.

More on Spatial Reference Systems

Oracle Spatial supports hundreds of SRSs, and it is very important to choose the right SRS for any given data set. The definition of an SRS can be easily obtained by looking at the **well-known text (WKT)** for that SRS. The WKTs for all the SRSs supplied as part of Oracle Spatial are available from the MDSYS.CS_SRS view. In addition to this view, there are several other metadata tables under MDSYS that contain more details on how these SRSs are defined. Oracle Spatial also supports the EPSG standard-based SRSs. SRS in Oracle Spatial is flexible and allows users to define new reference systems if they are not present in the supplied SRSs. We will revisit user-defined SRSs in the following chapters.

Creating spatial metadata

The tables used in our sample schema contain data that is geographically referenced. Spatial metadata can be created using Insert statements into the USER_SDO_GEOM_METADATA view. This view is defined as a public-writable view on top of MDSYS.SDO_GEOM_METADATA_TABLE that is used to store metadata for all the spatial columns in the database. Let us look at the metadata creation process for some of the spatial tables. Most of the tables used in the current schema have spatial data in the **California State Plane Zone 3 Coordinate System**. In Oracle Spatial, the corresponding SRID for this SRS is 2872. This coordinate system has foot as the unit of measurement and we will use 0.05 as our tolerance (that is, five-hundredths of a foot). The metadata is created using the Insert statement as shown in the following code:

```
Insert Into USER_SDO_GEOM_METDATA Values ('LAND_PARCELS', 'GEOM',
    SDO_DIM_ARRAY(SDO_DIM_ARRAY(SDO_DIM_ELEMENT('X', 5900000,
    6100000, 0.05), SDO_DIM_ELEMENT('Y',2000000, 2200000, 0.05)),
    2872);
```

The SDO_DIM_ELEMENT object is used to specify the lower and upper bounds for each dimension of the coordinate system along with the tolerance value. The metadata allows one entry for each dimension, even though it is very common to use the same tolerance value for the X and Y dimensions. When storing 3D data, it is very common to use a different tolerance value for the Z dimension.

The `BASE_ADDRESSES` table has geometries stored in two columns: `GEOMETRY` and `GEOD_GEOMETRY`. The `GEOMETRY` column has data in the 2872 SRID, while the `GEOD_GEOMETRY` column has data in longitude and latitude. As this is a geodetic system, the tolerance for such systems is required to be in meters. So, a tolerance of 0.05 means a tolerance of 5cm. For geodetic data, it is recommended that the tolerance should not be less than 5cm for all of the topology and distance-based operations.

```
Insert Into USER_SDO_GEOM_METADATA Values('BASE_ADDRESSES',
 'GEOD_GEOM', SDO_DIM_ARRAY((SDO_DIM_ELEMENT('Longitude',
 -122.51436, -122.36638, .05),
 SDO_DIM_ELEMENT('Latitude', 37.7081463, 37.8309382,
 .05)),
 8307);
```

As this is a geodetic system, the longitude range goes from -180 to 180 and the latitude range goes from -90 to 90. Even though it is normal practice to use these ranges for the metadata entry, many developers use the actual ranges spanned by the `SDO_GEOMETRY` object. Mapping tools and applications typically use this extent from the metadata to compute the initial extent of the data in each column of the spatial data.

Any application looking for all the spatial columns in the database should select the data from the `ALL_SDO_GEOM_METADATA` view. This will return one row for each column of spatial data in the database that is visible to the current user.

OGC-defined metadata views

Open Geospatial Consortium (OGC) defines a different set of standardized metadata views. OGC standard metadata can be defined using a new set of tables or views in Oracle Spatial. For a simple solution for the OGC metadata schema, we will show a view-based implementation using the Oracle Spatial metadata table. All Oracle supplied packages, functions, and types of Oracle Spatial are in the `MDSYS` schema. It is generally not recommended to create any user objects under this schema as it might cause problems during database upgrades. Oracle also supplies another predefined schema called `MDDATA` that can be used for Oracle Spatial-related user objects that are general purpose in nature. We use this `MDDATA` schema to create the OGC metadata views. This user comes locked and it is recommended that you do not unlock this user. But, it does require a few privileges to make the following code work, so grant those privileges as required.

Connect to the database as a user with `SYSDBA` privileges and execute all the following steps as the `MDDATA` user by changing the current schema to `MDDATA`. We need to grant an explicit `Select` privilege on `SDO_GEOM_METADATA_TABLE` to `MDDATA`.

```
Alter session set current_schema=MDDATA;
GRANT Select on MDSYS.SDO_GEOM_METADATA_TABLE to MDDATA;
```

The OGC standard requires the geometry type as part of the metadata view. But, this is not part of the MDSYS owned metadata view and has to be computed based on the geometry table information stored in the MDSYS table. So, first define a function that can compute the geometry type based on the rows in the spatial tables. Note that this function just looks at the first non-NULL geometry and returns the type of that geometry. Users can modify this to make it look at the whole table to decide on the geometry type, but it can be a very expensive operation.

```
Create Or Replace Function MDDATA.GET_GEOMETRY_TYPE (tsname
   varchar2, tname varchar2, cname varchar2) Return Number IS
gtype number;
Begin
  Begin
   execute immediate
     ' Select a.'||
     SYS.DBMS_ASSERT.ENQUOTE_NAME(cname, false)||
     '.sdo_gtype From '||
     SYS.DBMS_ASSERT.ENQUOTE_NAME(tsname, false)||'.'||
     SYS.DBMS_ASSERT.ENQUOTE_NAME(tname, false)||
     ' a Where a.'||
     SYS.DBMS_ASSERT.ENQUOTE_NAME(cname, false)||
     ' is not null and rownum < 2'
    Into gtype;
   Return gtype MOD 100;
   EXCEPTION When OTHERS Then
      Return 4;
  End;
End;
```

Notice all the uses of the ENQUOTE_NAME function from the SYS.DBMS_ASSERT package. This is used to avoid any possible SQL injection issues typically associated with functions that create SQL statements using the user supplied SQL. As we are creating a general purpose function that can be invoked by any user directly or indirectly, it is a good idea to protect the function from any possible SQL injection.

Next, we define an OGC metadata view to see all the rows from the MDSYS owned metadata table.

```
Create Or Replace View MDDATA.OGC_GEOMETRY_COLUMNS As
Select GM.SDO_OWNER        As F_TABLE_SCHEMA,
       GM.SDO_TABLE_NAME   As F_TABLE_NAME,
       GM.SDO_COLUMN_NAME  As F_GEOMETRY_COLUMN,
       Get_Geometry_Type(GM.sdo_owner,
                    GM.sdo_table_name,
```

```
                        GM.sdo_column_name)
                        As GEOMETRY_TYPE,
        (Select count(*)
           From Table(GM.SDO_DIMINFO)
        )                   As COORD_DIMENSION,
        GM.SDO_SRID          As SRID
    From MDSYS.SDO_GEOM_METADATA_TABLE GM;
```

And finally, we define a user view that will show all the geometry columns that are visible to the current user.

```
Create Or Replace View  GEOMETRY_COLUMNS  As
Select  b.F_TABLE_SCHEMA ,
        b.F_TABLE_NAME ,
        b.F_GEOMETRY_COLUMN,
        b.COORD_DIMENSION,
        b.SRID,
        b.GEOMETRY_TYPE
From MDDATA.OGC_GEOMETRY_COLUMNS b,
     ALL_OBJECTS a
Where  b.F_TABLE_NAME = a.OBJECT_NAME
  And  b.F_TABLE_SCHEMA = a.OWNER
  And  a.OBJECT_TYPE in ('TABLE', 'SYNONYM', 'VIEW');

Grant Select On MDDATA.GEOMETRY_COLUMNS to public;
Create PUBLIC SYNONYM GEOMETRY_COLUMNS FOR
MDDATA.GEOMETRY_COLUMNS;
```

Tolerance in Oracle Spatial

Tolerance is used in Oracle Spatial to associate a level of precision with the data and to check the validity of geometries among other things. Tolerance should be derived based on the resolution and accuracy of the data. If the devices or methods used to collect the spatial data are correct up to a five-meter resolution, the tolerance for that layer should be set to 5 meters. The actual tolerance value, inserted into the metadata view depends on the real-world tolerance value and the unit of measurement used in the coordinate system is used for the column of spatial data. For example, let the tolerance for the spatial data be 5 centimeters and the unit of measurement of the coordinate system used for the spatial column is feet. Then, the five-centimeter value should first be converted to feet (1 centimeter is 0.032 feet) — this comes out to be 0.164 feet. So, you use a value of 0.164 for tolerance in the metadata.

In practice, Oracle Spatial uses the following rules based on tolerance to determine if the geometry is valid or not. These are in addition to other topological consistency rules (as described by the OGC Simple Feature Specification) used to check the validity of geometries:

- If the distance between two consecutive vertices in the geometry is less than the tolerance value, the geometry is invalid. This rule applies to line-string and polygon type geometries.

- If the distance between a vertex and the nearest edge to that vertex in a polygon is less than the tolerance value, the geometry is invalid. This rule only applies to the polygon type geometries.

Managing homogeneous and heterogeneous data

If a spatial column in a table contains data of one single geometry type (for example, polygon or line-string, but not both), we can say that spatial data in that column is homogeneous. In other situations, a column may contain data from one or more geometry types (heterogeneous representation). For example, the spatial description of a rare and endangered flora object may normally be a single plant (a plant via a point), but in other situations, it may be an area (a patch via a polygon). Consider the CITY_FURNITURE table that is used for storing city assets like benches, trashcans, and streetlights. The geometry for the benches is represented using line-strings, while streetlights and trashcans are represented with points. It is perfectly correct, semantically, to store different types of observations within a single geometry column. However, while some mapping software systems can cope with multiple geometry types per SDO_GEOMETRY column, others, such as some traditional GIS packages, require homogeneity. We will describe how to achieve this next, when a column in the table has heterogeneous data.

We define three views on top of the CITY_FURNITURE table corresponding to each of the types of data stored in the table. This table has three classes of objects: benches, trashcans, and streetlights. After the views are defined, we also need to create metadata entries for these views in USER_SDO_GEOM_METADATA so that any GIS tool can discover these views as if they are tables.

Create the database views corresponding to each of the three types of data stored in the CITY_FURNITURE table.

```
-- DDL for View CITY_FURN_BENCHES
Create Or Replace FORCE VIEW
 CITY_FURN_BENCHES (FID, FEATTYPE, GEOM) As
  Select FID, FEATTYPE, GEOM From CITY_FURNITURE
```

```
        Where FEATTYPE='BENCH';
   -- DDL for View CITY_FURN_LIGHTS
   Create Or Replace FORCE VIEW
    CITY_FURN_LIGHTS (FID, FEATTYPE, GEOM) As
      Select FID, FEATTYPE, GEOM  From CITY_FURNITURE
   Where FEATTYPE='LIGHT';
   -- DDL for View CITY_FURN_TRASHCANS
   Create Or Replace FORCE VIEW
    CITY_FURN_TRASHCANS  (FID, FEATTYPE, GEOM) As
      Select FID, FEATTYPE, GEOM From CITY_FURNITURE
   Where FEATTYPE='TRASHCAN';
```

The preceding examples show how to use other relational attributes to create the
required views. Another way to do this is to constrain based on the SDO_GTYPE
attribute of the SDO_GEOMETRY column. The following example shows how to do this
for one of the preceding views, as the rest can be done with similar SQL:

```
   -- DDL for View CITY__FURN_BENCHES
   Create Or Replace FORCE VIEW
    CITY_FURN_BENCHES (FID, FEATTYPE, GEOM) AS
      Select FID, FEATTYPE, GEOM From CITY_FURNITURE A
   Where A.GEOM.SDO_GTYPE = 2002;
```

Now create the metadata for each of these views so that any GIS can access this as
if it is stored in a separate table. Note that these additional metadata entries are not
required for the correct usage of Oracle Spatial. They are created only to facilitate the
GIS tools that don't support heterogeneous data in spatial columns.

```
   Insert Into USER_SDO_GEOM_METADATA Values (
   'CITY_FURN_BENCHES', 'GEOM',
   SDO_DIM_ARRAY(SDO_DIM_ELEMENT('X', 5900000, 6100000, .05),
   SDO_DIM_ELEMENT('Y', 2000000, 2200000, .05)), 2872);

   Insert Into USER_SDO_GEOM_METADATA Values(
   'CITY_FURN_LIGHTS', 'GEOM',
   SDO_DIM_ARRAY(SDO_DIM_ELEMENT('X', 5900000, 6100000, .05),
   SDO_DIM_ELEMENT('Y', 2000000, 2200000, .05)), 2872);

   Insert Into USER_SDO_GEOM_METADATA Values(
   'CITY_FURN_TRASHCANS', 'GEOM',
   SDO_DIM_ARRAY(SDO_DIM_ELEMENT('X', 5900000, 6100000, .05),
   SDO_DIM_ELEMENT('Y', 2000000, 2200000, .05)), 2872);
```

How metadata is used

Applications typically look at the ALL_SDO_GEOM_METADATA view to see the spatial tables available in the database for a given user. If you select the data from this view, now you will see 11 rows returned: 8 rows corresponding to tables and 3 rows corresponding to the views defined in the CITY_FURNITURE table. From an application point of view, it does not make any difference whether this data is stored in a view or a table. It will all look the same to the application.

Sometimes it is useful to constrain the type of spatial data stored in the table to be homogeneous. For example, the ROAD_CLINES table should contain only linear geometries, as the roads are usually geometries of line type. This can be done by constraints that can be imposed by the spatial index defined in the ROAD_CLINES table. While creating the spatial index, provide the LAYER_GTYPE keyword and specify the type of data that will be stored in this table.

```
-- DDL for Index ROAD_CLINES_SIDX
  Create Index ROAD_CLINES_SIDX ON ROAD_CLINES (GEOM)
   INDEXTYPE IS MDSYS.SPATIAL_INDEX PARAMETERS
  ('LAYER_GTYPE=LINE');
```

Now, if you try to insert a row with a geometry that has a different SDO_GTYPE attribute than 2002, it will raise an error.

```
Insert Into ROAD_CLINES Values ( 198999, 402, 0, 0, 2300, 2498,
190201, 23564000,   23555000, 94107, 10,  'Y', 'DPW',
'Potrero Hill', '190201',  '03RD ST', '3', '3RD',
SDO_GEOMETRY(2001, 2872, NULL, SDO_ELEM_INFO_ARRAY(1, 1, 1),
SDO_ORDINATE_ARRAY( 6015763.86, 2104882.29)));

*
ERROR at line 1:
ORA-29875: failed in the execution of the ODCIINDEXInsert routine
ORA-13375: the layer is of type [2002] while geometry inserted has
type [2001]
ORA-06512: at "MDSYS.SDO_INDEX_METHOD_10I", line 720
ORA-06512: at "MDSYS.SDO_INDEX_METHOD_10I", line 225
```

The error message clearly indicates that the row that is currently being inserted has geometry with the wrong SDO_GTYPE attribute. This is the easiest way to strictly enforce the GTYPE constraints on the spatial data. However, this has the problem of rejecting the whole row when the geometry type does not match the LAYER_GTYPE keyword. And it is also not easy to log these cases as the error is thrown and the database moves on to process the next Insert statement. In some cases, the user might still want to insert the row into the table, but record the fact that there is invalid data in the row. Users can then come back and look at all the invalid entries and fix the issues. We will describe a few methods to do this logging and error processing later in this chapter.

Using database check constraints

Specifying the `layer_gtype` keyword is not the only way to constrain the type in a spatial layer. One can also use table level constraints to achieve the same result. With constraints, users get the additional benefit of specifying more complex constraints, such as allowing only points and lines in a layer. However, if only a single geometry type constraint is required, it is better to implement that constraint using the LAYER_ GTYPE method as this is more efficient than the check constraint. These constraints can also be enforced with database triggers, and these trigger-based constraints are discussed in a later section.

```
Alter Table CITY_FURNITURE ADD Constraint city_furniture_gtype_ck
CHECK ( geom.sdo_gtype in (2002, 2001) );
Insert Into CITY_FURNITURE Values (432432, 'BENCH',
SDO_GEOMETRY(2003,2872,NULL, SDO_ELEM_INFO_ARRAY(1, 1003, 3),
SDO_ORDINATE_ARRAY(6010548.53, 2091896.34, 6010550.45,2091890.11)));
```

This will fail with the following error:

```
ERROR at line 1:
ORA-02290: check Constraint (BOOK.CITY_FURNITURE_GTYPE_CK) violated
```

Similarly, this approach can be useful for an OBJTYPE column check in the CITY_ FURNITURE table.

```
Alter Table CITY_FURNITURE ADD Constraint city_furniture_type_ck
  CHECK ( feattype in ('BENCH','LIGHT','TRASHCAN') );

Insert Into  CITY_FURNITURE Values (432432, 'LIGHTS',
SDO_GEOMETRY(2001,2872, SDO_POINT_TYPE(6010548.53, 2091896.34,
  NULL), NULL, NULL));
ERROR at line 1:
ORA-02290: check Constraint (BOOK.CITY_FURNITURE_TYPE_CK) violated
```

Now these two constraints are checking two independent columns, but what we really need is a more complex check to ensure each value of OBJTYPE has the corresponding SDO_GEOMETRY with the right type. That is, we want to make sure that TRASHCAN and LIGHT types have a point geometry and BENCH has a line geometry.

```
Alter Table CITY_FURNITURE Drop Constraint  city_furniture_gtype_ck;
Alter Table CITY_FURNITURE Drop Constraint  city_furniture_type_ck;
```

```
Alter Table CITY_FURNITURE
  ADD Constraint city_furniture_objtype_geom_ck
  CHECK (
        ( ("FEATTYPE"='TRASHCAN' Or "FEATTYPE"='LIGHT')
          AND   "GEOM"."SDO_GTYPE"=2001
        )
        Or ("FEATTYPE"='BENCH' AND "GEOM"."SDO_GTYPE"=2002)
        /* Else Invalid combination */
        ) ;


Insert Into CITY_FURNITURE Values  (432432, 'BENCH',
SDO_GEOMETRY(2001,2872, SDO_POINT_TYPE(6010548.53, 2091896.34,
  NULL), NULL, NULL));

ERROR at line 1:
ORA-02290: check Constraint (BOOK.CITY_FURNITURE_TYPE_CK) violated
```

Multiple representations for the same objects

In some situations, it is beneficial to have multiple representations for the same geometric feature. For example, an address usually has a point representation for its location. If a footprint of a building is associated with the address, then that footprint will be represented as a polygon. In some cases, a building might have many different point locations associated with it. One point may allow a GIS application to draw an icon for the building depending on its function (for example, a fire station). Another point may allow the building to be labeled with its street address, and finally another one may show an alternate location that is used for main delivery or emergency services entry at the back of the building.

Similarly, a land parcel table can have an interior point of the parcel represented as point geometry in addition to the polygon representation. For a visualization application, it is sometimes useful to represent the land parcel as a point. When a map is displayed at a smaller scale (city level), the map will be cluttered if each land parcel is displayed as a polygon. In such cases, if land parcels are displayed as points with a suitable icon, the map will be less cluttered. When the map is displayed at a larger scale (street level), the same land parcel can be displayed as a polygon. Oracle Spatial allows such multiple representations by allowing multiple SDO_GEOMETRY columns in the same table.

We first start with the BUILDING_FOOTPRINTS table and alter it to add an additional SDO_GEOMETRY column to allow the street address to be represented at the center of the building via a point feature. We can use a spatial function that can compute a point inside a polygon automatically to populate this column.

Point representation for polygon objects

We first alter the table to add a new SDO_GEOMETRY column. For this, we pick ADDRESS_POINT as the column name.

```
Alter Table BUILDING_FOOTPRINT ADD (ADDRESS_POINT SDO_GEOMETRY);
```

We can then update the BUILDING_FOOTPRINT table and compute a value for the new column using the INTERIOR_POINT function in the SDO_UTIL package. This is a nice utility function that can compute a point that is interior to the polygon geometry. This function works even if the geometry has multiple rings or multiple elements. If the geometry has multiple rings, it will find the interior point inside the largest ring. The same rule applies when the geometry has multiple elements. It will find the interior point inside the largest ring of the polygon. Note that this is a fairly expensive operation and can take a few minutes on the large BUILDING_ FOOTPRINT table.

```
Update BUILDING_FOOT_PRINT SET ADDRESS_POINT =
sdo_util.interior_point(GEOM, 0.05);
```

This function takes the geometry as input along with the tolerance value associated with the geometry. The return value is a point geometry that is guaranteed to be inside the source polygon geometry.

Once this column is populated, a new metadata entry needs to be created for this and, if required, a spatial index should also be created.

```
Insert Into USER_SDO_GEOM_METADATA Values(
'BUILDING_FOOTPRINT', 'ADDRESS_POINT',
SDO_DIM_ARRAY(SDO_DIM_ELEMENT('X', 5900000, 6100000, .005),
SDO_DIM_ELEMENT('Y', 2000000, 2200000, 0.05)), 2872);

  Create Index BUILDING_FOOTPRINT_PT_SIDX ON
  BUILDING_FOOTPRINT (ADDRESS_POINT)
   INDEXTYPE IS MDSYS.SPATIAL_INDEX ;
```

Alternate representation in a different coordinate system

In some cases, the alternate representation for a feature might be a geometry with the same shape but in a different coordinate system or with reduced resolution. Data in different coordinate systems is useful in cases where the data has to be published over the web in a different coordinate system than what is stored in the database schema.

For the sample schema, the data is stored in the California state plane, but this data might be published into a state-level database where the coordinate system is Geodetic 8307. In such cases, another column of SDO_GEOMETRY can be added to each of the tables in the schema and set to use SRID 8307.

We take the LAND_PARCELS table and add another column to store the geometry with SRID 8307. We create the appropriate metadata for the geodetic system and create an index on the new spatial column.

```
Alter Table LAND_PARCELS  ADD (GEOD_GEOM SDO_GEOMETRY);

Update LAND_PARCLES SET GEOD_GEOM = SDO_CS.TRANSFORM(GEOM, 8307);
```

Now the GEOD_GEOM column has the geometry for the land parcels in the Geodetic system. We now create the metadata for the new spatial layer and create an index for it. Note that the tolerance now should be specified in meters, and we will use 0.05 meters for tolerance.

```
Insert Into USER_SDO_GEOM_METADATA Values(
'LAND_PARCELS', 'GEOD_GEOM',
SDO_DIM_ARRAY(SDO_DIM_ELEMENT('Longitude',  -180, 180, 0.05),
SDO_DIM_ELEMENT('Latitude',  -90, 90, 0.05)),  8307);

Create Index LAND_PARCELS_GEOD_SIDX ON LAND_PARCELS(GEOD_GEOM)
INDEXTYPE Is MDSYS.SPATIAL_INDEX;
```

Using generalized representation

In some cases, a generalized representation of the geometry in the same SRS is used for mapping applications. When the map scale is very small, it is better to use a generalized version of the geometry. This improves the appearance of the shape on the map and the performance as less data is transferred to the client. For this example, we will take the WATER_AREAS table and add another geometry column to store a generalized geometry.

```
Alter Table WATER_AREAS ADD (GEN_GEOM SDO_GEOMETRY);

Update WATER_AREAS SET GEN_GEOMETRY = SDO_UTIL.SIMPLIFY(GEOM, 500,
  0.05);
```

The SDO_UTIL.SIMPLIFY function takes an SDO_GEOMETRY column and simplifies it using the Douglas-Peucker algorithm. Here, a threshold of 500 feet is used to simplify the geometry. We can then create a metadata entry for this column and create a spatial index.

See the section on triggers to see how these additional columns can be kept in sync with the base column in the table. That is, whenever a new row is inserted or an existing geometry column is changed, the corresponding additional geometry column can be populated automatically using triggers.

Implementing spatial constraints in the database

After the schema is decided and the spatial layers are defined, it is useful to add spatial constraints to the database so that many of the data consistency checks for spatial data can be done in the database. These checks can be defined using the same mechanisms used to define the traditional relational database constraints. Typically, these constraints are defined and implemented in GIS applications, but a database is a common data store that is accessed by many GIS applications. So, it makes more sense to define the common data level constraints at the database level so that each application does not have to define these constraints separately.

A spatial constraint is a data consistency check that makes sure that the data stored in the spatial layers follows certain spatial rules. For example, it may be that a building footprint should always be contained within a land parcel or a road should never cross a land parcel. At the database level, there are several different ways to implement these constraints. The most common way is to define triggers on the tables and check for data consistency as soon as a new row is inserted. Then, based on the result of the check, the row of data can be rejected or accepted. This is called **synchronous trigger processing**. However, there may be cases where the checks can be done at a later time so that the processes inserting the new data are not blocked when checking for data consistency. When a bunch of data is inserted, a background process can go through the new data, run the data consistency checks, and generate a report with invalid rows that the data consistency checks useless. Because the processing is separated from the transaction, this is called **asynchronous trigger processing**.

Constraining the geometry type

We first revisit the previous problem of constraining the spatial layer to contain a single type of data. In the previous section, we showed how to implement this constraint using the spatial index. Now, if we want to constrain a table to contain only points and lines, the index-based constraint mechanism does not work. That was why we used a column or table constraint. But, it is also possible to use a trigger-based mechanism to implement such a constraint.

Consider the CITY_FURNITURE table. We mentioned that this table contains objects such as streetlights and benches. That is, this table can contain point and line type geometries. We define a simple trigger to enforce this constraint on this CITY_FURNITURE table.

```
Create Or Replace trigger CITY_FURNITURE_TYPE_CHK
before Insert or Update on CITY_FURNITURE
For Each Row
Begin
   If ( (:NEW.GEOM.SDO_GTYPE <>  2001) AND (:NEW.GEOM.SDO_GTYPE <>
2002)) Then
      RAISE_APPLICATION_ERROR(-20000, 'Geometry does not have the right
type for row with FID' || to_char(:NEW.FID)) ;
   End If;
End;
```

This trigger is defined as a BEFORE trigger. That means the trigger body is executed before the row is actually inserted into the table so that no table data will be changed if there is an error.

```
Insert Into CITY_FURNITURE Values (100, 'BENCH', SDO_GEOMETRY(2003,
2872, null,
SDO_ELEM_INFO_ARRAY(1, 1003, 1),
SDO_ORDINATE_ARRAY(0,0, 1,0, 1,1, 0,1, 0,0)));

ERROR at line 1:
ORA-20000: Geometry does not have the right type for row with FID 100
ORA-06512: at "BOOK.CITY_FURNITURE_TYPE_CONSTRAINT", line 3
ORA-04088: error during execution of trigger
'BOOK.CITY_FURNITURE_TYPE_CONSTRAINT'
```

As the error message shows, the row with the given FID does not have the right SDO_GTYPE for the geometry. Note that the error message is constructed to be very specific, and has enough information to find the correct row that caused the problem.

Downloading the example code

You can download the example code files for all Packt books you have purchased from your account at http://www.packtpub.com. If you purchased this book elsewhere, you can visit http://www.packtpub.com/supportand register to have the files e-mailed directly to you.

Implementing more complex triggers

With this trigger mechanism, we can do more complex checks to make sure that the data conforms to the requirements of the applications. Next, we look at an example to check the validity of the geometry data before it is stored in the table. First, we look at a synchronous trigger, and in the following section, we look at an asynchronous trigger to do the same check. If the table data is generated mostly from a user interface (for example, a desktop GIS), doing the checks with a synchronous trigger is better, as the user who is creating the data can be informed about the invalid data as soon as possible. If the data is usually created or loaded using a batch job, an asynchronous trigger is better suited for this situation. We will look at examples of asynchronous triggers in the following section:

```
Create Or Replace trigger LAND_PARCELS_VALIDATE_V1
after Insert or Update on LAND_PARCELS
For Each Row
Declare
    result            Varchar2(100);
Begin
  result := sdo_geom.validate_geometry_with_context(:NEW.GEOM,0.05);
  If (result <> 'TRUE') Then
    RAISE_APPLICATION_ERROR(-20000, 'Geometry not valid for row with
FID ' || to_char(:NEW.FID) || ' with error '||result) ;
  End If;
End;
```

Now let us insert a row with invalid geometry data into this table. Note that we altered the LAND_PARCELS table to add a new GEOD_GEOM column. So here, we just pass a NULL value for that argument.

```
Insert Into LAND_PARCELS Values(3326028, '0026T05AA', '0026T',
'055A', '2655', 'HYDE', '1 ST', 'HYDE', 'ST',   'O',
SDO_GEOMETRY(2003, 2872, NULL, SDO_ELEM_INFO_ARRAY(1, 1003, 1),
SDO_ORDINATE_ARRAY(6006801.83, 2121396.9,  6006890.33, 2121409.23,
6006878.19, 2121495.89,  6006788.97, 2121483.45 )), NULL);

ERROR at line 3:
ORA-20000: Geometry not valid for row with FID 3326028 with error
13348 [Element <1>] [Ring <1>]
ORA-06512: at "BOOK.LAND_PARCELS_VALIDATE_V1", line 6
ORA-04088: error during execution of trigger
  'BOOK.LAND_PARCELS_VALIDATE_V1'
```

Fixing invalid data using triggers

Oracle Spatial provides several utility functions that can be used to fix any invalid geometry data. For example, the SDO_UTIL.RECTIFY_GEOMETRY function can fix most of the common errors in the spatial data. These utility functions can be used with ease when the data is already stored in a table. But, the preceding trigger has the problem that when there is an error, it will immediately throw the error and will not let the row be inserted until the geometry is fixed. This makes it hard to use the SDO_UTIL functions to fix the invalid geometry data. We now look at a different trigger that lets the row be inserted into the table even when the geometry data is invalid. The trigger will make a note of the invalid row information in a journal table. After the data is created, users can go back to the journal table and fix the invalid geometry data using SDO_UTIL functions. For this, we first create the LANDPARCELS_INVALID table and use it as the journal table.

```
Create Table LAND_PARCELS_INVALID (geom SDO_GEOMETRY,
                                   FID Integer,
                                   result Varchar2(100));


Drop trigger land_parcels_validate_v2;


Create Or Replace trigger LAND_PARCELS_VALIDATE_V2
before Insert or Update on LAND_PARCELS
For Each Row
Declare
    result          Varchar2(100);
Begin

  result := sdo_geom.validate_geometry_with_context(:NEW.GEOM,
0.05);
  If (result <> 'TRUE') Then
      Insert Into LAND_PARCELS_INVALID Values(:NEW.GEOM, :NEW.FID,
result);
:NEW.GEOM := NULL;
  End If;
End;
```

Note that this is also defined as a BEFORE trigger. The geometry is checked for validity, and if it is invalid, a new row is created in the journal table with the geometry and FID values. Also note that the geometry value in the row for the LAND_PARCELS table is set to NULL. This is done so that if the geometry is invalid, then some other processes depending on the geometry data don't get any wrong data. Once the geometry is rectified, the row will be updated with the valid geometry value.

Once a batch of data is created, we go and update the invalid geometry data in the journal table.

```
Update LAND_PARCELS_INVALID SET GEOM =
  SDO_UTIL.RECTIFY_GEOMETRY(GEOM, 0.05);
Commit;
Select FID From LAND_PARCELS_INVALID
Where SDO_GEOM.VALIDATE_GEOMETRY_WITH_CONTEXT(GEOM, 0.05) <>
  'TRUE';
```

If the final statement in the preceding code does not return any rows, then we know that all the invalid data in the journal table is fixed. Now go and update the base table with this valid data.

```
Update LAND_PARCELS A SET A.GEOM = (Select B.GEOM From
  LAND_PARCELS_INVALID B  Where B.FID = A.FID)
Where A.FID IN (Select FID From LAND_PARCELS_INVALID);
```

This statement updates all the rows for which there are rows in the journal table.

Constraints with asynchronous triggers

Asynchronous triggers can be implemented in different ways, and one common practice is to use a queue. A queue is used to detect changes in the tables, and then based on the base table, specific checks can be done. As we are introducing a new concept of queues here, we will keep the spatial constraint part simple and do the validation checks using an asynchronous trigger.

Oracle provides PL/SQL procedures in the DBMS_AQ package for creating and using queues. In our example, a simple queue is used to insert information about new rows of data coming into the LAND_PARCELS table. Then, an asynchronous process can take entries out of the queue and process them one at a time.

Creating a queue

We first create a type to store the messages in the queue. This type is defined based on the information one wants to store in the queue. Make sure this type has enough fields so that the procedure processing the messages from the queue does not need to look for information elsewhere. The message type we use here has the Primary Key (PK) of the LAND_PARCELS table along with the value of the PK. Before we start, we should first drop the previous trigger that we have created on the LAND_PARCELS table.

```
Drop trigger land_parcels_validate_v1;
Drop trigger land_parcels_validate_v2;
```

```
-- Now create the type required for the messages
Create Type validate_q_message_type As Object(
        pk_column Varchar2(32),
         pk_value  Integer,
         table_name Varchar2(32),
         column_name Varchar2(32));
```

Let us create a queue called VALIDATE_QUEUE with a corresponding queue table, VALIDATE_Q_MESSAGE_TABLE. Once it is created, we need to start the queue so that it can be used.

```
-- First create the queue table
EXEC DBMS_AQADM.Create_QUEUE_TABLE( queue_table => 'validate_q_
message_table', queue_payload_type => 'validate_q_message_type');

-- Next create the queue
EXEC DBMS_AQADM.Create_QUEUE( queue_name => 'validate_queue', queue_
table => 'validate_q_message_table');

-- And finally start the queue

EXEC DBMS_AQADM.START_QUEUE( queue_name => 'validate_queue');
```

Now the queue is ready to receive messages.

Defining the trigger

Next, we code the trigger to look at the rows coming into the LAND_PARCELS table and create messages for the queue. We create one message in the queue for each row inserted or updated in the table. The message has information about the columns of the table that are required to access the new or changed geometry.

```
Create Or Replace trigger LAND_PARCELS_VALIDATE_V3
after Insert or Update on LAND_PARCELS
For Each Row
declare
    queue_options        DBMS_AQ.ENQUEUE_OPTIONS_T;
    message_properties   DBMS_AQ.MESSAGE_PROPERTIES_T;
    message_id           RAW(16);
    my_message           validate_q_message_type;
Begin
    my_message := validate_q_message_type('FID',
  :NEW.FID,'LAND_PARCELS','GEOM');
    DBMS_AQ.ENQUEUE(
        queue_name => 'validate_queue',
        enqueue_options => queue_options,
```

```
                  message_properties => message_properties,
                  payload => my_message,
                  msgid => message_id);
     End;
```

Now insert a row into the LAND_PARCELS table, and again we insert a row with an invalid geometry. Only this time, the row gets inserted into the table and no error is raised. But, a new message gets created in the message queue created by us.

```
Delete From LAND_PARCELS Where FID = 3326028;

Insert Into LAND_PARCELS values(3326028, '0026T05AA', '0026T',
   '055A', '2655',   'HYDE', '1 ST', 'HYDE', 'ST',  'O',
SDO_GEOMETRY(2003, 2872, NULL, SDO_ELEM_INFO_ARRAY(1, 1003, 1),
   SDO_ORDINATE_ARRAY(6006801.83, 2121396.9,  6006890.33,
2121409.23,  6006878.19, 2121495.89,  6006788.97, 2121483.45 )),
NULL);
```

Now look at the message queue to check if there are any messages. The easiest way to do this is to look at the queue table created for our message queue.

```
Select * From AQ$VALIDATE_Q_MESSAGE_TABLE;
```

This query should return one row corresponding to the message that our trigger just inserted. As you can see, the table is created using the name that we supplied in the Create_QUEUE_TABLE call with an additional AQ$ prefix.

Next, we will look at some PL/SQL that can be used to browse messages from the queue and to look at the values in the message.

```
Set SERVEROUTPUT ON;
Declare
     queue_options       DBMS_AQ.DEQUEUE_OPTIONS_T;
     message_properties  DBMS_AQ.MESSAGE_PROPERTIES_T;
     message_id          RAW(2000);
     my_message          validate_q_message_type;
Begin
     queue_options.dequeue_mode := DBMS_AQ.BROWSE;
     DBMS_AQ.DEQUEUE(
         queue_name => 'validate_queue',
         dequeue_options => queue_options,
         message_properties => message_properties,
         payload => my_message,
         msgid => message_id );
     Commit;
     DBMS_OUTPUT.PUT_LINE(
         'Dequeued PK Column: ' || my_message.pk_name);
```

```
DBMS_OUTPUT.PUT_LINE(
    'Dequeued PK value: ' || to_char(my_message.pk_value));
DBMS_OUTPUT.PUT_LINE(
    'Dequeued Table: ' || my_message.table_name);
DBMS_OUTPUT.PUT_LINE(
    'Dequeued Column: ' || my_message.column_name);
End;
/
```

This should print the following messages in SQLPLUS. Note that this queue does not do anything useful yet:

```
Dequeued PK Column: FID
Dequeued PK value: 3326028
Dequeued Table: LAND_PARCELS
Dequeued Column: GEOM
```

If you run the preceding PL/SQL code when the queue is empty, it will wait until some messages are inserted into the queue. So, it will wait for another transaction to insert some data into the LAND_PARCELS table and commit. For browsing the messages, the dequeue mode is set to DBMS_AQ.BROWSE. If you change the mode to DBMS_AQ.REMOVE, it will remove the message from the queue. Next, we will look at the code to remove the message from the queue, process it, and take an action depending on the validity of the geometry.

```
Set SERVEROUTPUT ON;
Declare
    queue_options         DBMS_AQ.DEQUEUE_OPTIONS_T;
    message_properties    DBMS_AQ.MESSAGE_PROPERTIES_T;
    message_id            RAW(2000);
    my_message            validate_q_message_type;
    tname                 Varchar2(32);
    cname                 Varchar2(32);
    pkname                Varchar2(32);
    result                Varchar2(100);
    geometry              SDO_GEOMETRY;
    rect_geom             SDO_GEOMETRY;
Begin
    queue_options.dequeue_mode := DBMS_AQ.REMOVE;
    DBMS_AQ.DEQUEUE( queue_name => 'validate_queue',
        dequeue_options => queue_options,
        message_properties => message_properties,
        payload => my_message, msgid => message_id );

    Execute IMMEDIATE ' Select '||
            SYS.DBMS_ASSERT.ENQUOTE_NAME(my_message.column_name)||
```

```
                    ', sdo_geom.validate_geometry_with_context(' ||
                    SYS.DBMS_ASSERT.ENQUOTE_NAME(my_message.column_name)||
                    ', 0.05) From ' ||
                    SYS.DBMS_ASSERT.ENQUOTE_NAME(my_message.table_name)||
                    ' Where ' ||
                    SYS.DBMS_ASSERT.ENQUOTE_NAME(my_message.pk_column)|| ' =
    :pkvalue '
        Into geometry, result USING my_message.pk_value ;
        If (result = 'TRUE') then
         return;
        else
         rect_geom := sdo_util.rectify_geometry(geometry, 0.05);
         result := sdo_geom.validate_geometry_with_context(rect_geom,
    0.05);
        If (result = 'TRUE') then
          EXECUTE IMMEDIATE ' Update '||
                    SYS.DBMS_ASSERT.ENQUOTE_NAME(my_message.table_name)||' set
    '||
                    SYS.DBMS_ASSERT.ENQUOTE_NAME(my_message.column_name)||
                    ' = :geometry Where '||
                    SYS.DBMS_ASSERT.ENQUOTE_NAME(my_message.pk_column)|| ' =
    :pkvalue '
          USING rect_geom, my_message.pk_value;
        Else
         RAISE_APPLICATION_ERROR(-20000, 'Geometry cannot be fixed for
    row with '||
                    my_message.pk_value|| ' in table' ||
                    my_message.table_name);
        End If;
        End If;
    dbms_output.put_line(result);
        Commit;
    End;
```

There are a few important things to note in this PL/SQL code. The message that
we have created in the queue is very generic so that the same code can be used to
process messages for many spatial tables. As the table name, column name, primary
key column, and Primary Key value are all retrieved from the message, the same
queue and dequeue mechanism can be used for a set of tables in the schema. This
code can be easily modified to be a procedure, and it can be executed in a loop while
waiting for messages in the queue. It can also be turned on when there is less activity
in the DB to reduce the load. You also need to think about what happens when the
geometry cannot be fixed using the rectify function. We left that part open in this code
example as it depends on other available tools for fixing the invalid geometry data.

Implementing rule-based constraints

So far, we have only looked at constraints that mainly deal with the validity of data or type of data present in our spatial tables. We will now look at more complex scenarios that define relationships between objects in different tables. These rules enforce data quality and make sure that the objects in the schema follow certain spatial relationship rules. An example of such a constraint is the one that specifies that the building footprints must be contained within the land parcels. Similarly, a land parcel must be contained inside a planning zone.

These constraints can be implemented using triggers (synchronous or asynchronous) as described in the previous sections. They can also be implemented via a spatial constraints table that would provide a flexible and generic data-driven approach. With this data-driven approach, a generic model can be used to enforce constraints between different spatial layers in our schema. A rules table is first created to store all the allowed spatial relationships between different spatial layers. Then, triggers are used on spatial layers to enforce these rules. The advantage of creating a rule table is that all the allowed rules are stored in one place instead of spreading them across the different triggers. This makes it easy to document the rules so that any GIS tool can easily look up the available rules.

Defining rules for the schema

We first look at the rules that will describe possible relationships between objects for this schema. These rules can be defined as MASK values and their combinations that are valid for the SDO_GEOM.RELATE function. We will use this function to enforce the following rules in our schema:

- **CONTAINS**: A land parcel may contain a building footprint
- **COVERS**: A planning neighborhood covers a land parcel
- **EQUAL**: A planning neighborhood can be equal to a land parcel
- **TOUCH**: A road segment can only touch another road segment
- **CONTAINS + COVERS**: A land parcel may contain and cover a building footprint

Next, we look at how to create a rule-based constraint between the BUILDING_FOOTPRINTS tables and LAND_PARCELS tables. The rule specifies that a building should be inside a land parcel (either completely inside or touching the boundary) And a building cannot exist without containing a land parcel. We first create a relationship table that specifies which land parcel contains which building footprint. Note that this information can always be obtained by executing a spatial query. But sometimes it is easier to persistently store this information as it is computed anyway to enforce the data consistency.

We will also enforce the constraint that the only possible values allowed for the SPATIAL_RELATIONSHIP column in this table are CONTAINS and COVERS. That is, a land parcel can only contain or cover a building footprint. We also want to enforce the UNIQUE constraint on the BUILDING_ID column, as each building footprint can only have one parent land parcel record in our schema.

```
Create Table Building_Land_Parcel
            (Land_parcel_id Varchar2(9),
             Building_id number,
             spatial_relationship Varchar2(100));

Alter Table Building_Land_Parcel add Constraint BLP_PK
  UNIQUE (Building_ID) enable;

Alter Table Building_Land_Parcel add Constraint BLP_FK_LP
   Foreign Key (Land_Parcel_ID) REFERENCES LANDPARCELS(BLKLOT) enable;

Alter Table Building_Land_Parcel add Constraint BLP_FK_BF
   Foreign Key (BUILDING_ID) REFERENCES BUILDING_FOOTPRINT(BUILDING_ID)
enable;

Alter Table BUILDING_LAND_PARCEL add Constraint BLP_Spatial_CHECK
CHECK ( spatial_relationship in ('COVERS', 'CONTAINS') );
```

Trigger for checking spatial relationships

Once we have the relationship table created, we next create a utility function that can be used to check for different spatial relationships between objects of different spatial layers. This is a generic procedure that can be used for many different spatial layer pairs with minor changes.

```
Create Or Replace PROCEDURE Check_Relation (tname varchar2,
                                            cname varchar2,
                                            pk_name varchar2,
                                            geometry SDO_GEOMETRY,
                                            mask  varchar2,
                                            pk_value OUT Integer,
                                            relation OUT varchar2) AS
stmt Varchar2(200);
rel_mask  Varchar2(200);
Begin
    rel_mask := 'MASK='||mask;
    stmt := ' Select '||
```

```
                SYS.DBMS_ASSERT.ENQUOTE_NAME(pk_name)||
                ', SDO_GEOM.RELATE('||SYS.DBMS_ASSERT.ENQUOTE_NAME(cname,
false)||
                ',  ''DETERMINE'', '||
                ':geometry,  0.05) From '||
                SYS.DBMS_ASSERT.ENQUOTE_NAME(tname, false)||
                ' Where SDO_RELATE('||
SYS.DBMS_ASSERT.ENQUOTE_NAME(cname)||
                ', :geometry, :rel) = ''TRUE'' ';
    Begin
     EXECUTE IMMEDIATE stmt Into  pk_value, relation
            USING geometry, geometry, rel_mask;

    EXCEPTION When NO_DATA_FOUND Then
      pk_value := NULL;
        When OTHERS Then
            raise;
    End;
End;
```

This procedure takes spatial layer information (table name, column name) to search for the given geometry with the specified spatial relationship. It returns the Primary Key of the spatial layer so that the row that satisfies the specified spatial relationship is identified easily for further processing. We use this information in the trigger to populate the rows in the relationship table that we have defined in the preceding code. It is very important to handle the exceptions in this procedure so that when the insert fails, the users will know exactly what failed. For this case, if valid values for table name and column name are passed in, the SDO_RELATE query can have three possible results:

- It finds exactly one row with the given mask
- It finds no rows
- It fails for some other reason

The first case is the valid case for our trigger, so we don't need to do error processing for this case. The second case means there is no corresponding land parcel that either contains or covers the given building footprint. We let the trigger handle this error, so in this procedure, we just pass a NULL value for FID. The third case means something is wrong with the parameters or the tables involved in the query. In this case, we raise the actual error so that the user can further process the error condition.

Next, we create a trigger on the BUILDING_FOOTPRINTS table to check each building footprint that is newly created or updated for containment with the LAND_PARCELS table.

```
Create Or Replace TRIGGER BF_LP_RELATION
after Insert or Update on BUILDING_FOOTPRINTS
FOr EACH ROW
Declare
    building_id number;
    FID Number;
    relation Varchar2(100);
Begin
    Check_Relation('LAND_PARCELS', 'GEOM', 'FID',
                        :NEW.geom, 'COVERS+CONTAINS', fid, relation );
    If ( (FID is NULL) Or
        (relation <> 'CONTAINS' AND relation <> 'COVERS') ) then
      RAISE_APPLICATION_ERROR(-20000, 'BuildingFootPrint with ID '||
      to_char(:NEW.fid)|| ' is not inside any landparcel');
    Else
      Insert Into Building_Land_Parcel Values(fid, :NEW.fid, relation);
    End If;
End;
```

This trigger first checks the relationship of the current geometry that is being inserted into the BUILDING_FOOTPRINT table. If it finds a land parcel that contains the footprint, this information is inserted into the relationship table. If there is an error (that is, no containing land parcel is identified), the trigger will raise an error and the insert into the footprint table fails. This can also be implemented as an asynchronous trigger so that the insert into the footprint table is allowed even if there is an error, but some error information is recorded in the relationship table so that it can be checked and fixed later.

Batch processing existing data

The trigger that we defined here will create the relationship for the new rows that are coming into the footprint table. In some cases, the data for these tables might have been populated using a batch process before this trigger is enabled. In such cases, how do we populate the relationship table? Next, we describe a process that looks at all the existing records in the footprint table and finds the land parcel that contains it. This can be a time-consuming process, as it has to check each footprint to find the corresponding land parcel.

```
Insert Into building_land_parcel
Select lp.fid, b.fid,
```

```
            sdo_geom.relate(lp.geom, 'determine', b.geom, 0.05)
    From BUILDING_FOOTPRINTS b, LAND_PARCELS lp
    Where SDO_RELATE(lp.geom, b.geom, 'mask=CONTAINS+COVERS') =
        'TRUE';
```

This SQL takes each building footprint and finds the corresponding land parcel. While this SQL is efficient in execution, there is no error reporting as part of this SQL. If there is a building footprint without a corresponding record from the LAND_PARCELS table, there won't be any record of that missing entry in the BUILDING_LAND_PARCEL table. The user should check the number of rows in the BUILDING_LAND_PARCEL table to make sure that there is one record corresponding to each building footprint in this table. This check is required only if the preceding SQL is used to populate this relationship table. In the normal processing of the building footprints, when new records arrive, the trigger that we have created on the BUILDING_FOOTPRINTS table will take care of these error conditions.

Summary

A spatial database application should be designed just like any other database application. It should follow the same standard data model practices like any other database application. In this chapter, we introduced a data model that can be used in a city-wide spatial data management system. This data model is used throughout the book to illustrate different aspects of spatial database application development. We discussed the use of triggers and queues to manage spatial data in the database. We also showed how to design database level constraints for spatial data management. We introduced the concepts of database triggers and queues and showed how they can be used for spatial data management in the database.

The next chapter will show different techniques for loading spatial data from different formats into Oracle Spatial tables. We will also describe some of the common spatial data interchange formats and show some PL/SQL examples for generating data in these formats.

2
Importing and Exporting Spatial Data

Once we have defined a data model, the next step is to load data into it. After the data is loaded, it needs to be checked for "cleanliness" before indexing and using it. There are many methods for loading data of different types and formats. In this chapter, we describe some of the most common formats, and how they can be loaded into the Oracle database using free tools, tools already available with Oracle, and tools from third-party vendors. In addition, we also discuss other issues relating to import performance and organization of data for efficient access by applications.

The goal of this chapter is to give you a complete overview of all aspects of data loading from tools, through physical loading techniques and data organization, data quality checking, and indexing.

- **Extract, transform, and load** (ETL) tools: GeoKettle, Oracle Spatial Shapefile loader, and Map Builder
- Using SQL, Application Express, and Excel
- Implementing theoretical storage resolution and minimal resolution used by functions
- Using ordinate resolution and the effect of rounding
- Using tolerance and precision
- Creating spatial autocorrelation via a Morton key
- Geometry validation and methods to clean imported data
- Coordinate system transformation techniques
- Spatial indexing
- Exporting formats, such as Shapefile, GML, WKT, and GeoJSON

Extract, transform, and load (ETL) tools

Spatial data is usually collected from many different sources, such as proprietary file systems, GIS tools, and third-party vendors. One of the biggest challenges for a **database administrator (DBA)** is to ingest all of these different formats of data, and present them in a uniform data model to their users. There are specific data loading tools (such as Shapefile, KML, and TAB file converters) to load the data from these specific file formats. In traditional database systems, an ETL tool is used to ingest data from different data sources. Now, there are several **graphical user interface (GUI)** driven ETL tools (both open source and third-party vendor provided) that can support many of the common spatial data formats. We look at one such ETL tool in detail to see how it can be used for the spatial data loading process.

An ETL tool supports extract, transform, and load operations. During the extract process, the tool can extract data from different source systems that have many different data organization structures. There are hundreds of known formats for spatial data, including different relational database models (Oracle, PostGIS, SQLServer, and so on) and popular GIS file formats, such as shape files, TAB files, and DGN files. A good ETL tool will be able to understand many of these formats for read and write purposes. That is, the ETL tool can read from any of these formats, and write to any system that stores the data in one of these formats. For Oracle users, it is very common to read the data from several different sources, and write to the Oracle Spatial database. These ETL tools can read the spatial data as well as the other attribute data, so a single tool can understand and manage all of these attributes of data.

During the transform process, a series of rules or procedures can be applied to the extracted data before loading it into the target system. Some data sources, if they are well known and clean, can be loaded directly into the target system with little or no data manipulation. Some data sources may require several data manipulation steps before loading the data into the target system. For example, the source data may be in a different coordinate system than the target; the spatial validation rules of a source may be different than the ones used in the target; the source might have additional columns of data that are not required in the target; or two different data sources may need to be combined before pushing the result into the target system.

The load process loads the processed data into the target system. And depending on the organization, this process can vary widely. Some systems may overwrite the existing data with new data every quarter. For example, a county might get the land parcel data from the respective city GIS departments every quarter. And the city provides the latest copy of all the land parcel data every quarter. In such cases, the county IT department can discard the existing data for that city, and load the new data. In some cases, the county might want to keep the older data set for historical purposes, and move it to a backup device before ingesting the new data. Whatever the required process, a good ETL tool provides the necessary UI and scripting capabilities to set up these workflows once, and execute them on a regular basis.

One of the popular open source ETL tools is GeoKettle, and it supports many functions of the ETL tool including support for many of the popular spatial formats. Another very popular open source tool is GDAL/OGR, which is available at `http://www.gdal.org`. The main drawback of GDAL is that it does not provide a GUI. On the other hand, it is very good for using with loader scripts. We will first look at GeoKettle as it has a good user interface to make the ETL process easy to use.

ETL processing with GeoKettle

GeoKettle is a free and open source metadata-driven spatial ETL tool. It supports integration of different spatial data sources for building data warehouses on different targets including database systems, GIS files, or Geospatial web services. GeoKettle can be downloaded from `http://sourceforge.net/projects/geokettle/` under the LGPL license for Linux, Mac, and Windows platforms. GeoKettle supports the extract operation from different databases (such as Oracle Spatial, PostGIS, MySQL, SQL Server, and IBM), different geo file formats (such as Shapefiles, GML, KML, and OGR), and OGC Web Services (CSW). GeoKettle supports different transformation functions, including coordinate system transformations, spatial buffers, centroids, area and length calculations, and geoprocessing operations such as geometry clipping, Delaunay triangulation, and polygon simplification. For load operations, it supports all the mentioned formats that are supported for the extract operation.

GeoKettle has a GUI (along with a command line tool) that is easy to use for defining the workflows. Users can select the source format (either files or databases), select a set of transformation operations that are chained together, and finally select a target system. Once this is defined, the workflow can be saved as a template and it can be used again. Batch jobs are created based on these workflows, and GeoKettle does the data processing and moves the data into the target system. Next, we describe this ETL process using an example to show how to load a Shapefile into Oracle Spatial along with a simple data transformation operation.

Starting the GeoKettle application will prompt the user with the dialog shown in the preceding figure. The user will be prompted to either enter the repository (that is similar to a project) information or skip the repository. Either option will take the user to the GeoKettle workbench. In the workbench, users can specify the input to be a Shapefile, with a transformation step to transform the geometries from WGS 84 to the **California State Plane Zone 3 Coordinate System**, and the output to be the Oracle Spatial table. This tool provides support for many complex data transformations including selecting specific rows from the input stream, and joining two or more input streams. The following figure shows a more complex data transformation process, where data is collected from two input sources: an HTTP client and a Shapefile. The data from the HTTP client can be filtered to look for any data related to San Francisco, and it can be joined with the data from the Shapefile. The result is then transformed into the required SRS, and inserted into an Oracle Spatial table.

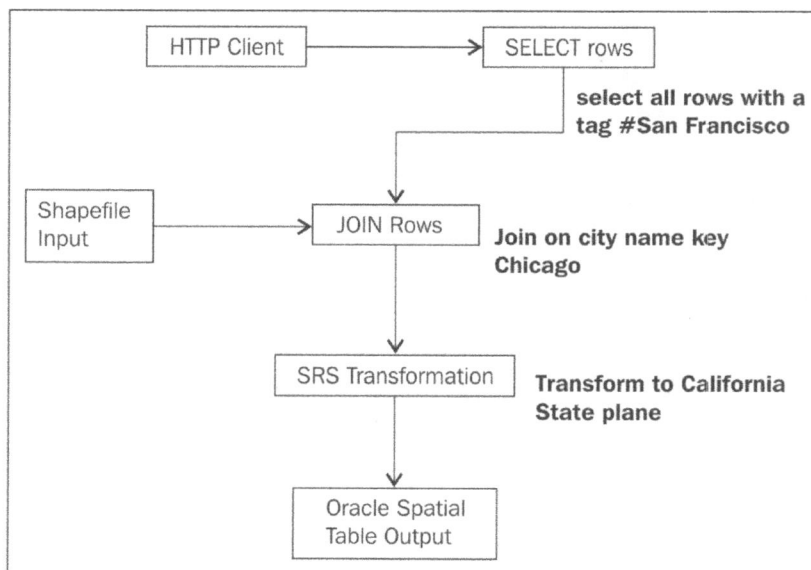

Loading Shapefiles

Shapefiles are one of the most commonly used file formats for exchanging vector spatial data in the industry. Oracle Spatial provides Java classes to load the data from Shapefiles into the database. Oracle Spatial also provides a GUI-based Shapefile, loading the tool as part of the Map Builder application. In this section, we describe both these approaches for loading the data from the Shapefiles.

Java Shapefile loader

The Java classes for the Shapefile loader are part of the `sdoutil.jar` file that can be found under the `$ORACLE_HOME/md/jlib` directory. This loader requires the `sdoapi.jar` and `sdoutil.jar` files to be in the `CLASSPATH` variable. We will show how to load this data using the `TIGER/Line` data published by the U.S. Census bureau at the following website: `http://www.census.gov/geo/www/tiger/tgrshp2012/tgrshp2012.html`.

We downloaded the block group data for the state of California (CA), and used that file in the following example:

```
-- this is the setup for UNIX/Linux machines
setenv clpath $ORACLE_HOME/jdbc/lib/ojdbc5.jar:$ORACLE_HOME/md/jlib/
sdoutl.jar:$ORACLE_HOME/md/jlib/sdoapi.jar


-- this is for Windows
> java -classpath %ORACLE_HOME%\jdbc\lib\ojdbc5.jar;%ORACLE_HOME%\md\
jlib\sdoutl.jar;%ORACLE_HOME%\md\jlib\sdoapi.jar oracle.spatial.util.
SampleShapefileToJGeomFeature -h localhost -p 1521 -s orcl -u book -d
<password-forboo-schema> -t ca_blocks -f tl_2012_06_bg -r 8307 -g geom


java -cp $CLASSPATH oracle.spatial.util.SampleShapefileToJGeomFeature -h
localhost  -p 1521 -s orcl -u book -d <password-for-book-schmea> -t ca_
blocks -f tl_2012_06_bg -r 8307 -g geom
```

Using Map Builder

Oracle Map Builder is a Java-based GUI tool that is mainly used as a map authoring tool along with Oracle MapViewer. It also has GUI driven support for loading Shapefiles into an Oracle database. Map Builder can load the Shapefiles, create the required geometry metadata, create spatial indexes, and get the data ready for use.

The following screenshot shows the menu-driven approach for loading Shapefiles into Oracle Spatial:

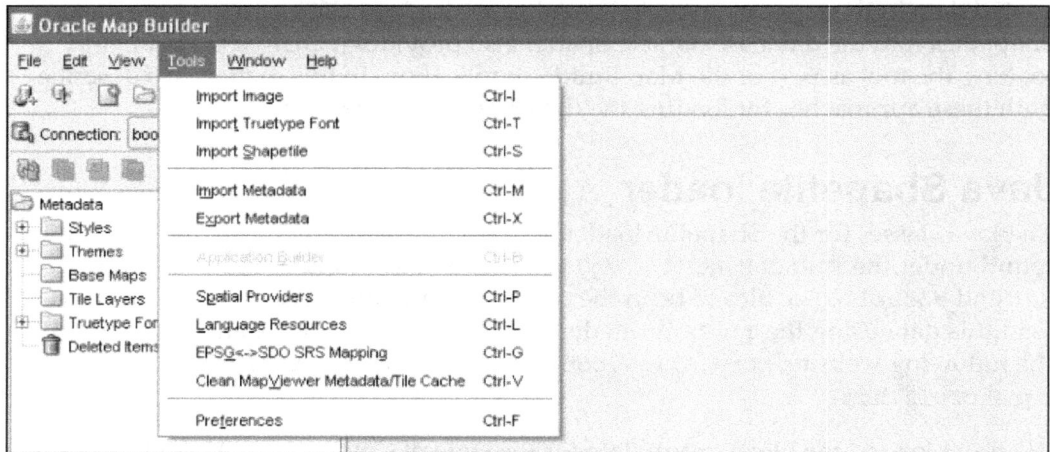

Map Builder has the option to load one Shapefile or multiple Shapefiles from a directory, which is useful for bulk loading of data. Users can specify the name of the table in the database, and this table will be created if it does not exist already. There is an option to append the data to an existing table.

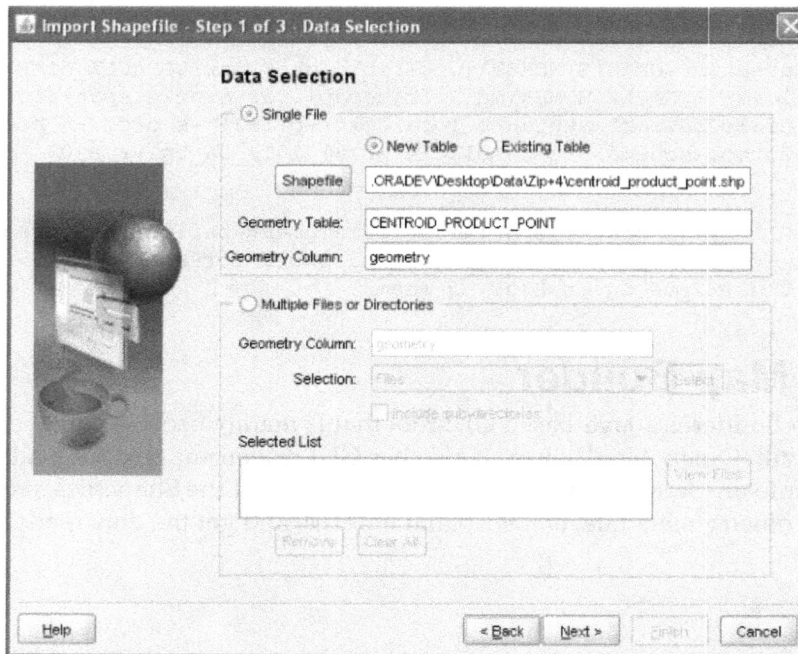

The Map Builder tool also has the option to create the metadata entry for the spatial table and create a spatial index on it. Users can also specify the target SRID if this information is missing from the input. If Map Builder is creating the table in the database, sometimes it renames the column names so that they do not conflict with Oracle key words. For example, if there is a field with the name LONG in the input Shapefile, it will be renamed LONG_MB to avoid any SQL error, as table columns cannot be named LONG.

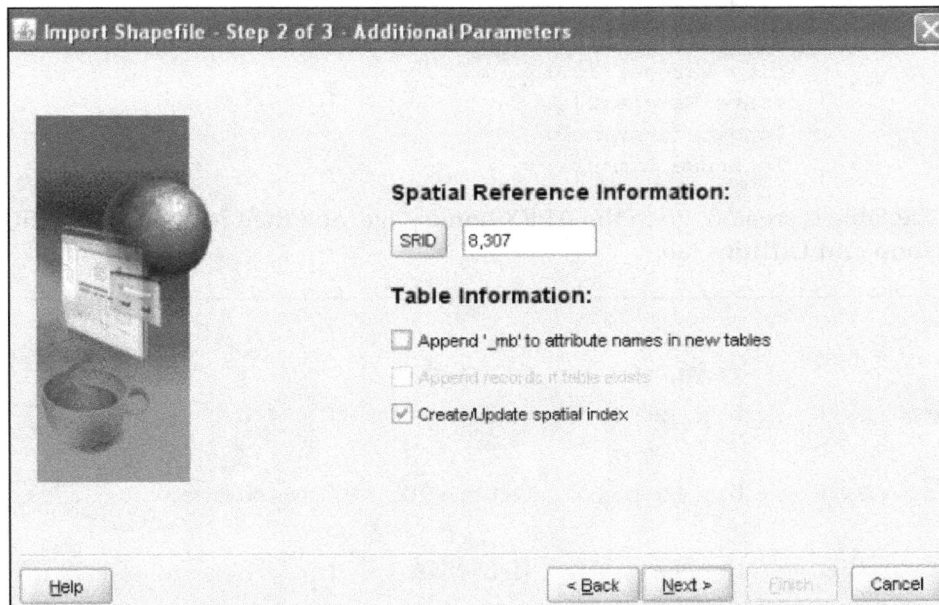

Using SQL, Application Express, and Excel

It is very common to exchange address data in the spreadsheet format between different groups or organizations. Often, this address contains the longitude/latitude data that the users might want to load into the Oracle Spatial tables. There are several tools available for loading the Excel data into the Oracle database, but **Application Express** (**APEX**) provides one of the easiest ways to do this job. In this section, we describe the procedure for loading the Excel data into an Oracle table that has two number columns to store the longitude and latitude data. We then show the use of function-based spatial indexing to run spatial queries against this data. The examples in this section are done with release of APEX 4.2. If users want to try Application Express features without installing their own version, they can try it by navigating to the following URL: http://apex.oracle.com.

After you create the required workspace and log into APEX, it takes you to the main screen. APEX does not read the full format of the Excel workbook format, so you need to store the data first as a **comma-separated values (CSV)** file. The sample Excel file that we are going to use in this example has six fields: name, street, city, state, longitude, and latitude. So, we first create a table to hold the data that we are going to read from the Excel spreadsheet. We can do this either via APEX, or directly on the database via SQLPLUS.

```
Create Table address_excel (name Varchar2(100),
            street Varchar2(200),
            city Varchar(100),
            state Varchar2(2),
            longitude number,
            latitude number);
```

Once the table is created, go to the APEX home page, and then navigate to the **SQL Workshop and Utilities** tab.

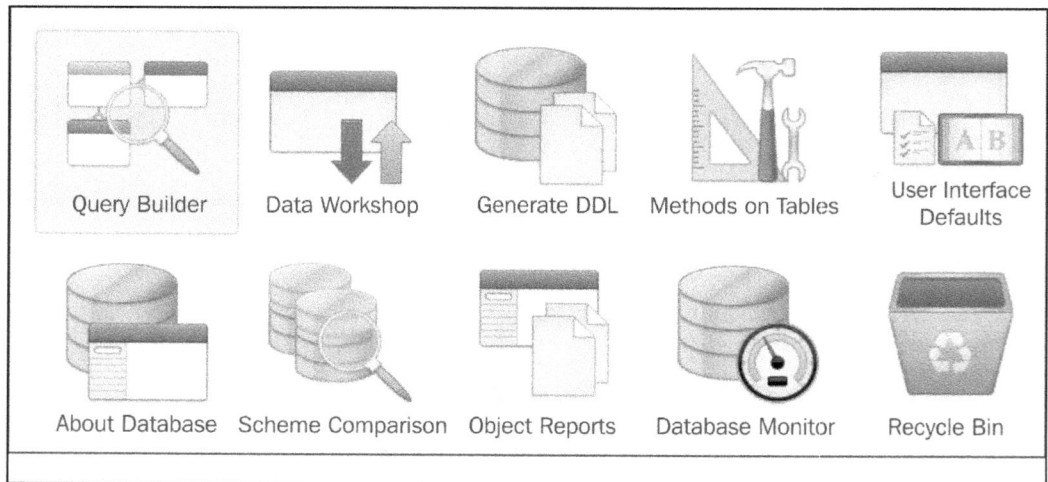

From here, click on **Data Workshop**, and it will bring up the page showing different options for **Data Load**. Select the **Spreadsheet** option, and it will bring you to the next screen.

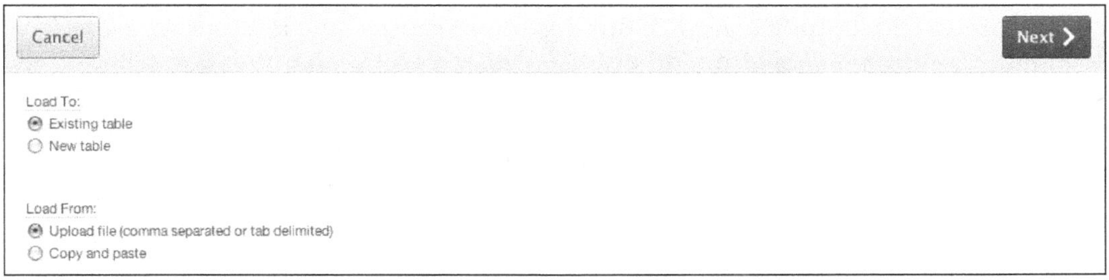

The next couple of screens take the user through the choice of the target table in the database and the source file to be used for the data. On the file upload screen, make sure to select the right option for the **First row contains column names** option, depending on the data in the CSV file. Once this is done, the following screen is displayed:

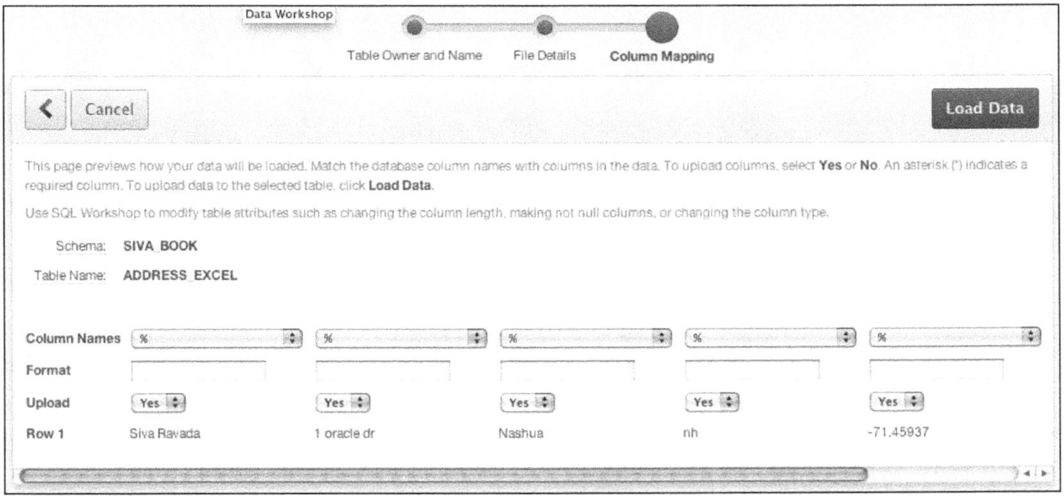

On the preceding screen, there are options to remap the columns in the CSV file to the columns defined in the table. If the order in the CSV file is different than what is defined in the table, they can be remapped at this stage. After this, click on the **load data** button, and the data will be loaded into the ADDRESS_EXCEL table.

Now, we have a table in the database with two number columns for storing the point data for each address. As the table does not have an SDO_GEOMETRY column, none of the spatial functions will work directly on this table. One option is to add an SDO_GEOMETRY column to this table, and populate that column based on the values from the longitude and latitude columns. But this might cause some issues if we want to load more data into a table from other CSV files. In practice, point data can be stored as two number columns, and can be spatially enabled using the function-based spatial index concept. This has the disadvantage of adding a little overhead (less than one percent) during queries although this is a very commonly used methodology. For this, we first define a function that takes two number values and returns an SDO_GEOMETRY object.

```
Create Or Replace Function get_Geom(longitude in number,
                                     latitude in number)
Return SDO_GEOMETRY DETERMINISTIC As
geom mdsys.SDO_GEOMETRY;
Begin
  -- assuming WGS 84 for the data
  -- notice that Longitude is X in Oracle Spatial
  geom := SDO_GEOMETRY(2001,8307,
         SDO_POINT_TYPE(longitude, latitude,null), NULL, NULL);
  Return geom;
End;
```

As this is longitude and latitude data, we construct a geometry with SRID 8307 which is a WGS 84 based Geodetic system. Then, we create the necessary metadata entry for the table and create the spatial index.

```
-- Create metadata
-- the Schema here is assumed to be BOOK
Insert Into USER_SDO_GEOM_METADATA VALUES(''ADDRESS_EXCEL'',
''BOOK.GET_GEOM(LONGITUDE,LATITUDE)'',
SDO_DIM_ARRAY(SDO_DIM_ELEMENT(''Longitude'', -180, 180, 0.05),
SDO_DIM_ELEMENT(''Latitude'', -90, 90, 0.05)), 8307);

-- Create the index
Create Index ADDRESS_EXCEL_SIDX on ADDRESS_EXCEL(BOOK.GET_
GEOM(LONGITUDE,LATITUDE))
indextype is mdsys.spatial_index;

-- run queries using the function-based index
Select  name, street, city, state
From address_excel a
```

```
Where SDO_ANYINTERACT(get_geom(longitude,latitude),
        SDO_GEOMETRY(2003,8307,NULL,
        SDO_ELEM_INFO_ARRAY(1,1003,3),
        SDO_ORDINATE_ARRAY(-122,37, -124, 38)) = ''TRUE'';
```

Note that the metadata entry has a different format than what we have seen before for the other tables in our schema. For the table name, we use the same format, but for the column name, we use the schema prefixed function name along with its parameters. This is the required format for the metadata entry for function-based spatial indexes to work. The `Create Index` statement also specifies the column name using the same format used for creating the metadata entry. After this, the table can be queried like any other indexed spatial table, but the column name is always referred to using the format used in the metadata.

CSV files as external tables

CSV files can also be loaded into the database using external tables. Oracle allows a user to access the external data by wrapping a table definition on top of the external data file, and using the SQL loader and DataPump technologies to convert the data on the fly. This creates a read-only table whose metadata is stored in the database, but whose data is stored outside the database. The table definition uses the physical properties clause to specify that the table be organized as an external structure. For example, the `Type` clause of `external_table_clause` allows the users to indicate the driver to be used to access the external table's data. The Oracle database provides two access drivers: `ORACLE_LOADER` and `ORACLE_DATAPUMP`.

The `ACCESS PARAMETERS` clause of `EXTERNAL_DATA_PROPERTIES` allows users to describe the physical layout of the file (the type of delimiter, number of fields, and so on). In the following example, we describe how to load the data from the CSV file that we used in our APEX example.

For this example to work, we first need to log in to the database using a DBA account, and create a directory object, and grant the required privileges to the BOOK user.

```
-- Create directory that points to actual data file
Create Or Replace DIRECTORY EXTERNAL_DATA_DIRECTORY As ''/tmp'';


-- If Create directory is done by DBA as SYSTEM then you would need to
-- grant the user doing the processing the ability to
-- read from that directory
GRANT READ,WRITE ON DIRECTORY EXTERNAL_DATA_DIRECTORY TO book;
```

Next we connect to the database as the BOOK user, and define the external table used to read the CSV file.

```
-- Create a new table to see the data from the CSV file
Create Table address_excel_external (
    name       Varchar2(100),
    street     Varchar2(200),
    city       Varchar2(100),
    state      Varchar2(2),
    longitude number,
    latitude  number)
ORGANIZATION EXTERNAL
(Type ORACLE_LOADER
DEFAULT DIRECTORY EXTERNAL_DATA_DIRECTORY
ACCESS PARAMETERS (
        RECORDS DELIMITED BY NEWLINE
        FIELDS TERMINATED BY '',''
        MISSING FIELD Values ARE NULL
        (name, street, city, state, longitude,latitude)
    )
LOCATION (EXTERNAL_DATA_DIRECTORY:''address.csv''))
REJECT LIMIT UNLIMITED;

Select * From address_excel_external;
```

The last SQL statement selects the rows from the table, and this statement should retrieve one row for each line in the CSV file. As this is an external table, it is a read-only table, and no data manipulation operations (delete or insert rows) are allowed on the table. Also no indexes can be created on the table. This is a different result than what we got while loading the CSV data into the database using APEX. We can get the same effect by creating a normal table, and inserting rows from this external table into the normal database table. In a workflow where new updates come in on a regular basis via CSV files, this external table setup is required once. After the initial setup, the normal database table is populated by inserting rows from the external table whenever a new file with the CSV data is collected.

Storage resolution versus resolution used by functions

When data is collected from external data sources, it is very important to understand different accuracy and storage characteristics of the data. Coordinates of the geometry are stored in VARRAYS of numbers in Oracle Spatial. The Number data type in the Oracle database stores fixed and floating-point numbers. Numbers of virtually any magnitude can be stored, with up to 38 digits of precision. This means() that any coordinate in the Oracle Spatial format can have up to 38 digits in decimal representation. This level of precision is not usually required for many applications, so it is important to understand how the number of digits used for each coordinate affects the storage and performance of the spatial applications.

Precision and accuracy

Oracle Spatial also has the concept of tolerance for geometry data. As described in *Chapter 1, Defining a Data Model for Spatial Data Storage*, tolerance is used to distinguish unique coordinates in geometry. But there is no direct relationship between the tolerance used in Oracle Spatial, and the number of digits used for each coordinate. The number of digits used for each coordinate reflects the accuracy of a single coordinate. The higher the number of digits used to represent a coordinate, the more accurate is its position with respect to the coordinate system. For example, a **Global Positioning System (GPS)** is used to measure the location of a spatial feature along the equator. The distance between two points along the equator that are one degree apart is about 70 miles. Suppose the coordinates measured by the GPS are represented as an integer pair (that is, only as the integer part of longitude and latitude). The measurement of the GPS is only accurate up to 70 miles, that is, the GPS measurements stored only as integers cannot distinguish between the two points that are 70 miles apart. If we add two digits after the decimal to the measurements, then the accuracy goes up to 0.7 miles. So, the higher the number of digits present in each coordinate, the more accurate the measurement is with respect to the actual position of the spatial feature. If a GPS can only measure with an accuracy of 1 mile, then there is no point in storing more than two digits after the decimal for each coordinate that is obtained using this GPS. Hence, it is important to understand how the data is collected when deciding on the number of digits to be used for each coordinate.

Storage cost for each coordinate

We first run some simple SQL statements to compute the storage cost for each number stored using the Oracle Number type. As Oracle numbers are stored as decimal numbers, the storage cost mainly depends on the number of digits in each number. For example, 123456 takes up as much space as 123.45 or 1.2345. In the Number format, one byte is used to store two digits of the number in addition to some header information that takes one byte. So, the approximate space used for each number is *(1 + (number of digits in the number))/2*.

```
SQL> Select dump(12345), vsize(12345) From dual;

DUMP(12345)                VSIZE(12345)
------------------------   ------------
Typ=2 Len=4: 195,2,24,46              4

Select dump(1234.5), vsize(1234.5) From dual;

DUMP(1234.5)               VSIZE(1234.5)
------------------------   -------------
Typ=2 Len=4: 194,13,35,51             4


SQL> Select dump(1.2345), vsize(1.2345) From dual;

DUMP(1.2345)               VSIZE(1.2345)
------------------------   -------------
Typ=2 Len=4: 193,2,24,46              4
```

When these numbers are stored in a VARRAY of numbers, there is an additional overhead for the headers of VARRAY. But, most of the storage cost is derived from the number of digits in each number. We now describe a function to compute the storage cost for the geometry in terms of number of bytes used to store all the numbers in the SDO_GEOMETRY object.

```
Create Or Replace function geometry_space(
 geometry IN SDO_GEOMETRY,
digits_after_decimal in Number default NULL)
Return Number is
tsize Number;
gsize Number;
Begin
```

```
If (geometry is NULL) Then
  Return 1;
End If;
Select vsize(geometry.sdo_gtype) Into gsize From dual;

If (geometry.sdo_srid is not NULL) Then
    Select vsize(geometry.sdo_srid) Into tsize From dual;
Else
  -- one byte cost if the attribute is NULL
  tsize := 1;
End If;
gsize := gsize+tsize;

-- if the POINT_TYPE is not NULL
If (geometry.sdo_point is NOT NULL) Then
  If (digits_after_decimal is NULL) Then
    Select nvl(vsize(geometry.sdo_point.X), 1)
                Into tsize From dual;
    gsize := tsize + gsize;
    Select nvl(vsize(geometry.sdo_point.Y), 1)
                Into tsize From dual;
    gsize := tsize + gsize;
    Select nvl(vsize(geometry.sdo_point.Z), 1)
                Into tsize From dual;
    gsize := tsize + gsize;
  Else
    Select nvl(vsize(round(geometry.sdo_point.X, digits_after_
decimal)), 1)
            Into tsize From dual;
    gsize := tsize + gsize;
    Select nvl(vsize(round(geometry.sdo_point.Y, digits_after_
decimal)), 1)
            Into tsize From dual;
    gsize := tsize + gsize;
    Select nvl(vsize(round(geometry.sdo_point.Z, digits_after_
decimal)), 1)
            Into tsize From dual;
    gsize := tsize + gsize;
  End If;

End If;
```

```
   -- if Elem Info is null, assume the ordinates are also NULL
   If (geometry.sdo_elem_info is NULL) Then
      Return gsize/1000/1000;
   End If;


   Select sum(vsize(column_value))
    Into tsize
   From Table(geometry.sdo_elem_info) a;
   gsize := gsize+tsize;

   If (digits_after_decimal is NULL) Then
      Select sum(vsize(a.X)+vsize(a.Y)+nvl(vsize(a.Z),0)+nvl(vsize(a
.W),0))
      Into tsize
      From Table (SDO_UTIL.GetVertices(geometry)) a;
   Else
      Select sum(vsize(round(a.X, digits_after_decimal))+
                 vsize(round(a.Y, digits_after_decimal))+
                 nvl(vsize(round(a.Z, digits_after_decimal)),0)+
                 nvl(vsize(round(a.W, digits_after_decimal)),0))
      Into tsize
      From Table (SDO_UTIL.GetVertices(geometry)) a;
   End If;

   gsize := gsize+tsize;
   Return gsize/1000/1000;

End;


Select sum(geometry_space(geom))
From land_parcels;

SUM(GEOMETRY_SPACE(GEOM))
-------------------------
               27.545218
```

This function computes the storage cost for a given storage, based on the number of digits used in each coordinate and the element information arrays: GTYPE and SRID. Using this function, we can see that the geometry data in the LAND_PARCELS table takes up 27.5 MB of space.

The LAND_PARCELS table has data in the **California State Plane Coordinate System** with the unit of measurement in feet. So, storing each coordinate with more than three digits after the decimal point() (that is, a thousandth of a foot) is not going to help much with the accuracy of the measurements. So we can check how much space we can save if we only use three digits after the decimal for each coordinate by calling the above function with a parameter value of 3. As you can see from the result, the storage space would be reduced by about 30 percent for the geometry if we reduce the number of digits after the decimal to 3. In the following chapters, we will also show a function that can be used to reduce the number of digits after the decimal.

```
Select sum(geometry_space(geom, 3))
From land_parcels;

SUM(GEOMETRY_SPACE(GEOM))
-------------------------
                21.582843
```

Storage cost for Oracle Spatial tables

When the geometry is actually stored in a table, the storage cost will be slightly different from the cost computed by this function. This is due to the overhead associated with the storage of VARRAYs and the storage model used for the VARRAYs. When a VARRAY is less than 4 KB in size, it is stored inline in the Oracle row storage. If the VARRAY is more than 4 KB in size, the data is stored as an out of line **Large Object (LOB)**. The actual storage for this comes from the space allocated for storing LOB segments. Depending on how large the VARRAYs are, the storage cost for each geometry varies when the geometry is stored in a table. Next, we describe a function that can be used to compute the storage space used by a table storing SDO_GEOMETRY data. This is done by the actual number of bytes allocated for storing the row data of a given table.

```
Create Or Replace Function geom_table_space_mbytes(tname in Varchar2)
Return Number Is
tsize number;
Begin

  Select sum(bytes) Into tsize
  From
  (Select bytes
   From user_segments
   Where  segment_name = tname
   UNION ALL
```

```
      Select bytes
      From user_lobs l,
           user_segments s
      Where l.TABLE_NAME = tname
        and s.segment_name = l.segment_name);

    tsize := tsize/1000/1000;
    Return tsize;
  End;



  Select geom_table_space_mbytes(''LAND_PARCELS'')
  From dual;

  GEOM_TABLE_SPACE_MBYTES(''LAND_PARCELS'')
  ---------------------------------------
                            129.10592
```

Effect of tolerance on performance of Oracle Spatial functions

In general, tolerance does not have much impact on the performance of most spatial operations. But in some cases, unnecessarily small values of tolerance can cause severe performance problems. As mentioned in *Chapter 1, Defining a Data Model for Spatial Data Storage*, tolerance is used to distinguish between duplicate vertices in geometries and decide when the vertices snap to the edges when they are close to each other. When comparing two geometries (as in doing a relate operation or a union/intersection operation), tolerance plays an important role in deciding the number of edges and vertices in the topology that is computed for the two input geometries. If the tolerance is unnecessarily small, two vertices that can be snapped into a single vertex will remain as distinct vertices. This can increase the size of the topology that is computed, resulting in performance degradation. When the two input geometries are not well aligned with each other on the boundary, the neighboring line segments from each geometry might be nearly parallel to each other or nearly coincident with each other. If the tolerance is unnecessarily small, instead of treating two edges as the same or as parallel edges, Oracle Spatial might compute several intersection points that are not required in the resulting geometry.

This is illustrated in the following figure with two polygons that are next to each other. If both the polygons are valid at five centimter tolerance, then an SDO_UNION operation between these two polygons with a tolerance of five centimter will produce a shape that has no gaps between the two input geometries, that is, the resulting geometry will be a single polygon with one ring. This also means that the number of intersection points created while doing the SDO_UNION operation is also reduced as many of the line segments along the shared boundary end up being parallel to each other with a five centimter tolerance. On the other hand, an SDO_UNION operation with a one centimter tolerance can produce a resulting shape that has holes along the shared boundary. This example also shows another hidden capability of Oracle Spatial; that is, the ability to snap geometries together even when their boundaries do not align well with each other. For example, consider two input geometries that are next to each other and are valid at a tolerance of one centimter. However, there are some parts of the geometry that are more than one centimter away from each other along the shared boundary. If the greatest distance between the two geometries is five centimter along the shared boundary, then an SDO_UNION operation with a tolerance of five centimter can snap the shared boundary together. Note that this snapping result can be obtained only if both the polygons are valid at a higher tolerance of five centimter which is used for the SDO_UNION operation.

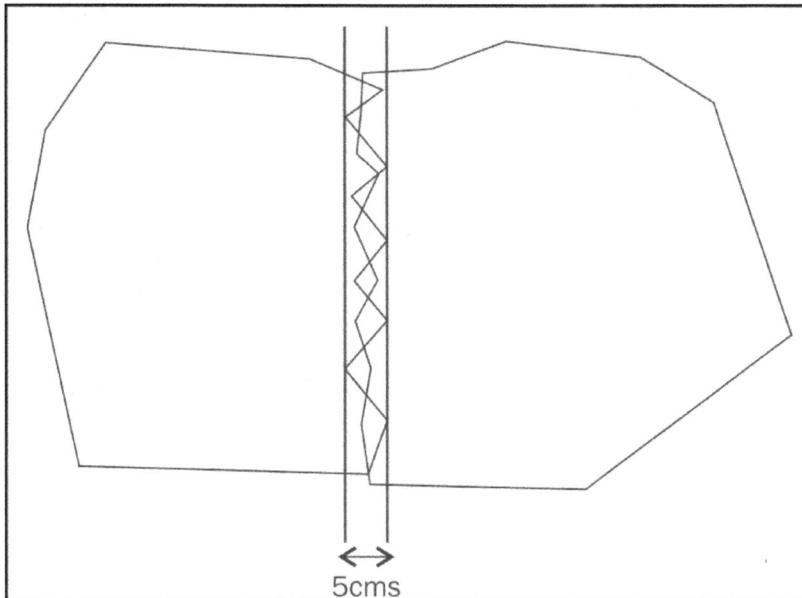

5cms

Creating spatial autocorrelation via clustering

When spatial data is loaded into a table, data is usually organized depending on the order of the incoming rows. This ordering of the data on disk has a direct impact on the performance of the spatial queries. Consider a simple SDO_ANYINTERACT query that retrieves all the data inside a rectangular box. After the spatial query is performed, the rows that satisfy the result are retrieved from the table. If all of these rows are spread over different data blocks, the cost of the query increases, as many blocks have to be fetched to form the result set. If data in the blocks can be organized in such a way to minimize the number of blocks fetched for each query, the query performance would improve. This improvement will be greater for queries that fetch a large number of rows for each query. For example, in web mapping applications, small scale maps show less detail and large scale maps show more detail. As the scale goes from small to large, more and more layers of data are added to the map. Consider the layer that stores local roads. This table typically has a large number of rows in urban areas. In medium scale maps when this layer first starts appearing in the map, a web mapping application query will cover a large area resulting in the fetching of a large number of rows from this local road's table. If the number of rows fetched for each query is large (in the order of tens of thousands of rows), and the rows are widely spread out on the disk, the I/O cost for each query would be high. Spatial clustering of data minimizes this query overhead. The idea here is to try and place the spatial data that is close together in space (called spatial autocorrelation) and close together on the disk.

We want to store the data in the table such that the data is sorted using a spatial key, and stored in the table using this sorted order. Note that if the data is one-dimensional, such as numbers or characters, a simple sorting of the data gives a linear order that can be used to cluster the data in the table. As spatial data is multi-dimensional, a simple sort of data does not give spatial clustering of the data.

The **space-filling curve** concept is designed to address this clustering problem for multi-dimensional data.

Space-filling curves

In mathematical analysis, a space-filling curve is a curve whose range contains the entire two-dimensional unit square. In practical terms, this is a curve that connects every single point in a gridded two-dimensional space. Imagine a gridded two-dimensional space with equally spaced grids along X and Y. A space-filling curve will pass through each of these grids without any self-intersections. If we start at one end of the curve and move along the curve, the path will touch each of the squares once and only once. This gives us a linear ordering of all the squares in the two-dimensional space, and ordering can then be used to sort the squares in the two-dimensional space.

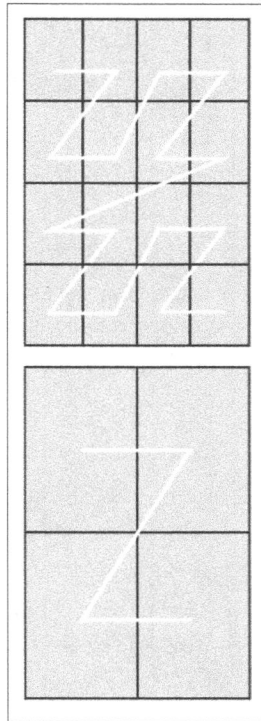

This space-filling curve concept can be used to linearly order any arbitrary shapes in a two-dimensional space. For this process to work, we need to know the full extent of the space taken by the spatial objects. We take the target two-dimensional space, and decide on the size of the grid cell. The smaller the grid cell, the better the linear ordering of the data in that space. Note that an arbitrary shape such as a line or a polygon can span multiple cells in the grid space. So for these shapes, we first reduce them to a point by using the centroid or some other representative point corresponding to the original shape. Once we have reduced the candidate data set to points in the two-dimensional space, each point can only be in one grid cell.

We then find the linear ordering of the grid cells and associate that order with the corresponding two-dimensional shape. These shapes are then sorted and stored using this linear order to improve the clustering of those close to each other in the two-dimensional space, so that they can be close to each other even on disk when storing them in the database tables.

In our examples, we will use Morton code, as the space-filling curve used in the program to generate the codes is very simple and efficient to implement in PL/SQL.

```
Create Or Replace Function Morton (p_col Natural, p_row Natural)
Return INTEGER
As
     v_row          Natural := ABS(p_row);
     v_col          Natural := ABS(p_col);
     v_key          Natural := 0;
     v_level        BINARY_INTEGER := 0;
     v_left_bit     BINARY_INTEGER;
     v_right_bit    BINARY_INTEGER;
     v_quadrant     BINARY_INTEGER;

     /* sub-routine to left shift the bits of a number */
     Function Left_Shift( p_val Natural, p_shift Natural)
     Return PLS_Integer
     As
     Begin
        Return trunc(p_val * power(2,p_shift));
     End;

Begin
     while ((v_row>0) Or (v_col>0)) Loop
        /* split off the row (left_bit) and column (right_bit) bits and
then combine them to form a bit-pair representing the quadrant */
        v_left_bit   := MOD(v_row,2);
        v_right_bit  := MOD(v_col,2);
        v_quadrant   := v_right_bit + ( 2 * v_left_bit );
        v_key        := v_key + Left_Shift( v_quadrant,( 2 * v_level ));
        /* row, column, and level are then modified before the loop
continues */
        v_row := trunc( v_row / 2 );
        v_col := trunc( v_col / 2 );
        v_level := v_level + 1;
      End Loop;
     Return (v_key);
End Morton;
```

The preceding function takes a Y value and an X value and returns the corresponding Morton code. To explain how this function works, we first create a test table and populate it with point data. Using this data, we will show how to generate the Morton code, and explain how to determine the number of cells in the grid space. As mentioned in the preceding code, the higher the number of cells in the space, the better the clustering will be for the data stored in the tables.

```
Drop    Table MORTONED_P;
Drop    Table MORTON_P;
Create Table MORTON_P (
   id Integer,
   morton_key Integer,
   geom mdsys.SDO_GEOMETRY );

-- Now Create the points and write then to the table
--
Insert Into MORTON_P
Select rownum, NULL,
mdsys.SDO_GEOMETRY(2001,NULL,
      MDSYS.SDO_POINT_TYPE(
            ROUND(dbms_random.value(350000  - ( 10000 / 2 ),
                                    350000  + ( 10000 / 2 )),2),
            ROUND(dbms_random.value(5400000 - ( 10000 / 2 ),
                                    5400000 + ( 10000 / 2 )),2),
                  NULL), NULL,NULL)
      From DUAL
   CONNECT BY LEVEL <= 1000;
Commit;
```

The preceding SQL statement generates data in the range 345000-355000 along X and 5395000-5405000 along Y in the two-dimensional space. Note the use of DUAL and CONNECT by clause to generate the required number of rows (1000 in this case) from a single SQL statement. This is a very useful trick when generating rows using a single SQL statement and DUAL.

Now let us see how we can divide this space into a 20 x 20 grid. Note that the range for X as well as Y is 10,000. If we want to divide this space into 20 cells along X and Y, we divide the numbers in the space by 500. We can run the following SQL statement to see the number of grid cells that we get with this approach. If we want to divide the space into 200 cells, we divide the X and Y values by 50 instead of 500.

```
Select min(morton_gx), max(morton_gx),
       min(morton_gy), max(morton_gy)
From (
Select id,
        a.geom.sdo_point.y As y,
        a.geom.sdo_point.x As x,
        FLOOR(a.geom.sdo_point.y/500) - MIN(FLOOR(a.geom.sdo_
point.y/500)) OVER (Order By FLOOR(a.geom.sdo_point.y/500)) As
morton_gx,
        FLOOR(a.geom.sdo_point.x/500) - MIN(FLOOR(a.geom.sdo_
point.x/500)) OVER (Order By FLOOR(a.geom.sdo_point.x/500)) As
morton_gy
    From MORTON_P a );

MIN(MORTON_GX) MAX(MORTON_GX) MIN(MORTON_GY) MAX(MORTON_GY)
-------------- -------------- -------------- --------------
             0             19              0             19
```

Next, we use the Morton function to create the key for all the rows in the MORTON_P table, and then create a new table that has the rows ordered based on the Morton key.

```
Update MORTON_P A
    SET a.morton_key = Morton(
    (Select y From
       (Select id, FLOOR(b.geom.sdo_point.y/500) - MIN(FLOOR(b.geom.
sdo_point.y/500)) OVER (Order By FLOOR(b.geom.sdo_point.y/500)) y From
Morton_P b) M Where m.id = a.ID ),
    (Select x From
       (Select id, FLOOR(c.geom.sdo_point.x/500) - MIN(FLOOR(c.geom.
sdo_point.x/500)) OVER (Order By FLOOR(c.geom.sdo_point.x/500)) x From
Morton_P c) M Where m.id = a.ID ));


 Create Table mortoned_p
 as
 Select rownum As id, morton_key, geom
  From (Select morton_key, geom
          From morton_p mp
         Order By mp.morton_key desc
     );
```

The above process has some extra steps to create the Morton key column in the base table that is not really necessary in many cases. So we can directly create the new table using this SQL:

```
Drop Table mortoned_p;
Create Table mortoned_p As Select * From (
with min_x As (Select min(floor(a.geom.sdo_point.x/500)) x From
morton_p a),
     min_y As (Select min(floor(a.geom.sdo_point.y/500)) y From
morton_p a)
Select id,  morton(
    (FLOOR(a.geom.sdo_point.y/500) - (Select y From min_y)),
    (FLOOR(a.geom.sdo_point.x/500) - (Select x From min_x))) morton_
key, geom
From morton_p a) Order By morton_key;
```

Next, we will show how this can be used to order the data in one of the tables in our sample schema that has polygon data. The LAND_PARCELS data has polygon features, so we will use a simple function to find a representative point for the polygon. We can then use that point to cluster the data using the Morton function we have created. The data in the LAND_PARCELS table spans a square of 60,000 x 60,000 units with extents of (5970000, 6030000) along the X axis and (2080000, 2140000) along the Y axis. We can divide this space into 36 tiles for the Morton key generation. The following SQL shows the steps required to create a new table, LAND_PARCELS_ CLUSTERED, that has the data clustered by the Morton key values:

```
Create Table LAND_PARCELS_CLUSTERED As Select * From (
Select FID, MAPBLKLOT, BLKLOT, BLOCK_NUM, LOT_NUM, FROM_ST, TO_ST,
       STREET, ST_TYPE, ODD_EVEN, GEOM,
       Morton ( FLOOR((a.g.sdo_point.y -2080000 )/10000),
                FLOOR((a.g.sdo_point.x - 5970000)/10000) ) MORTON_KEY
   From
    (Select c.*, sdo_geom.sdo_pointonsurface(geom, 0.005) g
     From LAND_PARCELS c) a
     Order By morton_key );
```

Geometry validation and methods to clean imported data

When data is loaded from external data sources, it is often possible that the spatial data is invalid based on the validation rules in Oracle Spatial. Sometimes these errors are due to variations in how the validation rules are defined between different systems. For example, in Oracle Spatial, polygons should be ordered counter-clockwise for exterior rings and clockwise for interior rings. Such errors can be easily fixed using the utility functions provided by Oracle Spatial. Sometimes these errors are due to inherently bad data that has not been fixed or validated. Such errors are harder to fix, as there might not be a way to fix the geometries without drastically modifying the data. In such cases, users need to manually inspect the data and decide on the appropriate fix.

The SDO_GEOM.Validate_Geometry_With_Context function gives very specific information about what is wrong with the invalid geometries. The errors associated with the orientation of polygon rings 13366 (invalid combination of interior/exterior rings), 13367 (wrong orientation for interior/exterior rings), and 13368 (simple polygon has more than one exterior ring) can be fixed with the SDO_UTIL.Rectify_Geometry function. This function will check for the orientation of each ring in the polygon and rearrange them to make it a valid geometry. The error 13356 associated with the duplicate vertices can happen in two cases. In the first case, two consecutive vertices are duplicates of each other as they have exactly the same coordinates. In the second case, two vertices are so close to each other that the distance between then is less than the value of the tolerance. In both cases, SDO_UTIL.Rectify_Geometry can fix the geometry by removing these duplicate vertices. But in the second case, removing the duplicate vertices changes the shape of the geometry by moving some edges. And when this happens, the resulting geometry might become invalid as an edge might have moved too close to some other vertex. Some geometry might be invalid with 13356, and when the duplicate vertices are removed, now it might be invalid with the error 13349 (edges cross other edges).

The 13349 errors can happen in two distinct cases. In the first case, two edges of a polygonal ring cross each other as shown in the following figure. In the second case, two consecutive edges of a geometry form a very narrow angle such that the distance of the end points from an edge to the other edge is less than the tolerance value. In the first case, the geometry is fixed by splitting the ring into two rings at the intersection point of the two crossing edges. In the second case, one of the two edges is snapped to the other edge leaving a dangling edge. If the geometry is a polygonal geometry, this dangling edge is removed from the final geometry producing a valid polygonal geometry. The SDO_UTIL.Rectify function can fix both these types of 13349 errors.

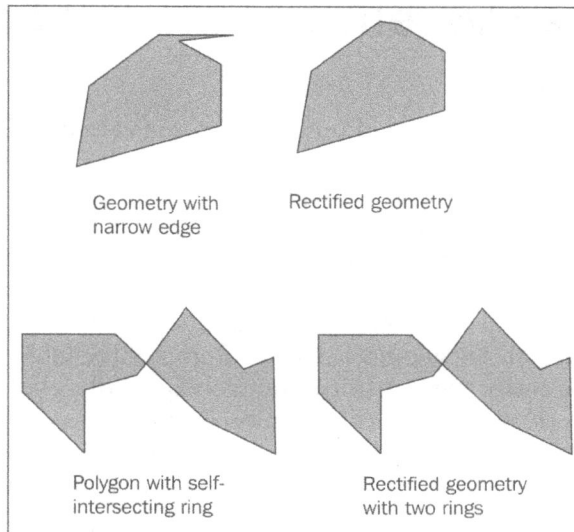

Geometry with narrow edge

Rectified geometry

Polygon with self-intersecting ring

Rectified geometry with two rings

Two other types of common errors are 13350 and 13351. Both these errors happen only for polygons with multiple rings. The 13350 errors happen when two inner rings, or two exterior rings share a part or all of an edge. Dissolving the shared edge and combining the two rings into one ring fixes these errors. That is, the two inner rings will be combined into a single inner ring, or two exterior rings will be combined into a single exterior ring. The 13351 errors happen when an inner ring shares an edge with an exterior ring (either from the same polygon element or from a different polygon element). In many of these cases with 13351 error cases, there is no single way to fix the invalid geometry. The figure shows the case with 13351 error and the two different ways to fix this problem:

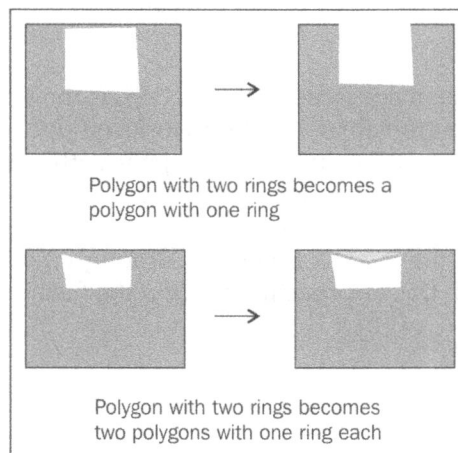

Polygon with two rings becomes a polygon with one ring

Polygon with two rings becomes two polygons with one ring each

Coordinate system transformation techniques

As we have seen in the first section of this chapter, it is very common to do coordinate system transformations as part of the data loading process. Many of the external data loading tools provide these transformations, but sometimes it is convenient and more efficient to do these transformations in the database. Oracle Spatial provides two distinct methods for doing coordinate system transformations. The SDO_CS.TRANSFORM function works on one row at a time to transform a geometry, while the SDO_CS.TRANSFORM_LAYER function works on a whole table of data and transforms all the geometries in a column of the table. Both methods support different use cases: the transform function is mainly used in dynamic query situations, while the TRANSFORM_LAYER function is mainly used in bulk update or bulk loading situations. As we are describing data loading techniques in this chapter, we will describe how both these methods can be used in bulk data loading situations.

For this example, we use the ROAD_CLINES table. Assume that we are bulk loading road data into this table, but the input data has SRID 2872, but we want to have the data in the WGS 84 Geodetic coordinate system (with SRID 8307). Let's assume that the data is loaded into the ROAD_CLINES table using an ETL tool, but without any coordinate system transformation. We will then move this data into a new table called ROAD_CLINES_GEO where the geometries are stored with SRID 8307. This can be achieved with a single SQL statement as follows:

```
Create Table road_clines_geo  As Select
a.fid, a.road_id, a.lf_fadd, a.lf_toadd, a.rt_fadd, a.rt_toadd,
a.cnn, a.f_node_cnn, a.t_node_cnn, a.zip_code, a.district,
a.accepted, a.jurisdiction, nhood, a.cnntext, a.streetname,
a.classcode, a.street_gc,
sdo_cs.transform(a.geom, 8307) geod_geom
From road_clines a;
```

While this works well, one major issue with this SQL is the performance of the transform function. As the transform function works on one row at a time, the transformation context is built for each row independently, thus increasing the Create Table statement's execution cost. If the table has thousands of rows, this cost can easily add up to a very large value. The alternative way to achieve this same result is to use the TRANSFORM_LAYER function. Although it takes multiple steps to get the desired result with this approach, it is often the faster way to get the result.

```
-- first create the new table as Select * From the old table
Create Table road_clines_geo As Select * From road_clines;

-- create the metadata entry for this newly created table
-- use the SRID 2872 as that is the current SRID for this table

Insert Into user_sdo_geom_metadata Values (''ROAD_CLINES_GEODETIC'',
''GEOM'',
SDO_DIM_ARRAY(SDO_DIM_ELEMENT(''X'', 5900000, 6100000, .05),
SDO_DIM_ELEMENT(''Y'', 2000000, 2200000, .05)), 2872 );

-- execute the transform_layer call
EXEC sdo_cs.transform_layer(''ROAD_CLINES_GEODETIC'', ''GEOM'',
''ROAD_CLINES_CS'', 8307);

-- create an index the sdo_rowid column of the result table
Create index rid_idx on road_clines_cs(sdo_rowid);

-- now update the new table with the transformed geometries
Update road_clines_geo a set geom =
   (Select geometry From road_clines_cs b Where b.sdo_rowid = a.rowid);

-- update the metadata to reflect the new SRID
Update user_sdo_geom_metadata set diminfo =
SDO_DIM_ARRAY(SDO_DIM_ELEMENT(''Longitude'',-122.51471, -122.35631,
.05),
SDO_DIM_ELEMENT(''Latitude'', 37.7079188, 37.8332976, .05)),
 srid = 8307
Where table_name = ''ROAD_CLINES_GEODETIC'';
```

The transform layer call creates a result table with the columns: GEOMETRY (to store the transformed geometry), and SDO_ROWID (stores the row ID from the source table). This SDO_ROWID column can then be used to join back to the source table when we want to update the source table with the transformed geometry data. For this we create a **B-tree index** on the SDO_ROWID column to help in improving the performance of the Update statement on the source table. In the end, we update the metadata to reflect the new SRID for the data in the source table.

Spatial indexing

Spatial indexing is one of the most important concepts in Oracle Spatial. Using the index can dramatically reduce the query cost for many operations. As Oracle uses an R-tree index, there are actually no maintenance operations required once the index is built. However, there are several important considerations while building the index to get the best possible performance for queries. In the following section, we describe some of the most commonly used parameters in the Create Index statement.

Layer GTYPE for point data

As we mentioned in *Chapter 1, Defining a Data Model for Spatial Data Storage*, the LAYER_GTYPE parameter can be used to enforce type consistency across all the rows in the table. For point data, this parameter can also be used to improve the performance for spatial index-based queries against these tables. When an operator like SDO_RELATE is evaluated, the R-tree index is used to do the primary filtering. In some cases, the index nodes can be used to perform secondary filtering as well; for example, when the index node is completely inside the query polygon, the data entries in those index nodes will also be completely inside the query polygon. In many cases, the actual data geometry has to be fetched from the table to perform the full geometry-to-geometry comparison with the query geometry. This means that the data geometry has to be fetched from the rows of the table. And as we have discussed before, if the data is not clustered based on the spatial location, this can result in many random I/O operations. For point data, we can avoid this problem by specifying the LAYER_GTYPE parameter. With this parameter specified, Spatial knows that the data geometry is a point, which means the index entry is the exact copy of the data geometry, so there is no need to fetch the data geometry from the base table. Note that this optimization only works for point data, and for all other cases the data geometry has to be fetched from the base table. The following SQL shows how to use the LAYER_GTYPE parameter for a table with point data:

```
Create Index BASE_ADDRESSES_SIDX ON BASE_ADDRESSES (GEOM)
INDEXTYPE IS MDSYS.SPATIAL_INDEX PARAMETERS
(''LAYER_GTYPE=POINT'');
```

The Work tablespace

During the R-tree index creation process, many temporary tables called work tables are created to store temporary results. As many tables are created and deleted, it will end up defragmenting the tablespace used for these tables. For this reason, Oracle Spatial allows the user to specify a different tablespace to create these worktables. This is specified as a parameter in the `Create Index` statement as shown in the following SQL:

```
Drop Index ROAD_CLINES_SIDX;
Create Index ROAD_CLINES_SIDX on ROAD_CLINES(GEOM)
indextype is mdsys.spatial_index
parameters (''work_tablespace=USERS'');
```

DML batch size

The `SDO_DML_BATCH_SIZE` parameter can improve application performance, because Spatial can pre-allocate the system resources to perform multiple index updates more efficiently than successive single index updates. However, to gain the performance benefit, users must not perform commit operations after each insert operation, or at intervals less than or equal to the `SDO_DML_BATCH_SIZE` value. It is generally not recommended to use a value greater than 10,000 (ten thousand) for the batch size, as the cost of the additional memory and other resources required will probably outweigh any marginal performance increase resulting from such a value.

When spatial tables are updated, the corresponding spatial index is also updated. Updating the spatial index requires updates to the R-tree index leaf nodes. Sometimes these changes can propagate to the root of the tree. Each update to the index node will also result in redo generation as the R-tree index is stored in a normal Oracle table. Thus, if these changes to the spatial table are propagated to the index for every single row, it is very likely the same index nodes get updated multiple times. If these changes to the index can be done in batches, the number of updates to the index can be substantially reduced. Oracle Spatial uses a temporary table to log the changes to the spatial table until the number of updates reaches the number specified by the `SDO_DML_BATCH_SIZE` parameter. Once this number is reached, the index is updated using a batch update algorithm.

Non-leaf index table

The R-tree index is a tree structure with leaf and non-leaf nodes. Performance of the spatial queries will greatly improve if the whole R-tree index table can be cached in memory. But in practice, systems usually don't have enough resources to cache the whole index table of the spatial index. The same higher-level nodes of the index (non-leaf) are accessed more often than the lower-level leaf nodes of the index. So if these non-leaf level nodes that tend to be smaller in number than the leaf-level nodes are cached, spatial queries can still benefit. For this reason, Oracle Spatial provides a mechanism to split the index table into two tables. Specifying SDO_NON_LEAF_TABLE=TRUE creates two index tables: one table with the MDNT prefix to store the non-leaf level nodes and one table with the MDRT prefix to store the leaf-level nodes. In this case, users must also cause Oracle to buffer the MDNT_...$ table in the KEEP buffer pool, for example, by using Alter Table and specifying STORAGE (BUFFER_POOL KEEP). The following SQL example shows how to create the index with the separate non-leaf index table:

```
Create Index land_parcels_sidx on land_parcels(geom)
indextype is mdsys.spatial_index
parameters(''sdo_non_leaf_tbl=true'');

-- find the MDNT table for the index
Select sdo_index_table, replace(sdo_index_table, ''MDRT'', ''MDNT'')
From user_sdo_index_info
Where index_name = ''LAND_PARCELS_SIDX'';

-- alter the table to specify buffer pool keep
Alter Table MDNT_165DB$ storage (buffer_pool keep);
```

Exporting formats – GML, WKT, and GeoJSON

Oracle Spatial provides different types of converters to convert the geometry data to GML (both Version 2.1 and Version 3.1.1), KML, and the well-known text and binary (WKT and WKB) representations. These converters are provided as PL/SQL functions and Java APIs. In this section, we will show examples of using these converters to generate GML and KML documents. The WKT and WKB examples will be very similar and are left as an exercise for the reader.

We first look at the KML converter using the following SQL example:

```
Select SDO_UTIL.To_Kmlgeometry(geom) gml_feature
From land_parcels Where fid= 16;

GML_FEATURE
-------------------------------------------------------------------
<Polygon><extrude>0</extrude><tessellate>0</tessellate><altitudeMo
de>relativeToGround</altitudeMode><outerBoundaryIs><LinearRing>
<coordinates>6006491.42346619,2121789.41749893
6006423.35503277,2121779.7587256 6006442.27297394,2121643.56575651
6006510.34140736,2121653.22420177
6006491.42346619,2121789.41749893
</coordinates></LinearRing></outerBoundaryIs></Polygon>
```

Similarly, we can use the GML converter to generate the GML representation for this geometry.

```
Select SDO_UTIL.to_gmlgeometry(geom) gml_feature
From land_parcels Where fid=16;

GML_FEATURE
-------------------------------------------------------------------
<gml:Polygon srsName=""SDO:2872""
xmlns:gml=""http://www.opengis.net/gml""><gml:outerBoundaryIs><gml
:LinearRing> <gml:coordinates decimal=""."" cs="","" ts=""
"">6006491.42346619,2121789.41749893
6006423.35503277,2121779.7587256 6006442.27297394,2121643.56575651
6006510.34140736,2121653.22420177
6006491.42346619,2121789.41749893
</gml:coordinates></gml:LinearRing></gml:outerBoundaryIs></gml:Pol
ygon>
```

In many applications, it is useful to generate the XML representation for the whole row instead of just the geometry column. In Oracle, XDB provides many useful functions to convert the row data into an equivalent XML representation. The following example shows the conversion of a row of the LAND_PARCELS data to the following XML format:

```
Select ''<LAND_PARCEL_FEATURE>'' ||
    XMLFOREST(fid, mapblklot, blklot, block_num, lot_num,
From_st, to_ST,
       street, st_type, odd_even ) ||
SDO_UTIL.to_gmlgeometry(geom) ||
    ''</LAND_PARCEL_FEATURE>'' land_parcel_feature
LAND_PARCEL_FEATURE
-------------------------------------------------------------
<LAND_PARCEL> <FID>16</FID><MAPBLKLOT>0025012</
MAPBLKLOT><BLKLOT>0025012</BLKLOT
> <BLOCK_NUM>0025</BLOCK_NUM><LOT_NUM>012</LOT_NUM><FROM_ST>799
</FROM_ST><TO_ST>799</TO_ST><STREET>BEACH</STREET><ST_TYPE>ST</ST_
TYPE><ODD_EVEN>O</ODD_EVEN><gml:Polygon srsName=""SDO:2872""
xmlns:gml=""http://www.opengis.net/gml""><gml:outerBoundaryIs><gml
:LinearRing> <gml:coordinates decimal="".""  cs="","" ts="" "">
6006491.42346619,2121789.417498936006423.35503277, 2121779.7587256
6006442.27297394,2121643.56575651 6006510.34140736,
2121653.22420177 6006491.42346619,2121789.41749893
</gml:coordinates></gml:LinearRing></gml:outerBoundaryIs>
</gml:Polygon> </LAND_PARCEL>
```

We can use the XMLFOREST function to convert all the columns of the table into XML format, and generate one XML document for each row. If we want to generate a single document for a set of rows, we can use the XMLAGG function to combine the rows to form a single XML document.

Generating a GeoJSON format

JavaScript Object Notation (JSON) is a lightweight data interchange format used in many web applications. GeoJSON is an extension to represent the spatial data, and is one of the most popular formats for JavaScript-based mapping applications. This format can be used for encoding a variety of geographic data structures. An object in this format may represent some geometry, a feature, or a collection of features. GeoJSON supports the following geometry types: Point, linestring, Polygon, MultiPoint, MultiLineString, MultiPolygon, and GeometryCollection. Features in GeoJSON contain a geometry object and additional properties, and a feature collection represents a list of features. In this section, we describe a procedure for generating the GeoJSON format from the SDO_GEOMETRY data. As the procedure for converting the geometry to GeoJSON is fairly long, we will not list the code in this book, but it can be accessed from the sample code website used for this book. This procedure called sdo2geojson can be used to convert each instance of the geometry to the GeoJSON format, as shown in the following example:

```
Select sdo2geojson(geom) From land_parcels Where fid = 16;

SDO2GEOJSON(GEOM)
----------------------------------------------------------------
{""type"":""Polygon"",""coordinates"":[[[6006491.42,2121789.42],[60064
23.36,2121779.76],[6006442.27,2121643.57],[6006510.34,2121653.
22],[6006491.42,2121789.42]]]}
```

Sometimes, it is useful to take a set of rows from a table and generate one single GeoJSON document containing all the features. The following procedure can be used to take a ref cursor as input and generate a GeoJSON document containing all the geometries selected by the ref cursor. Using a ref cursor as an input argument simplifies the coding of these functions, as we don't have to worry about constructing the SQL statements from the user input. Depending on the complexity of the predicates in the where clause, constructing SQL statements can be very cumbersome. Using a ref cursor shifts this burden to the user, and the function developer can focus on the main logic and correctness of the function.

```
Create Or Replace Function sdo2geojson_document(r_cur SYS_REFCURSOR)
Return CLOB is
geom MDSYS.SDO_GEOMETRY;
js_geom clob;
j_geom clob;
start_tag Varchar2(100);
end_tag   Varchar2(100);
Begin
  DBMS_LOB.createtemporary (lob_loc => js_geom, cache => TRUE);
  start_tag := ''{ ""type"": ""FeatureCollection"", ""features"": ['';
  DBMS_LOB.write(lob_loc => js_geom, amount => LENGTH(start_tag),
                   offset => DBMS_LOB.GETLENGTH(js_geom)+1,
                   buffer => start_tag );

  Loop
    FETCH r_cur Into geom;
    EXIT When r_cur%NOTFOUND;
    j_geom := sdo2geojson(geom);

    DBMS_LOB.write(lob_loc => js_geom, amount => LENGTH (j_geom),
                     offset => DBMS_LOB.GETLENGTH(js_geom)+1,
                     buffer => j_geom );
  End Loop;
  end_tag := ''] }'';
  DBMS_LOB.write(lob_loc => js_geom, amount => LENGTH (end_tag),
                   offset => DBMS_LOB.GETLENGTH(js_geom)+1,
                   buffer => end_tag );

  Return js_geom;
End;
```

A user can now invoke this function to create a CLOB data type with the GeoJSON data for a set of rows from any table in the database. The following code example shows the usage of this function:

```
Declare
cur SYS_REFCURSOR;
json_geom clob;
Begin
    open cur for ''Select geom From land_parcels Where fid in (16,15)
'';
    json_geom := sdo2geojson_document(cur);
    dbms_output.put_line(json_geom);
End;
/

{ ""type"": ""FeatureCollection"", ""features"": [{""type"":""Polygon"
","",coordinates"":[[[6007197.92,2121691.11],[6007194.18,2121718.01],[6
007189.99,2121748.06],[6007187.23,2121747.68],[6007167.3,2121744.89],[
6007186.26,2121608.7],[6007208.96,2121611.88],[6007197.92,2121691.11]]
]}{""type"":""Polygon"",""coordinates"":[[[6006491.42,2121789.42],[600
6423.36,2121779.76],[6006442.27,2121643.57],[6006510.34,2121653.22],[6
006491.42,2121789.42]]]}] }
```

Summary

Spatial data is usually available in many different file formats as many GIS tools use their own proprietary formats for managing this data, so it is very important to understand how to load all these different formats of spatial data into Oracle Spatial. Many ETL tools and data loaders support Oracle Spatial as both input and output format. In this chapter, we looked at different methods for loading the spatial data into Oracle Spatial including simple CSV files to the most common file formats. We also show how to load and generate data in formats that are very common in web mapping applications. This chapter also described issues related to the storage of geometry data in Oracle Spatial. Sometimes it is beneficial to understand the data layout in Oracle Spatial tables, and this was explained with concepts like spatial clustering.

Once the data is loaded into the database, it is now ready for use with applications. Oracle provides many other database features that can be used to manage the application development process more efficiently. In the next chapter, we will have a look at some of the important features of Oracle that are commonly used in spatial applications.

3
Using Database Features in Spatial Applications

The goal of this chapter is to introduce some of the standard features that the Oracle database makes available to solve some common data processing, auditing, version management, and quality issues related to the maintenance of the spatial data. These database features allow developers to keep the data processing operations in the database instead of doing them in the application code. The topics covered in this chapter are:

- Row-level and statement-level triggers
- Avoiding the mutating table problem
- Using materialized views
- Logging changes independently of an application
- Flashback queries
- AWR reports and ADDM
- Database replay
- Workspace Manager
- SecureFile compression

Using row-level and statement-level triggers

Database triggers can be defined as row-level or statement-level triggers. All the triggers we have seen so far in *Chapter 1, Defining a Data Model for Spatial Data Storage,* are defined as row-level triggers; that means the trigger is executed for each row of the table that is inserted, deleted, or updated. In some cases, we do not want the trigger to execute for each row, but we want it to execute for the whole SQL statement. For example, consider the following two SQL statements:

```
Insert Into LAND_PARCELS Values(3326028, '0026T05AA', '0026T', '055A',
  '2655',    'HYDE', '1 ST', 'HYDE', 'ST',   'O',
SDO_GEOMETRY(2003, 2872, NULL, SDO_ELEM_INFO_ARRAY(1, 1003, 1),
SDO_ORDINATE_ARRAY(6006801.83, 2121396.9,  6006890.33, 2121409.23,
6006878.19, 2121495.89,  6006788.97, 2121483.45 )), NULL);

Insert Into LAND_PARCELS Select * From TEMP_LAND_PARCELS;
```

Let's assume that the TEMP_LAND_PARCELS table has ten rows. The first statement inserts one row into the LAND_PARCELS table, while the second statement inserts ten rows into the LAND_PARCELS table. If the Insert trigger on this table is defined as a row-level trigger, then the trigger will be executed ten times for the second statement. If the trigger is defined as a statement-level trigger, then it will be executed just once for both the statements. So, the application developer has to carefully choose which triggers will be the statement-level triggers and which will be the row-level triggers. By default, Oracle database triggers are created as statement-level triggers unless the keyword For Each ROW is specified in the definition of the trigger.

From the examples listed in *Chapter 1, Defining a Data Model for Spatial Data Storage,* we can see that we want to define the row-level triggers when we want to take an action for each row of the table that is inserted, deleted, or modified. But the statement-level triggers can be used in conjunction with the row-level triggers to do batch operations more efficiently. For example, consider the trigger LAND_PARCELS_VALIDATE_V2 defined in *Chapter 1, Defining a Data Model for Spatial Data Storage.* This trigger fires once for each row inserted into the LAND_PARCELS table and moves the invalid data into a staging table. After the inserts are done, we showed the steps required to rectify this invalid data and move it back to the LAND_PARCELS table. Now, instead of doing these steps manually, they can be done in a statement-level trigger. If the statements are inserting many rows with each SQL statement, then a statement-level trigger will be useful. If the SQL statements only insert one row at a time, then a statement-level trigger is not advisable. There are other uses for statement-level triggers, such as recording the number of changes done to each table (for auditing purposes) and to avoid the mutating-table problem, which we'll discuss next.

Avoiding the mutating table problem

It is very common for developers to run into the mutating table problem while using database triggers. While executing a trigger, if you notice an error like this: ORA-04091: table is mutating, the trigger/function may not see it, which implies that you have run into the mutating table problem. The main reason for this error is the inconsistent view of the data read by the trigger. The error is commonly encountered when a row-level trigger tries to access the same table on which the trigger is defined. Since the trigger is invoked during an update, insert, or delete operation on the table and the code inside the trigger tries to read the table, it might not see a consistent view of the table data. Since the table is changing while the trigger is executing, the table is said to be mutating. This mutation does not occur if a single row is changed with a SQL statement. This typically only happens when a set of rows are changed with a single SQL DML statement.

We will illustrate this problem with a trigger on the ROAD_CLINES table. This table has address ranges for the left side and the right side of the road segments stored in the table. Ideally, no two road segments corresponding to a road should have overlapping address ranges. We first define a trigger to check the incoming rows to see if there are any other road segments corresponding to the same road with overlapping address ranges.

```
Create Or Replace Trigger ROAD_CLINES_ADDR_CHECK_V1
Before Isert Or Update On ROAD_CLINES
For Each Row
Declare
cnt number;
Begin

  Select count(*) Into cnt From road_clines
  Where road_id = :NEW.road_id
    And ( (:NEW.lf_fadd between lf_fadd  and lf_toadd) Or
          (:NEW.lf_toadd between lf_fadd  and lf_toadd) Or
          (:NEW.rt_fadd between rt_fadd  and rt_toadd) Or
          (:NEW.rt_toadd between rt_fadd  and rt_toadd) );

  If (cnt > 0) Then
   RAISE_APPLICATION_ERROR(-20000, 'ROAD_CLINE with FID ' || to_
char(:NEW.FID) || ' has overlapping address ranges with other rows ')
;
  End If;

End;
```

This trigger checks the address ranges of the incoming data and raises an error if the new address ranges overlap with any of the existing road segments with the same road_id as the incoming row. Now, if we try to insert a new row into the ROAD_CLINES table with the overlapping address ranges with road_id=2747, we'll see the error raised in our trigger.

```
Insert Into road_clines_stage Values (155420, 2747,
1510,1569,1500,1598, 857000, 23679000,23744000,'94107','10', 'Y',
'DPW',  'Potrero Hill',
   '857000', '18TH ST', 5, '18TH',
SDO_GEOMETRY(2002, 2872, NULL, SDO_ELEM_INFO_ARRAY(1, 2, 1), SDO_
ORDINATE_ARRAY(6013195.42, 2105673.63, 6012915.75, 2105658.06)));

ERROR at line 4:
ORA-20000: ROAD_CLINE with FID 155420 has overlapping address ranges
with other
rows
ORA-06512: at "BOOK.ROAD_CLINES_ADDR_CHECK_V1", line 13
ORA-04088: error during execution of trigger 'BOOK.ROAD_CLINES_ADDR_
CHECK_V1'
```

We did not see the mutating table error, but we see that the error we wanted to raise from the trigger body is the error that is raised. What happened? As we mentioned earlier, the mutating table error is raised only when more than one row is inserted into the table. Since the previous Insert statement only inserted one new row, the table is not mutating when the trigger is executed. Now, let's create a temporary table to collect the new rows and then do a bulk insert into the ROAD_CLINES table. Note that we are trying to simulate a bulk insert using the staging table, but in practice, the application logic will be batching several inserts into one SQL insert to reduce the database overhead.

```
Create Table road_clines_temp As Select * From road_clines
Where 1 = 2;

Insert Into road_clines_temp Values (155420, 2747, 1510, 1569, 1500,
1598, 857000,23679000,23744000,'94107','10', 'Y', 'DPW',  'Potrero
Hill','857000', '18TH ST', 5, '18TH',
SDO_GEOMETRY(2002, 2872, NULL, SDO_ELEM_INFO_ARRAY(1, 2, 1), SDO_
ORDINATE_ARRAY(6013195.42, 2105673.63, 6012915.75, 2105658.06)));

Insert Into road_clines_temp Values (155421, 2747,
1569,1579,1500,1598, 857000,23679000,23744000,'94107','10', 'Y',
'DPW',  'Potrero Hill', '857000', '18TH ST', 5, '18TH',
SDO_GEOMETRY(2002, 2872, NULL, SDO_ELEM_INFO_ARRAY(1, 2, 1), SDO_
ORDINATE_ARRAY(6013195.42, 2105673.63, 6012915.75, 2105658.06)));
Insert Into road_clines Select * From road_clines_temp;
```

```
ERROR at line 1:
ORA-04091: table BOOK.ROAD_CLINES is mutating, trigger/function may
not see it
ORA-06512: at "BOOK.ROAD_CLINES_ADDR_CHECK_V1", line 5
ORA-04088: error during execution of trigger 'BOOK.ROAD_CLINES_ADDR_
CHECK_V1'
```

As we tried to insert two rows into the ROAD_CLINES table, we see the mutating table error, and the trigger sees inconsistent data between the inserts of the two rows. This error can be avoided by changing the way we define our trigger. There are two possible workarounds to the mutating table error: first, change the trigger in to a statement-level trigger, and second, move some of the logic out of the trigger so that the table is not queried from inside the trigger. In the next example, we describe this workaround using a statement-level trigger to manage the inserts/updates to the ROAD_CLINES table.

For this example to work, we first alter the ROAD_CLINES table so that we can tag each row with a status code. We add a column called ACCEPTED that can have either 'Y' or 'N' as possible values. When the data is first inserted into this table, the data is set to 'N' for the ACCEPTED column. A batch process can periodically check this table and verify all the rows with ACCEPTED='N' and set the value to 'Y' if the address ranges do not violate the consistency rule.

```
-- First add constraint to make sure only 'N' and 'Y' are allowed
Alter Table ROAD_CLINES ADD Constraint ACCEPTED_CHK
  Check (ACCEPTED in ('Y', 'N'));

-- Trigger to check the for the address range consistency
Create Or Replace Trigger ROAD_CLINES_ADDR_CHECK_V2
Before Insert Or  Update on ROAD_CLINES
For Each Row
Declare
cnt number;
Begin
  If Inserting Then
    -- change the ACCEPTED to 'N'
    :NEW.ACCEPTED := 'N';
  ElsIf Updating Then
   If ( (:NEW.lf_fadd <> :OLD.lf_fadd) Or (:NEW.lf_toadd <> :OLD.lf_
toadd)
       Or
       (:NEW.rt_fadd <> :OLD.rt_fadd) Or (:NEW.rt_toadd <> :OLD.rt_
toadd) ) then
       :NEW.ACCEPTED := 'N';
   End If;
  End If;
End;
```

In this trigger, we take two different actions depending on the type of DML performed on the table. On all `Insert` operations, the `ACCEPTED` field is automatically set to `'N'`. For any `Update` operation that changes the address range values, the `ACCEPTED` field is set to `'N'`. This makes sure that the address range values are always consistent; if the `ACCEPTED` field is set to `'Y'`, then a batch program can periodically check the address ranges and fix the data and set the `ACCEPTED` filed to `'Y'`. Note that this update to set the `ACCEPTED` field to `'Y'` will go through only if none of the address ranges are modified in the same SQL statement.

Understanding materialized views

A view in the database presents a logical view of a table that is different from the actual physical structure of the table. We have used different views in our sample schema to make subsets of rows from the `CITY_FURNITURE` table look like tables. A view can also be defined on the result set of a query between two or more tables, for example, we can define a view to look at all the `LAND_PARCELS` that are affected by the maintenance work done to a `sidewalk`. For this, we define the view `land_parcel_sidewalk` as follows:

```
Create View land_parcel_sidewalk As
Select a.fid lp_fid, b.fid sw_fid
From land_parcels a, sidewalks b
Where sdo_anyinteract(a.geom, b.geom) = 'TRUE';

-- lets find the land_parcels affected by a sidewalk
Select * From land_parcel_sidewalk
Where sw_fid = 6882;
```

When the query is executed to find a `land_parcel` corresponding to the sidewalk with `sw_fid=6882`, a spatial query is executed using the two base tables involved in the view definition. Sometimes, these spatial queries can be computationally expensive and may consume many CPU cycles. If these queries are executed many times, it will waste resources, as the same computationally expensive spatial query is executed many times. Oracle database provides a feature called materialized views to avoid such repeated executions of the same query. With a materialized view, the query result can be materialized as a regular database table. Since the materialized view is treated as a regular table, indexes can also be built on it. We'll now describe how a materialized view can be defined instead of the normal view shown earlier:

```
Drop View land_parcel_sidewalk;
-- Create the view as materialized view
```

```
Create Materialized View land_parcel_sidewalk As
Select a.fid lp_fid, b.fid sw_fid
From land_parcels a, sidewalks b
Where sdo_anyinteract(a.geom, b.geom) = 'TRUE';

-- Create an index on sw_fid column
Create Index land_parcel_sidewalk_idx On land_parcel_sidewalk(sw_fid);

-- run the query
Select * From land_parcel_sidewalk
Where sw_fid = 6882;
```

This now works as expected, but what happens if the LAND_PARCELS table or the SIDEWALKS table is updated? The materialized view is a snapshot of the result of the query, and if one of the underlying tables changes after the view is created, then this materialized view is not automatically updated to reflect those changes. Next, we delete a row from the LAND_PARCELS table to show that the materialized view does not change its rows to reflect the new state of the underlying query result:

```
Select * From land_parcel_sidewalk
Where sw_fid = 6084;
```

LP_FID	SW_FID
130335	6084
202990	6084
130119	6084
130118	6084
204046	6084
130348	6084

```
Delete From land_parcels Where fid = 130335;

Select * From land_parcel_sidewalk
Where sw_fid = 6084;
```

LP_FID	SW_FID
130335	6084
202990	6084
130119	6084
130118	6084
204046	6084
130348	6084

```
Rollback;
```

As seen from the result of the query after the delete from the LAND_PARCELS table, the materialized view still has the result snapshot since the deletion was performed during the view creation. Oracle provides multiple options to refresh the materialized view to reflect the latest state of the view based on the current state of the underlying tables. When creating an Oracle materialized view, developers have the option of specifying whether the refresh occurs manually (ON DEMAND) or automatically (On Commit). If the data in the underlying tables is changed via batch jobs, then it is better to refresh the materialized views manually. Since the refresh of a materialized view will incur a cost, doing it on demand will reduce the refresh overhead by not updating the view for each update that happens on the base tables. In our schema, the typical workflow will now use batch jobs to update the LAND_PARCELS table; we will look at ways to refresh the materialized view on demand.

When the underlying tables of a materialized view are updated, one option is to do a complete refresh of the materialized view. With this option, the query defining the view is executed using the latest snapshot of the underlying tables. As one can see, this kind of refresh can use many database resources if the refresh is done whenever incremental changes are made to the base tables. Oracle also provides a fast refresh option that only uses the changes since the last refresh. When the changes are made to the underlying tables, the database stores rows to describe these changes in a log (called the materialized view log), and when users ask for a refresh of the materialized view, these change logs are used to quickly refresh the materialized view. To enable the fast refresh of our materialized view, we need to create materialized view logs for each of the two tables. Unfortunately, when the query contains a complex predicate such as the spatial predicate, Oracle will not allow the materialized view to be fast refresh enabled. So, we will show the example with the complete refresh option, but if the join predicate is based on a simple relational operator, then the fast refresh will work as expected:

```
Drop Materialized View land_parcel_sidewalk;
-- Create the MV with the refresh option ON DEMAND
Create  Materialized View land_parcel_sidewalk
Refresh Force  On Demand As
Select a.fid lp_fid, b.fid sw_fid
From land_parcels a, sidewalks b
Where sdo_anyinteract(a.geom, b.geom) = 'TRUE';
```

```
-- now delete a row from the LAND_PARCELS table
Delete From land_parcels Where fid = 130335;

-- refresh the MV
EXEC dbms_mview.refresh('LAND_PARCEL_SIDEWALK', 'C');

Select From the MV to see If the Delete Row is present

    LP_FID      SW_FID
---------- ----------
    202990        6084
    130119        6084
    130118        6084
    204046        6084
    130348        6084
```

As can be seen from the result of the last query, the materialized view is refreshed to reflect the fact that fid=130335 is deleted from the LAND_PARCELS table. As we mentioned already, the refresh should be done sparingly to avoid the expensive join operation between the base tables. Since the data for the LAND_PARCELS table comes in batches, a good option will be to run the refresh operation once a day to reflect the changes in the underlying base tables.

Materialized view restrictions with geometry columns

Materialized views have several restrictions if the view definition contains spatial columns. We have already mentioned the restriction with FAST REFRESH if the Where clause contains an Oracle Spatial operator. There are two other restrictions that developers commonly encounter with materialized views and spatial tables. The first is a view that has a Select clause with a geometry column and is defined as a table JOIN. The second is a view that has an Oracle Spatial function such as SDO_AREA in the Select clause. In both cases, the materialized view can be defined, but the FAST REFRESH option will be disabled for the view. In summary, if FAST REFRESH is not required, then materialized views offer benefits for the Oracle Spatial tables. However, if materialized views are used due to the benefits of FAST REFRESH, then there are very few cases where these views are supported for the Oracle Spatial tables.

Logging changes independently of applications

Some spatial applications require that the old and new values in the schema be tracked so that the changes can be traced back in case of errors. For example, new surveying measurements from the field can be used to update the geometries for the existing land parcels, but the older values need to be tracked for a certain amount of time in the database in case of disputes. Oracle database provides an auditing technology called database auditing that can be used to monitor and track all database level actions. This auditing can be done for individual actions such as DML SQL statements or DDL SQL statements, or any other database level action. Database auditing is typically done to account for actions taken at the schema, table, or an other object level to monitor actions of specific users for accounting purposes, investigate suspicious activity, monitor and gather statistics for specific database operations, and for compliance purposes. Therefore, Oracle auditing provides a very broad technology for tracking changes at the database level. This may be overkill for applications that only want to track changes to certain columns in a specific table. And if the application wants to have control over how these changes are tracked and acted upon, auditing might not provide the flexibility required for these applications. In this section, we'll describe the use of triggers to log changes to the tables at the database level, and depending on the data model, these triggers can be easily modified to log only those changes that are relevant for the data model.

We first look at a trigger to log the changes applied to the geometry column of the LAND_PARCELS table. We want to monitor all the updates to this table and only log the changes if the geometry column has changed. The purpose of this initial approach is to show the flexibility of the triggers in logging the changes with very fine-grained control on what is logged.

```
Create Table LAND_PARCELS_LOG
(  FID   Number(38)
   GEOM_OLD SDO_GEOMETRY,
   GEOM_NEW SDO_GEOMETRY,
   LAST_MODIFIED_DATE timestamp (6) DEFAULT sysdate,
   LAST_MODIFIED_USER Varchar2(64)
);

Create Or Replace Trigger LP_LOG_TRIGGER
After Update On LAND_PARCELS
REFERENCING NEW As NEW OLD As OLD
For Each Row
Declare
```

```
user_name Varchar2(62);
relation Varchar2(32);
Begin

  relation := sdo_geom.relate(:NEW.geom, 'EQUAL', :OLD.geom, 0.05);
  If (relation = 'TRUE') Then
   -- no change in the geometry value; return
    Return;
  End If;

 Select user Into user_name From dual;

Insert Into LAND_PARCELS_LOG VALUES
    (:new.FID, :OLD.GEOM, :NEW.GEOM, SYSDATE, user_name);
End LP_LOG_TRIGGER;
```

The previous trigger can be easily modified to track changes to more columns other than the geometry column. One important aspect to consider here is the LAST_ MODIFIED_USER field in the logging table. Here we added code to get the currently logged user from the database, and use that as the user modifying the table. In many cases with a three-tier architecture, the application user is different from the database user, and it is very common to use one database user for many different application users. In such cases, the application should set some context or pass the application user name information to the database so that it can be logged in the log tables. How this context is passed to the database is completely up to the application, and we leave it to the user to decide on how this is done.

Next, we develop a generic package-based mechanism that will make it easy to track changes for different tables and different columns. For this, we will use a generic table structure that stores all the column types as Varchar2 values. In this version of the package, a separate logging table is used for each application table.

```
-- Create a generic log table for logging the changes to LAND_PARCELS
Create Table LAND_PARCELS_LOG (fid Number(38),
                               column_name Varchar2(32),
                               old_value   Varchar2(4000),
                               new_value   Varchar2(4000),
                               LAST_MODIFIED_DATE timestamp(6),
                               LAST_MODIFIED_USER Varchar2(64));

-- this package has one log procedure for each of the data types
-- we encounter in the LAND_PARCELS table
-- If there are more types, each type needs a new procedure
```

```
Create Or Replace package audit_book_tables is
procedure log_values (fid           IN number,
                    column_name IN varchar2,
                    new_val      IN varchar2, old_val IN varchar2);
procedure log_values (fid           IN number,
                    column_name IN varchar2,
                    new_val      IN number, old_val IN number);
procedure log_values (fid           IN number,
                    column_name IN varchar2,
                    new_val      IN SDO_GEOMETRY, old_val IN SDO_
GEOMETRY);
End audit_book_tables;

-- Create the package body
Create Or Replace package body audit_book_tables is
  procedure log_values (fid           IN number,
                    column_name IN varchar2,
                    new_val      IN varchar2, old_val IN varchar2)
Is
Begin
  If ( (new_val is not NULL and old_val is not NULL
                                and new_val <>   old_val) Or
        ((new_val is NULL OR old_val is NULL) And
                      NOT (new_val is NULL and old_val is NULL) ) )
    Then
      Insert Into LAND_PARCELS_LOG
      Values(fid, column_name, old_val, new_val,SYSDATE, USER);
  End If;
End;
procedure log_values (fid           IN number,
                    column_name IN varchar2,
                    new_val      IN number, old_val IN number) IS
Begin
  If ( (new_val is not NULL and old_val is not NULL
                              and new_val <> old_val) Or
        ((new_val is NULL OR old_val is NULL) and
                      NOT (new_val is NULL and old_val is NULL) ) )
Then
      Insert Into LAND_PARCELS_LOG values(fid, column_name,
        to_char(old_val),to_char(new_val), SYSDATE, USER);
  End If;
End ;
```

```
procedure log_values (fid         IN number,
                     column_name IN varchar2,
                     new_val      IN SDO_GEOMETRY, old_val IN SDO_
GEOMETRY) Is
relation Varchar2(32);
Begin
  relation := sdo_geom.relate(new_val, 'EQUAL', old_val, 0.05);
  If ( (new_val is not NULL and old_val is not NULL
                          and relation='TRUE') Or
      ((new_val is NULL OR old_val is NULL) and
                    NOT (new_val is NULL and old_val is NULL) ) )
    Then
      Insert Into LAND_PARCELS_LOG values(fid, column_name,
        sdo_util.TO_GMLGEOMETRY(old_val),
        sdo_util.TO_GMLGEOMETRY(new_val),SYSDATE, USER);
  End If;
End;
End;
```

Next, we define the trigger on LAND_PARCELS to log the changes to the table. In the trigger body, we only need to add one call to the above packaged procedures for each column we want to log.

```
-- this trigger will log changes to 3 columns of LAND_PARCELS
Create Or Replace Trigger LP_LOG_TRIGGER
After Update On LAND_PARCELS
REFERENCING NEW As NEW OLD As OLD
For Each Row
Declare
user_name Varchar2(62);
relation Varchar2(32);
Begin
  -- log changes to MAPBLKLOT column
  audit_book_tables.log_values(:NEW.fid, 'MAPBLKLOT', :NEW.MAPBLKLOT,
          :OLD.MAPBLKLOT);
  -- log changes to LOT_NUM column
  audit_book_tables.log_values(:NEW.fid, 'LOT_NUM', :NEW.LOT_NUM,
          :OLD.LOT_NUM);
  -- log changes to GEOM column
  audit_book_tables.log_values(:NEW.fid, 'GEOM', :NEW.GEOM,
          :OLD.GEOM);
End LP_LOG_TRIGGER;
```

This package assumes that the geometry column can be converted to a `Varchar2` value of less than 4000 bytes. If the geometry data is larger than this when converted to a `varchar2` value, a separate table may be used to store the changes to the geometry column. Once the logging schema is defined and the triggers are created, applications don't need to worry about tracking these changes, and these changes will be visible to the application since the changes are stored in a regular database table.

Flashback queries

When applications are updating the database tables, some inadvertent action may cause data loss. Some bug in the application logic may cause the deletion of some data that is not supposed to be deleted. Oracle database provides very robust data recovery and backup features, but these are mostly used in case of media or hardware failures. The data loss caused by application or normal user actions often does not need the traditional data recovery mechanism provided by the database. the flashback feature of Oracle provides an easier to manage mechanism for recovering the data loss at the row-level or at the table-level. The main features of the flashback query are as follows:

- **Flashback Database**: This restores the entire database to a specific point-in-time using Oracle-optimized flashback logs, rather than via backups and forward recovery.

- **Flashback Transaction**: This undoes the effects of a single transaction, and optionally, all of its dependent transactions via a single PL/SQL operation or by using an Enterprise Manager wizard.

- **Flashback Transaction Query**: This sees all the changes made by a specific transaction, which is useful when an erroneous transaction changed data in multiple rows or tables.

- **Flashback Table**: This easily recovers tables to a specific point-in-time, which is useful when a logical corruption is limited to one or a set of tables instead of the entire database.

- **Flashback Drop**: This recovers an accidentally dropped table. It restores the dropped table and all of its indexes, constraints, and triggers from the **Recycle Bin** (a logical container of all dropped objects).

- **Flashback Query**: This queries any data at some point-in-time in the past. This powerful feature can be used to view and logically reconstruct corrupted data that may have been deleted or changed inadvertently.

- **Flashback Versions Query**: This retrieves the different versions of a row across a specified time interval instead of a single point-in-time.

In this section, we only look at the table, drop, query, and versions query. The flashback database is not an application-level operation, so we will not discuss it here. The flashback transaction and flashback transaction query have limited support for a table with SDO_GEOMETRY columns. When the DML operations such as Insert, Update, or Delete are executed, Oracle writes data to an undo tablespace that is used for transaction rollbacks and for guaranteeing read consistency. The flashback feature depends on the amount of history stored in the undo tablespace. The bigger the undo tablespace, the further back in time one can go to recover the data.

Flashback table

The flashback table feature enables the DBAs to restore a table to its state as of a previous point in time. This feature provides a fast and online solution for recovering the contents of a table that has been accidentally deleted by a user or an application. This eliminates the need for a more complex database recovery operation. For this feature to work, the tables have to be enabled for row movement. All the application tables in our schema are created with no row movement enabled. So, the first operation we need to do is to enable row movement for the tables.

```
Alter Table building_land_parcel enable row movement;
Alter Table building_footprints enable row movement;
Alter Table land_parcels enable row movement;
```

Next, we do some transactions on the land_parcels, building_footprints, and the building_land_parcel tables and delete some rows. Then, we will show how to back out of these transactions and recover the lost data.

```
-- first check to see if the data exists for the rows plan to delete
Select * From building_land_parcel Where building_id = 85034;
Select * From building_footprints Where fid = 85034;
Select * From land_parcels Where fid = 840;

Alter session set nls_date_format = 'DD-MON-YYYY HH24:MI:SS';
Select sysdate From dual;

Delete From building_land_parcel Where building_id = 85034;
Commit;
Delete From building_footprints Where fid = 85034;
Commit;
Delete From land_parcels Where fid = 840;
Commit;
```

```
-- connect as the DBA user to the DB
connect / As sysdba

-- use the sysdate obtained above to generate a TIMESTAMP
flashback table book.land_parcels to
TIMESTAMP to_timestamp('05-JAN-2013 13:58:37' , 'DD-MON-YYYY
HH24:MI:SS');
Select fid From book.land_parcels Where fid = 840;

flashback table book.building_footprints to
TIMESTAMP to_timestamp('05-JAN-2013 13:58:37' , 'DD-MON-YYYY
HH24:MI:SS');
Select fid From book.building_footprints Where fid = 85034;

flashback table book.building_land_parcel to
TIMESTAMP to_timestamp('05-JAN-2013 13:58:37' , 'DD-MON-YYYY
HH24:MI:SS');
Select building_id From book.building_land_parcel Where building_id =
85034;
```

Now, when we execute the query to select the rows from these tables, we can find them easily, as the flashback restored the tables to their previous state as of the given timestamp value.

Flashback drop

The flashback drop feature reverses the effects of a Drop Table operation, and it can be used to recover the table after an accidental drop. As with other flashback operations, this is substantially faster than doing a database level recovery operation. When a user drops a table, the database does not immediately remove the table completely from the database. The table is renamed, and along with any associated objects, such as triggers and constraints, the table is moved to the **Recycle Bin** of the database. The flashback drop operation can recover the table from this **Recycle Bin**. For this feature to work, the value of the RECYCLEBIN parameter must be set to ON for the database.

```
-- make sure you do not need anything from the recyclebin before
-- emptying it
purge recyclebin;
Select * From user_recyclebin;
-- drop the sidewalks table as it does not have any other dependent
-- tables
Drop Table sidewalks ;
```

```
-- see what is in the recyclebin
Select * From user_recyclebin;

-- recover the table
flashback table sidewalks to before drop;

purge recyclebin;
```

This method can be used to recover tables, but not indexes or views. If an index is dropped by itself, it cannot be recovered using the flashback feature. But, if a table with indexes is dropped and recovered, the corresponding indexes are recovered along with the table. Note that this feature does not work with Oracle Spatial indexes; spatial indexes cannot be recovered if the corresponding table is recovered.

Flashback query

The flashback query feature can retrieve data from a table as it existed at an earlier point-in-time. The query explicitly refers to the past time via a timestamp or the Oracle **System Change Number** (**SCN**). The query will return the committed data from the specified pointin-time from the table. This feature is commonly used to recover the data lost due to user or application errors or to undo unwanted changes. This can also be used to compare the state of the table between two different timestamps. The users can also perform historical queries to find the state of the table at a given time in the past. The flashback query-based recovery operations can be done by a normal application user without the need for a DBA. In this section, we show how to use this feature using the road_clines table:

```
-- find the current SCN and the TIMESTAMP
Select current_scn, SYSTIMESTAMP
From v$database;
CURRENT_SCN SYSTIMESTAMP
-------------------------------------------------------------------
     820646 05-JAN-13 02.42.23.947242 PM -05:00

-- execute a Spatial query to find a row From ROAC_CLINES
Select fid From  road_clines
Where sdo_anyinteract(geom, SDO_GEOMETRY(2003,2872,null,
sdo_elem_info_array(1,1003,3),
sdo_ordinate_array(6011019, 2106642, 6011029, 2107298))) = 'TRUE';

     FID
---------
   15877
```

```
Delete From road_clines Where fid = 15877;

Commit;

Select fid From  road_clines
Where sdo_anyinteract(geom, SDO_GEOMETRY(2003,2872,null,
sdo_elem_info_array(1,1003,3),
sdo_ordinate_array(6011019, 2106642, 6011029, 2107298))) = 'TRUE';

no rows selected

-- now execute the query, but use the previous SCN
Select fid From  road_clines
AS OF SCN 820646
Where sdo_anyinteract(geom, SDO_GEOMETRY(2003,2872,null,
sdo_elem_info_array(1,1003,3),
sdo_ordinate_array(6011019, 2106642, 6011029, 2107298))) = 'TRUE';

    FID
---------
   15877

-- same query, with TIMESTAMP format
Select fid From  road_clines
AS OF TIMESTAMP to_timestamp('05-JAN-13 02.42.23.000000 PM')
Where sdo_anyinteract(geom, SDO_GEOMETRY(2003,2872,null,
sdo_elem_info_array(1,1003,3),
sdo_ordinate_array(6011019, 2106642, 6011029, 2107298))) = 'TRUE';
```

The next step is to recover the deleted row from the road_clines table. For this, we can select the row with the known FID that we want to recover and insert it into the main table:

```
-- find the row with FID=15877 From the previous timestamp
Insert Into road_clines
Select * From  road_clines
AS OF TIMESTAMP to_timestamp('05-JAN-13 02.42.23.000000 PM')
Where fid = 15877;
```

With this AS OF syntax, it is easy to find rows in the table for any given time period for example, we can use this feature to easily find the new rows that are inserted into the road_clines table in the last 60 minutes. For this, we define the following view that will list all the FIDs that are inserted into the table:

```
Drop View road_clines_60_minutes;
Create View road_clines_60_minutes As
  Select fid From road_clines
  minus
  Select fid From road_clines
    AS OF TIMESTAMP (SYSTIMESTAMP - INTERVAL '60' MINUTE);
```

Similarly, it is easy to define the views that can show all the FIDs that are deleted in the last 60 minutes by switching the order of SQL statements in the previous view definition. Therefore, it is easy to provide historical context for any of the tables in the database with the flashback query feature.

Flashback versions query

Oracle creates a new version of a row whenever a row is updated and a Commit statement is executed. A row can be updated multiple times in a transaction, but only one new version of a row is created when the Commit is executed. The users can use the flashback version query to retrieve the different versions of a specific row during a given time interval or during the lifetime of the row:

```
-- set the date format for display
Alter session set nls_date_format = 'DD-MON-YYYY HH24:MI:SS';
Select sysdate From dual;

-- look at the value for the row with FID=3000
Select * From roads Where fid = 3000;
   FID    STREET_NAME   CLASSCODE
----------------------------------------------------------------------

   3000   DUNNES ALY    5

-- make some updates to the row
Update roads Set street_name = 'DUNES ALY' Where fid = 3000;
Commit;

Update roads Set classcode = 6 Where fid = 3000;
Commit;

Update roads Set street_name = 'DUNNES ALLY' Where fid = 3000;
Commit;
```

```
-- retrieve different version of the row from the beginning timestamp
Select  versions_starttime, versions_endtime, versions_operation,
      fid, street_name, classcode
  From ROADS
  VERSIONS BETWEEN TIMESTAMP
      TO_TIMESTAMP('05-JAN-2013 17:34:29', 'DD-MON-YYYY HH24:MI:SS')
  And TO_TIMESTAMP(sysdate, 'DD-MON-YYYY HH24:MI:SS')
  Where fid = 3000
  Order By versions_xid;

VERSIONS_STARTTIME VERSIONS_ENDTIME VERSIONS_OPERATION
   FID      STREET_NAME  CLASSCODE
------------------------------------------------------------------


05-JAN-13 05.40.22 PM  05-JAN-13 05.40.40 PM    U
   3000    DUNES ALY      5

05-JAN-13 05.40.40 PM  05-JAN-13 05.41.16 PM    U
   3000    DUNES ALY      6

05-JAN-13 05.41.16 PM                           U
   3000    DUNNES ALLY    6
```

As seen from the last row in the result, it only has a start time and no end time since that is the current value for the row.

AWR reports

Oracle provides powerful tools for diagnosing database performance bottlenecks, such as long running queries and disk I/O waits. These are **Advanced Workload Repository (AWR)** and **Automated Database Diagnostic Monitor (ADDM)** tools. These are very useful tools for understanding how the application level APIs affects how the DB performs. Often times, some application API calls will take more time to complete while a seemingly similar application API call takes much less time to complete, for example, a user's application might be drawing different layers of spatial data on a map. From the application's point of view, there might not be much difference between the different layers of data, but on the database, the underlying queries against the different layers might perform differently. In this section, we describe how to use database supplied tools to collect and analyze information regarding the application workloads.

While analyzing bottlenecks for an application, it is good practice to isolate the workload on the database specific to the application workload you want to analyze. Before an application workload can be analyzed, a snapshot of the current state of the database is captured. Then, the application workload is executed and another snapshot of the database is captured. An AWR report is then generated on the captured workload and analyzed further.

```
-- Step 1: login to the DB as a DBA
--         Create a before snapshot of the DB
--
   EXEC DBMS_WORKLOAD_REPOSITORY.Create_SnapShot;
```

Now, run the application load test to capture the workload we want to analyze for bottlenecks. Once the workload is completed, we need to take the after snapshot of the database.

```
-- Step 2: connect to the DB as DBA again do the after snapshot
   EXEC DBMS_WORKLOAD_REPOSITORY.Create_Snapshot;
```

Using these two snapshots, we can now create analysis reports using the AWR and ADDM scripts. Note that these reports cover all database activities for all sessions during the capture period. So, if there are other sessions active on the database during this time, these need to be excluded from the analysis. While running the following script, users need to answer several questions, so the whole session is captured in the following example and the user's responses are highlighted in bold:

```
-- Step 3: Generate the AWR report
-- run this script while connected to the DB as a DBA
@ORACLE_HOME/rdbms/admin/awrrpti.sql


Specify the Report Type
~~~~~~~~~~~~~~~~~~~~~~~~~
Would you like an HTML report, or a plain text report?
Enter 'html' for an HTML report, or 'text' for plain text
Defaults to 'html'
Enter value for report_type: html


Type Specified:  html


Instances in this Workload Repository schema
~~~~~~~~~~~~~~~~~~~~~~~~~~~~~~~~~~~~~~~~~~~~~~~
```

```
     DB Id       Inst Num DB Name      Instance     Host
------------ -------- ------------ ------------ ------------
* 679018072        1 DO11203      book         host1

Enter value for dbid: 679018072
Using 679018072 for database Id
Enter value for inst_num: 1
Using 1 for instance number


Specify the number of days of snapshots to choose from
~~~~~~~~~~~~~~~~~~~~~~~~~~~~~~~~~~~~~~~~~~~~~~~~~~~~~~~~~~~
Entering the number of days (n) will result in the most recent
(n) days of snapshots being listed.  Pressing <return> without
specifying a number lists all completed snapshots.


Enter value for num_days: 1


Listing the last day's Completed Snapshots
                                                        Snap
Instance      DB Name      Snap Id    Snap Started     Level
------------ ------------ --------- ----------------- -----
do11203      DO11203           39 06 Jan 2013 00:00      1
                               40 06 Jan 2013 02:01      1
                               41 06 Jan 2013 12:09      1
                               42 06 Jan 2013 12:10      1



Specify the Begin and End Snapshot Ids
~~~~~~~~~~~~~~~~~~~~~~~~~~~~~~~~~~~~~~~~~
Enter value for begin_snap: 41
Begin Snapshot Id specified: 41

Enter value for end_snap: 42
End   Snapshot Id specified: 42

Specify the Report Name
~~~~~~~~~~~~~~~~~~~~~~~~~
The default report file name is awrrpt_1_51_52.html.  To use this
name,
press <return> to continue, otherwise enter an alternative.

Enter value for report_name: book_awr.html
```

After the last step, `report book_awr.html` will be generated in the current directory. Go to a browser and open up the HTML file to see the content of the report. This report will have the analysis of the workload by I/O waits, by SQL statements, and a host of other database parameters. One of the important sections is the SQL statements section that lists all the SQL executed in the workload, and these SQL statements are ranked using different types of parameters such as elapsed time, CPU time, and I/O wait time.

While AWR is very useful for providing users with a detailed breakdown of the database events and timings, it is complemented by the adviser tool, ADDM, that automatically analyzes the same set of stats between the two snapshots and can provide concrete advice on how to fix the bottlenecks in the database.

```
-- Step 4: generate the ADDM report
--   connect to the DB as DBA run the following script
ORACLE_HOME/rdbms/admin/addmrpt.sql
```

While users can always use the Enterprise Manager to view the ADDM report, one can also generate a text file by running the above script. The script will prompt for the snapshot to use (as in Step 3) and in the end it will generate a text file in the current directory. This text file will have concrete suggestions on database configuration changes, schema changes, and possible application changes.

Database replay

Many users often find unintentional changes to their applications when a DB is upgraded to a newer version or a patch set is applied. Organizations usually have a test system and a production system so that the changes to the DB can be first tested on the test system before applying them to the production system. But often times it is hard to run the same production application load on the test system to completely test in a production-like environment. Database replay provides DBAs with the ability to systematically re-run accurate production workloads in test environments. It allows the DBAs to capture full database workloads from the production systems including all concurrency, dependencies, and timing scenarios and then replay them on the test system. This is done by capturing the SQL level operation on the production system and replaying the same set of SQL statements on the test system. With database replay, the DBAs can capture a workload very easily with a set of scripts or clicks (via Enterprise Manager) and replay the workload on the test system with minimum effort.

Database replay consists of four main steps:

- **Workload Capture**: When this is enabled, all SQL operations (DDL, DML, and so on) executed on the database are captured on the filesystem
- **Workload Processing**: During this phase, the workload captured previously is processed and the replay files and necessary metadata files are created
- **Workload Replay**: A client program called replay client processes the replay files and submits calls to the test DB instance
- **Analysis and Reporting**: The reports are generated at the end of the replay process with extensive information about any errors or differences in expected results as well as differences in performance

Workload capture

When workload capture is enabled, all SQL commands executed in the Oracle database are tracked and stored in a set of binary files called capture files. This process captures DML commands on the DDL along with all the data required to be able to completely replay them on the second system. Users need to specify the location for the capture files and the workload capture start and end time. So, it is recommended that the capture should be done at a time when the database workload is typical for the production system.

```
--
Prompt !!!
Prompt !!! Please read and understand the script before proceeding
Prompt !!!

--
-- 1. Create directory in your file system to store the capture files
--      ie. in Linux: mkdir /scratch/capture

-- 2. Create directory Object and point it to the EMPTY directory you
--      created
Create or Replace directory "book_cap_dirobj" As '/scratch/capture';

--
-- 3. (Optional) Create capture filter
--   These are used to INCLUDE/EXCLUDE the workload
-- Create filters for user "BOOK"

EXEC dbms_workload_capture.add_filter('book_filter', 'USER', 'BOOK');
-- To see what filters are in place
```

```
Select type, name, attribute, status, value From dba_workload_filters;


--
-- 4. Start Capture
-- we want to capture ONLY user "BOOK" workload and exclude
-- everything else

EXEC dbms_workload_capture.start_capture('book_capture', 'book_cap_
dirobj', NULL, 'EXCLUDE');


--
-- 5. (Optional) Check status of the capture
--
Select name, directory, status, start_time, end_time,
        duration_secs, errors
From dba_workload_captures;
```

All the steps in this script are self-explanatory except for Step 3 and how the filter is used in Step 4. In a database replay, a filter can be used to filter out all the SQL corresponding to a specific schema, or the same filter can be used to include only the SQL corresponding to a specific schema. So, in this case, we defined the filter for the BOOK schema, and we use the EXCLUDE option while calling the start_capture procedure to let the relay know that we want to exclude all the SQL from the workload that does not belong to the BOOK schema. After Step 5, the database is ready to capture the workload for the BOOK schema. The users can now run the application against the database as it is done during the normal operation of the application. After a suitable amount of time, the capture process has to be stopped.

```
--
-- 1. finish current capture
--
EXEC dbms_workload_capture.finish_capture;


--
-- 2. (Optional) Export AWR snapshots automatically taken at the
-- start and end of the capture.
-- So you could use it later to analyze the system
--

Declare
  capture_id number;
Begin
  Select max(id) Into capture_id
  From dba_workload_captures
```

```
    Where status = 'COMPLETED';

    dbms_workload_capture.export_awr(capture_id);
End;
```

At the end of this capture process, a set of capture files will be created under the directory created in Step 1. The capture files first have to be copied to the replay (the target test system) before they can be processed.

Workload processing

After the workload is captured, the information in the capture files has to be processed to transform the captured data into the replay files, and the necessary metadata needed for replaying the workload is also created as a part of this processing. Once the replay files are generated, the captured workload can be replayed repeatedly on a system or different systems. The following steps assume that the capture files are moved to the replay system:

```
--
Prompt !!!
Prompt !!! Please read and understand the script before proceed. !!!
Prompt !!!

-- NOTE: All of steps from this point can be performed using
-- Enterprise Manager GUI

--
-- 1. Create directory object
--
Create or Replace directory "book_cap_dirobj" As '/scratch/capture';


--
-- 2. Process the capture: Create necessary meta-data to replay
--    Need to be done once per capture per replay database version
--
EXEC dbms_workload_replay.process_capture('book_cap_dirobj');


--
-- 3. Initialize replay
--
```

```
EXEC dbms_workload_replay.initialize_replay('book_capture', 'book_cap_
dirobj');

--
-- 4. (Optional) List connection used in the workload
--
Select conn_id As connection_id, capture_conn, replay_conn
From dba_workload_connection_map
Where replay_id=<replay_id>;


--
-- 5. (Optional) Remap the connection
--   If your capture and replay system need different connection
--   string to connect, you need to remap the connection so replay
--   session can connect to the correct database.
--
-- For example if connection_id 1 needs to be remapped

EXEC dbms_workload_replay.remap_connection(1, '(DESCRIPTION=(ADDRESS=
(PROTOCOL=TCP)(HOST=sample_host)(PORT=1234))
   (CONNECT_DATA=(SERVICE_NAME=sample_sid)))');

--
-- 6. Prepare replay: put DB state into REPLAY mode
--
EXEC dbms_workload_replay.prepare_replay(synchronization => TRUE,
connect_time_scale => 100, think_time_scale => 100, think_time_auto_
correct => TRUE);
--
```

Step 4 needs to be carried out, if the connection string is different (which would be the most common case as the replay is done on a different system) for the replay system. In Step 6, the replay mode can be used to describe how the replay is played on the target system, for example, the synchronization parameter specifies whether or not the commit order is preserved on the replay system. The users can also specify how to adjust the time between calls if the execution time on the replay system is different from the execution time on the capture system. Once this setup is done, we start the replay client that will replay the workload on the target system.

Workload replay

Workload replay is done via the replay client that will replay all the SQL commands from the replay files. This replay is done with the exact same timing and concurrency as in the capture system. Depending on the workload, the users may need one or more replay clients to accurately replay the workload. Since the entire workload is replayed including DML and SQL queries, it is very important that the data on the replay system be identical to the production system (that is, the data in the production system as of the beginning of the process). This will avoid any errors due to missing data or schema objects and will generate an accurate analysis for the workload.

```
--
-- 1. We will need to start workload replay client which is
       reponsible to
--   parse and send the workload to the server.
-- Number of clients that needs to be started depends on the workload
-- captured.
-- WRC comes with a mode to recommend how many clients
     you need to start
-- This command is executed from the UNIX/Windows shell
$ORACLE_HOME/bin/wrc <username>/<password> mode=calibrate
     replaydir=/scrath/capture'

--
-- 2. Start Replay
-- Now connect to the DB as a DBA and start the replay
--
EXEC dbms_workload_replay.start_replay;
--
```

These steps should be done as a DBA user, and the **Workload Replay Client** (**WRC**) client will replay the whole workload on the target replay system. The replay will not start until the start_replay command is executed on the database system.

Analysis and reporting

When the replay is done, it will generate a set of reports used to analyze how the workload performed on the test system compared to what is seen on the production system. Note that when the capture process is started on the production system, it automatically generates the AWR reports for the workload on the production system. These reports can now be compared with the AWR reports on the replay system to identify any differences and potential bottlenecks.

Workspace Manager

Spatial applications often have the need to work with multiple versions of the data, for example, a city planning application might provide design choices for adding new sub-divisions to the city. At the same time, a reviewer should be able to look at all the possible designs and choose the best design. This requires access to multiple versions of the data. Some of the common reasons for multiversioning of data are concurrency (multiple users working on different designs), history, and what-if scenario creation. The concurrency here is different from the concurrency provided by the database. Another requirement is to keep the changes available beyond the lifetime of the database transaction, especially for what-if scenarios, for example, with the database transaction model, one application user cannot make changes and make them visible to some other users without committing the changes so that they are visible to all the users on the database. This concept is called a long transaction, and the Workspace Manager feature provides these long transactions at the database level. The versioning of history allows earlier versions of the data to be kept indefinitely. This is useful for users to go back to see how the database looked with the changes at a particular point in time. As we have seen earlier, the flashback query also provides similar history information, but it depends on the amount of undo space used in the database. With the Workspace Manager, the history can be kept for an indefinite amount of time.

The Workspace Manager provides a set of PL/SQL APIs to the version-enabled user tables. When a table is version-enabled, the table is renamed to a system-extended name derived from the original table name, and a view with the same name as the table is created with triggers. This ensures that all DML references to the original table still continue to work and thus the applications do not need to be aware that the underlying table is version-enabled. The Workspace Manager uses the concept of a workspace to logically group a collection of changes to a set of version-enabled tables. It provides session context that can be used to switch to any workspace to see the changes made in that particular workspace. The top most workspace is called LIVE and it is the default workspace for user activity and the production version of the data. The workspaces can be arranged in hierarchies, that is, a workspace can in turn have other child workspaces, and the child workspaces inherit all the changes done in the parent workspace at the time of the child workspace's creation. The LIVE workspace is a parent workspace for any top level workspace that is created, and a workspace can have sibling workspaces. These sibling workspaces allow a single row to have multiple versions at the same time. Unlike some other application-based versioning technologies, the Workspace Manager creates a new version for a row only if the row has changed in a workspace. Therefore, the storage for the tables is minimized in the Workspace Manager as a copy of the row is created only if a row is updated in a workspace.

Version-enabling tables

The Workspace Manager can version-enable one or more user tables in the database. When a table is version-enabled, all rows in that table can support multiple versions of data. The versioned rows are stored in the same table as the original rows. The Workspace Manager implements these capabilities by renaming the TABLE_NAME to TABLE_NAME_LT, adding a few columns to the table to store versioning metadata, creating a view on the version-enabled table using the original TABLE_NAME, and defining INSTEAD OF triggers on the view for the SQL DML operations. If after some time the user no longer needs a table to be version-enabled, then they can disable versioning for the table. In the following sections, we describe how the ROAD_CLINES table is version-enabled to support the city-planning application and show some of the important features of the Workspace Manager with examples. It is important that the table that is version-enabled has a Primary Key defined on it, otherwise the Workspace Manager will raise an error.

```
-- call enableVersioning on ROAD_CLINES
-- the parameter NONE is specified here to not keep history
-- (like timestamp, etc) of the rows that are changing
-- note however the values for the rows are still kept
EXEC dbms_wm.enableVersioning('ROAD_CLINES','NONE');

-- see how many objects are now created when table is versioned
Select count(*) From tab Where tname like 'ROAD_CLINES%';

  COUNT(*)
----------
        15
-- now describe the table with _LT suffix to see additional columns
-- created to track version information

desc  ROAD_CLINES_LT

SQL> desc  ROAD_CLINES_LT
 Name                                      Null?    Type
 ----------------------------------------- -------- ------------------
 ------------
 FID                                                Number(38)
 ROAD_ID                                            Number
 LF_FADD                                            Number
 LF_TOADD                                           Number
 RT_FADD                                            Number
 RT_TOADD                                           Number
 CNN                                                Number
```

F_NODE_CNN	Number(38)
T_NODE_CNN	Number(38)
ZIP_CODE	Varchar2(9)
DISTRICT	Varchar2(3)
ACCEPTED	Varchar2(1)
JURISDICTION	Varchar2(4)
NHOOD	Varchar2(30)
CNNTEXT	Varchar2(10)
STREETNAME	Varchar2(36)
CLASSCODE	Varchar2(1)
STREET_GC	Varchar2(29)
GEOM	SDO_GEOMETRY
VERSION	Number(38)
NEXTVER	Varchar2(500)
DELSTATUS	Number(38)
LTLOCK	Varchar2(150)

Once the table is versioned, applications can continue to use it just like a normal table, only when a session wants to create a workspace and change some rows in the table, the DML is different.

Creating and using workspaces

In this example, we create two workspaces and show how the new rows or updates to the rows across the different workspaces interact with each other. Note that any new session is in the LIVE workspace by default. This workspace is always there and cannot be created or dropped by users.

```
-- Create workspace called SITE1
EXEC dbms_wm.CreateWorkspace('SITE1') ;

-- goto the new workspace; all SQL will be executed in this workspace
EXEC dbms_wm.GotoWorkspace('SITE1') ;

Insert Into road_clines Values (137901, 1518, 50, 55, 51, 56,
    4366000, 32960000, 26126000, '94114', '08','N', NULL,
'Twin Peaks', '4366000', 'COPPER ALY', '0', 'COPPER',
SDO_GEOMETRY(2002, 2872, NULL, SDO_ELEM_INFO_ARRAY(1, 2, 1),
SDO_ORDINATE_ARRAY( 5999730.36, 2103832.58, 5999750, 2103845)));
Commit;

-- this will confirm that there is one row with the new FID
Select count(*) From ROAD_CLINES Where FID = 137901;
```

```
EXEC dbms_wm.GotoWorkspace('LIVE') ;
-- this will show that there is no row with the new FID
Select count(*) From ROAD_CLINES Where FID = 137901;
```

As we can see from the last step, the newly created row only exists in the workspace SITE1. Unless the session switches to that specific workspace, the new row is not visible until it is merged into the LIVE data. Now we can create a new workspace and create a row with the same FID but with a different geometry. This will not be permitted in a normal table as two rows cannot have the same FID (as it is the PK), but in the version-enabled tables, two rows in two different workspaces can have the same FID.

```
-- Create a new workspace
EXEC dbms_wm.CreateWorkspace('SITE2') ;

EXEC dbms_wm.GotoWorkspace('SITE2') ;

Insert Into road_clines Values (137901, 1518, 50, 57, 51, 56,
    4366000, 32960000, 26126000, '94114', '08','N', NULL,
'Twin Peaks', '4366000', 'COPPER ALY', '0', 'COPPER',
SDO_GEOMETRY(2002, 2872, NULL, SDO_ELEM_INFO_ARRAY(1, 2, 1),
SDO_ORDINATE_ARRAY( 5999730.36, 2103832.58, 5999755, 2103849)));

Commit;

-- this will confirm that there is one row with the new FID
Select count(*) From ROAD_CLINES Where FID = 137901;
```

Now we want to merge these changes back to the LIVE workspace so that normal sessions can see this new row that is created. This is called merging the changes to the LIVE workspace and is done with another PL/SQL procedure call. Assume that the application session working with the SITE2 workspace finished the work and wants to merge the changes to the LIVE workspace.

```
-- you need to be in LIVE workspace to execute merge commands
EXEC dbms_wm.GotoWorkspace('LIVE') ;
-- specify to keep the workspace in tact after the merge
EXEC dbms_wm.MergeWorkspace('SITE2', remove_workspace=>FALSE) ;

-- this query show the current workspace as LIVE
Select DBMS_WM.GetWorkSpace From DUAL;

-- now check for the new row in the LIVE workspace and it shows up
Select count(*) From ROAD_CLINES Where FID = 137901;
```

At this point, the second application session working in the other workspace finishes work and wants to merge the changes to the LIVE workspace, but it will run into an error since SITE1 has completed the merge first.

```
-- goto the LIVE workspace
EXEC dbms_wm.GotoWorkspace('LIVE') ;
-- this will throw an error as shown
EXEC dbms_wm.MergeWorkspace('SITE1', remove_workspace=>FALSE) ;

ERROR at line 1:
ORA-20055: conflicts detected for workspace: 'SITE1' in table:
'BOOK.ROAD_CLINES'
ORA-06512: at "WMSYS.LT", line 6030
ORA-06512: at line 1
```

Conflict resolution

Since there can be only one row with FID=137901, this version of the row is not allowed in the LIVE workspace. At this time, the second workspace can either abandon the changes or try to consolidate the changes with the other workspace and resolve the differences. In such cases, some human interaction might be required to resolve the differences between the two workspaces.

```
-- goto the SITE1 workspace
EXEC dbms_wm.GotoWorkspace('SITE1') ;
-- this will show the actual conflicting values for the row
EXEC DBMS_WM.SetConflictWorkspace ('SITE1');
-- this will show two rows: one from LIVE and one from SITE1
Select * From road_clines_conf;

-- we need to resolve the rows that have the conflict
-- and this is done the doing a Resolve
-- after the commit, the values in SITE1 for the new row will
-- will take precedence over what is in LIVE if we merge
EXEC DBMS_WM.BeginResolve ('SITE1');
EXEC DBMS_WM.ResolveConflicts ('SITE1', 'ROAD_CLINES',
'FID = 137901', 'child');
Commit;
EXEC DBMS_WM.CommitResolve ('SITE1');

-- now we decided that SITE1 values are correct for all columns
-- except LF_TOADD column
-- For LF_TOADD, we update to reflect the values used by SITE2
```

```
Update ROAD_CLINES Set LF_TOADD = 57
Where fid = 137901;

-- now merge the changes from workspace SITE1
EXEC dbms_wm.GotoWorkspace('LIVE') ;
EXEC dbms_wm.MergeWorkspace('SITE1', remove_workspace=>FALSE) ;

-- now remove both the workspaces
EXEC dbms_wm.removeWorkspace('SITE1') ;
EXEC dbms_wm.removeWorkspace('SITE2') ;
```

In this example, after reviewing the changes from the LIVE workspace, we have decided that the changes from the SITE1 workspace are correct except for the LF_TOADD column. For this column, the change done in the SITE2 workspace is correct. So, we update the row in SITE1 to the value used in SITE2.

The Workspace Manager provides the view table_name_conf to show the conflicts between different workspaces. This view can be used to peek at possible conflicts with other workspaces and to resolve the differences as early as possible to avoid building up many conflicts that will be harder to resolve during the merge to the LIVE workspace.

Workspace locking

The Workspace Manager provides exclusive and shared version locks in addition to locks provided by the regular Oracle transactions. You can enable locking on a workspace for a user session, on the specified rows, or for some combination of the two. These locks are primarily intended to eliminate the row conflicts between a parent workspace and a child workspace. The workspace level locking locks any row changed in the workspace. The session level locking locks any row changed by the session regardless of the workspace. The row-level locking locks particular rows and can ensure all rows that must be updated are available for update.

Workspace-exclusive locks and version-exclusive locks are the forms of exclusive locking that control which users can and cannot change data values, but (unlike exclusive locking) they do not prevent conflicts from occurring. The workspace-exclusive locks lock rows so that only the user that set the lock can change the values in the current workspace; however, other users in other workspaces can change the values. The version-exclusive locks lock rows so that only the user that set the lock can change the values (and that user can be in any workspace); no other users (in any workspace) can change the values.

DDL operations on version-enabled tables

One of the big differences one notices with version-enabled tables is during the DDL operations on the version-enabled tables. If we want to add a new column to the ROAD_CLINES table to store the number of lines, then the standard DDL command to alter the table will raise an error. Once the table is version-enabled, the Workspace Manager PL/SQL APIs are required to execute any DDL commands against that table.

```
-- try the alter with the normal DDL command to see the error
Alter Table ROAD_CLINES add (lanes number);

ERROR at line 1:
ORA-00942: table or view does not exist

-- use the PL/SQL api
-- EXEC DBMS_WM.BeginDDL('ROAD_CLINES');
-- notices the special table name used in the DDL command
-- with a _LTS suffix
Alter Table ROAD_CLINES_LTS ADD (lanes number);

-- now commit the DDL changes
EXEC DBMS_WM.CommitDDL('ROAD_CLINES');
```

Two main points need to be highlighted in the previous example. The Alter Table command is executed against a special table name that has the _LTS suffix. Unlike the normal DDL commands that are auto committed, these DDL commands are not committed until the PL/SQL API is called to commit the DDL changes.

Valid-time support

Sometimes applications need to store data with an associated time range that indicates when the data is valid; that is, each row is valid only for a certain time range. The Workspace Manager supports this valid-time concept by adding a column to store the valid time range for each row in the table. The valid-time can span the past, present, or the future. This is useful in the city-planning application where a new set of roads will become valid after a certain date (after the construction is completed). The rows can be entered into the table at design time, but are not made valid for everyone until a future time when the construction is completed. A session level valid-time can be set with a PL/SQL API call to see which rows are valid in the table for any specified time range. In the following example, we show how the valid-time can be used with the ROAD_CLINES table:

```
-- reset the table back to the original state
EXEC dbms_wm.disableVersioning('ROAD_CLINES');
Alter table road_clines Drop (lanes);
Delete road_clines Where FID=137901;

-- now enable versioning with the valid-time option
EXEC dbms_wm.enableVersioning('ROAD_CLINES', 'VIEW_WO_OVERWRITE',
FALSE, TRUE);

EXEC dbms_wm.CreateWorkspace('SITE1') ;

EXEC dbms_wm.GotoWorkspace('SITE1') ;

-- insert a row into the table with a valid-time range
Insert Into road_clines Values (137901, 1518, 50, 57, 51, 56,
    4366000, 32960000, 26126000, '94114', '08','N', NULL,
'Twin Peaks', '4366000', 'COPPER ALY', '0', 'COPPER',
SDO_GEOMETRY(2002, 2872, NULL, SDO_ELEM_INFO_ARRAY(1, 2, 1),
SDO_ORDINATE_ARRAY( 5999730.36, 2103832.58, 5999755, 2103849)),
 WMSYS.WM_PERIOD(TO_DATE('01-07-2013', 'MM-DD-YYYY'),
                 TO_DATE('12-31-9999', 'MM-DD-YYYY')));

Commit;

-- this will show  the row in the table
Select * From ROAD_CLINES Where FID=137901;

-- for the session set a valid time range
EXEC DBMS_WM.SetValidTime(TO_DATE('01-04-2013', 'MM-DD-YYYY'), TO_
DATE('01-06-2013', 'MM-DD-YYYY'));

-- now the select will return no rows
Select * From ROAD_CLINES Where FID=137901;

-- reset the table back to the original state
EXEC dbms_wm.disableVersioning('ROAD_CLINES');
```

As seen in this example, a valid-time range can be set for the session, and this time range can then be used for all DML statements in the session. In addition, the Workspace Manager also provides relationship operators that accept two time period parameters to apply them as time filters for the query. These operators, such as WM_OVERLAPS and WM_CONTAINS, can be used to check if any rows in the table satisfy the specified predicate on the time ranges. For example, the following query will show all the rows from the ROADS_CLINES table having a valid-time contained in the specified time range:

```
Select * From ROAD_CLINES e
Where WM_CONTAINS(e.wm_valid,
    wm_period(TO_DATE('01-01-1995', 'MM-DD-YYYY'),
              TO_DATE('01-06-2013', 'MM-DD-YYYY'))) = 1;
```

Other features of the Workspace Manager

The Workspace Manager provides many more useful functions for spatial applications. We only covered an introduction to this feature in this chapter due to lack of space. The other important features, such as save points and the history of changes, are also very useful for spatial applications.

SecureFiles compression

Oracle SecureFiles is a new feature in Oracle 11*g* that offers the best solution for storing **Large Object (LOB)** data in a database. This feature also provides native compression for the data stored in the lob columns. The SDO_GEOMETRY column has two VARRAYs that can be optionally stored as lobs in the database. The default storage for VARRAYs uses the old style lobs, but by specifying storage parameters for these VARRAYs, one can use the new SecureFiles storage for the VARRAYs. However, this is not always recommended, as there are cases where the default storage for the VARRAYs will be better than the SecureFiles storage option.

VARRAYs by default are stored inline with the rest of the row data if the size of the VARRAY is less than 4000 bytes. When the size of the VARRAY goes beyond the 4000 bytes, the data is stored in a lob in the LOB segment storage. If the users specify storage parameters for the VARRAY, then they are always stored as lobs; inline lobs if the size is less than 4000 bytes and in LOB segments when the size is larger. However, this behavior can be overridden with the ENABLE STORAGE IN ROW parameter while creating the tables. Traditional lobs have more overhead than the normal storage for VARRAYs (for the case when the VARRAY is less than 4000 bytes), so specifying the storage for VARRAYs of the geometry columns is not recommended before 10*g*. But with the SecureFiles storage option, it is sometimes useful to use this new storage option, especially when the compression can be used. The following examples show how to use SecureFiles compression for tables with SDO_GEOMETRY columns:

```
-- create a new table to show the SecureFiles option
Create Table MAP_WORLD_SEC (COUNTRY Varchar2(50),
                            CAPITAL Varchar2(30),
                            GEOMETRY MDSYS.SDO_GEOMETRY )
VARRAY "GEOMETRY"."SDO_ELEM_INFO" STORE AS
SECUREFILE LOB (CACHE COMPRESS HIGH ENABLE STORAGE IN ROW)
VARRAY "GEOMETRY"."SDO_ORDINATES" STORE AS
SECUREFILE LOB (CACHE COMPRESS HIGH ENABLE STORAGE IN ROW);
```

By default, the SecureFiles lobs are created without any compression, and there are different levels of compression available with this feature; the more the compression, the more expensive it will be to select the row data from the table. For example, it usually costs about 20 percent more time to read the geometry data from the table with CACHE COMPRESS HIGH compared to the uncompressed lobs. But at the same time, the storage for the compressed lobs will be about 40 percent of the space taken by the uncompressed lobs. So, this is a trade-off that should be carefully evaluated before deciding on a specific compression option for the geometry data.

Summary

It is very common for spatial applications to try to do many of the data management tasks that can be done in the database more efficiently. In this chapter, we have discussed many common database features useful for the spatial applications. We have shown examples of when these features can be useful for spatial applications along with several code examples. The data consistency check is a common task that is traditionally done in a GIS, but these can be done at the database using triggers as shown in this chapter. We learned how to use materialized views to reduce the cost associated with the repeated query operations. We explained the database replay used to replay the workload captured on a production system on other systems. The versioning of spatial objects is another common operation that is usually managed by a GIS. Oracle provides the Workspace Manager technology to manage the different versions of spatial data, and we also described the storage options for the SDO_GEOMETRY data to reduce the space overhead.

In the next chapter, we'll describe some data replication features provided by Oracle. Data replication is useful when data from a production database is moved to a backup system, or when more than one system is used in production to improve the availability of data.

4
Replicating Geometries

Replication is the process of copying and maintaining data across the different databases in a distributed system. The goal of this chapter is to present a few methods for replicating geometry data. Some of the traditional Oracle replication technologies do not directly support replication of tables with SDO_GEOMETRY data. The examples given here show alternative ways of replicating tables with the geometry data. Replication does not always mean replicating the same data in different databases. In some cases, it also means copying the data from one database into a different database in a different form, for example, data in **Online Transaction Processing (OLTP)** databases can be converted to **Online Analytical Processing (OLAP)** databases by replicating or combining data from different OLTP tables into a single table in the OLAP system. We will show how to do this conversion of data from a transactional OLTP database to a publication or OLAP database. Starting with the 12cR1 release, the logical standby support for SDO_GEOMETRY data types is introduced, so we will look at how this feature can be used to replicate the geometry data. In this chapter, we will cover the following topics:

- Introducing different types of replication
- Materialized view based replication
- Streams based replication
- Physical and logical standby
- OLTP and OLAP databases

Introducing different types of replication

An organization might want to serve users in different geographies from a central database. If the datacenter is located in the US, and some of the users are located in Asia, the network latency for the database queries generated in Asia might be longer than the expected response time for the application. In such cases, the organization might want to maintain another local database in a datacenter in Asia. Once a second database is added to the system, these two databases have to be maintained in sync with each other so that the users in Asia see the same results as the users in the US. Another example is the duplication of state level data sets in the local state plane projection at a state level database, while the national database has the same data in a geodetic system. One way to achieve this data duplication is to use a replication strategy to push the changes from one database to the other database. In this distributed system, there are different scenarios that can be supported depending on the application's requirements. In a simple master-replication strategy, changes to the database are pushed to one designated database called the **master database**. Once the changes are committed to the master database, it can push these changes to the other secondary databases in the system. These secondary databases are called the **slave databases**. In some scenarios, the distributed system might want to accept changes in all the databases. In such cases, a multi-master replication strategy is used to maintain all the databases in sync with each other using some of the advanced replication strategies of Oracle. As can be expected, the multi-master replication strategies are more complex than the master-slave replication strategies.

The examples in this chapter are based on a typical application workflow often used in the geographic databases. In a real-world system, there can be multiple hierarchical levels, but we will keep our example to a simple system with two levels. In this system, a county (district) level agency is responsible for maintaining the master map for each county (district). At the next level down, the cities are responsible for collecting and maintaining the map data for each city. The city pushes its map data to the county agency at regular intervals. The county collects data from each city, aggregates and aligns the map data, and publishes the resulting map. The individual cities can then get the master map back from the county to use as a reference map. In such a system, the information has to flow from the city to the county and then from the county to the cities. A master-slave replication strategy can be used in this scenario to push the data back and forth from the city and county level databases. Note that the replication strategy can be set at different granularities, namely database, schema, or table. Depending on the application's requirements, a suitable level of replication can be chosen. The following figure shows a replication process at the table level with a master-slave replication strategy. A replication group can be the whole database, or a schema, or one of more tables in a schema.

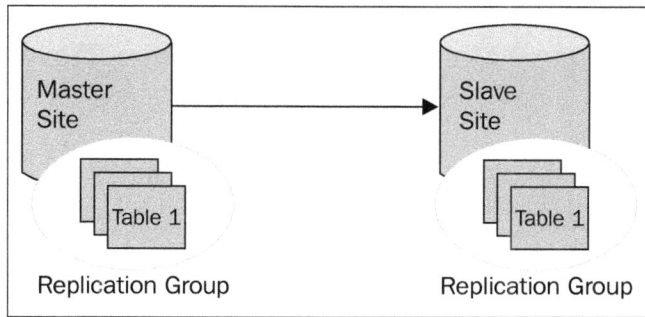

Replicating data with materialized views

Materialized view based replication is very useful when the changes at the master site do not have to be immediately available at the secondary sites. This is also useful when the connectivity between the master and slave sites is not continuous. When only a subset of tables or portions of the tables are required at the secondary sites in an asynchronous fashion, materialized view based replication is one of the most common ways to achieve it.

As we discussed in the previous chapter, a materialized view is a replica of a table that can be refreshed at regular intervals or on demand. During a refresh, only the final values of the changed rows are pulled down, no matter how many updates were applied to the master table. This results in the reduced amount of data being transferred to the remote site.

Let's use the LAND_PARCELS table in our example to show the replication process using the materialized views. First, the database administrators at the city and county agencies have to agree on a protocol for sharing the data between the databases. Once this is done, the database links have to be established at both the sites so that the users from one database can access the data from the other database. If users are familiar with the Oracle database network concepts, the following example will be easy to understand. If not, they should refer to these concepts in the Oracle User guides before proceeding with the rest of the example.

In the county database, the administrator first creates a TNSNAMES entry to connect to a city database. A typical TNSNAMES entry in the tnsnames.ora file looks like the following:

```
City1 =
(DESCRIPTION=
  (ADDRESS_LIST =
   (ADDRESS=(PROTOCOL=tcp)(HOST=host1.city1.com) (PORT=26091)))
  (CONNECT_DATA=
     (SERVICE_NAME=db1.city1.com)))
```

With this network entry defined, the users from the county database can connect to the city database using the following SQL command:

```
sqlplus book/book@city1
```

Once it is confirmed that the connection can be established between these two databases, a database link can be created using the following SQL. Note that the Oracle Enterprise Manager has GUI tools that provide more helpful screens for doing these steps, but we will show the SQL examples here, as the Enterprise Manager steps are self-explanatory.

```
Drop database link city1_db;
Create DATABASE LINK city1_db
   CONNECT TO book IDENTIFIED BY book
   USING 'city1';
```

Note that the USING clause should specify the same name used to identify the database in the tnsnames.ora entry. If the database has set the GLOBAL_NAMES initialization parameter value to TRUE, this database link creation process will be different. If these instructions do not work, please check this parameter and follow the appropriate instructions specified in the Oracle User Guide. The database link here specified the user name and password for the remote database. With this type of link, remote users can only access the specified schema and are not able to access other objects in other schemas on the remote database. The city level DBAs can use this model to limit the remote access to the data in their database. They can create a new schema that has read-only access to the tables they want to publish to the remote site. Then they share this username and password to this new schema with the remote site administrators, and using the database link to this new schema, the remote site users only have read-only access to those tables that are visible to this new schema. After the database link is established, the users from the county database can connect to the city database and look at the contents of the tables.

```
Select * From land_parcels@remotedb Where fid = 5256;
```

Now, the users at the county database can access the cities' LAND_PARCELS table. So, users might wonder "what is the need for replication, if we can already see the table from the county database". Note that the data is still remote, so every time the county database needs to look at the data from the LAND_PARCELS table, the data has to be sent over the network from the remote database. Also, the types of queries that can be performed on a remote database are limited, and if this data has to be joined with other tables in the local database, the query performance will suffer. For these reasons, the county DBAs would like to have a copy of the LAND_PARCELS tables in the local database. We can use materialized views to solve this local copy problem.

```
Drop materialized View land_parcels_city1;
Create  materialized View land_parcels_city1
Refresh force  on demand as
Select * From land_parcels@remoteDB;

Select count(*) From land_parcels_city1;

  COUNT(*)
----------
    154202
```

Once the materialized view is created, it can be refreshed on demand by the country DBAs. This can be done periodically or on a request from the city DBAs. Using this methodology, the materialized view refresh can be done to copy only those rows that are changed from the last refresh using the FAST REFRESH option. First, go to the city database and delete one row from the LAND_PARCELS table and after the refresh we should see 154201 rows from the county database.

```
EXEC dbms_mview.refresh('LAND_PARCELS_CITY1', 'F');
Begin dbms_mview.refresh('LAND_PARCELS_CITY1', 'F'); End;

ERROR at line 1:
ORA-12004: REFRESH FAST cannot be used for materialized view
"BOOK_C"."LAND_PARCELS_CITY1"
ORA-06512: at "SYS.DBMS_SNAPSHOT", line 2563
ORA-06512: at "SYS.DBMS_SNAPSHOT", line 2776
ORA-06512: at "SYS.DBMS_SNAPSHOT", line 2745
ORA-06512: at line 1
```

This refresh call raises an oracle error, so what happened? We can get more information about the cause of this error by using the EXPLAIN_MVIEW procedure. But first, the script utlxmv.sql should be loaded into the database to create the required tables to run the EXPLAIN_MVIEW procedure.

```
@$ORACLE_HOME/rdbms/admin/utlxmv.sql

EXEC DBMS_MVIEW.EXPLAIN_MVIEW ( 'LAND_PARCELS_CITY1' );

Select possible, related_text, msgtxt
From mv_capabilities_table
Where capability_name = 'REFRESH_FAST_AFTER_INSERT';

POSSIBLE
-
RELATED_TEXT
----------------------------------------------------------------------
MSGTXT
----------------------------------------------------------------------
N
BOOK.LAND_PARCELS
the detail table does not have a materialized view log
```

In the city database, the table has to be enabled with the fast refresh view option by creating a materialized view log for the LAND_PARCELS table. This log keeps track of the changes to the table and only those changes that need to be pushed to the materialized view are used during the refresh operation.

```
-- connect to the city database first and do this
Create materialized View log on land_parcels;

Delete From LAND_PARCELS Where fid = 5246;

-- connect to the county database now and try the refresh operation
EXEC dbms_mview.refresh('LAND_PARCELS_CITY1', 'F');

PL/SQL procedure successfully completed.

Select count(*) From land_parcels_city1;

  COUNT(*)
----------
    154201
```

Now, as one can see, the county database shows `154201` rows in the table. Using this mechanism, a subset of tables can be easily replicated at multiple sites, and this is the most flexible way to replicate the spatial tables. In some situations, the replicated data might be an aggregated view of the data in the city database; for example, in the city database, the road centerline data is stored at a very granular scale. But in the county database, they need only one geometry for each street, if they plan to use this data for the base map of the county. In such cases, a materialized view can be created that aggregates the data from the `road_clines` table.

```
Create materialized View road_clines_aggr
As Select road_id, zip_code, district, streetname,
sdo_aggr_concat_lines(geometry) geometry
From road_clines@remoteDB
Group by road_id, zip_code, district, streetname;

Select count(*) From road_clines_aggr;

   COUNT(*)
   --------
       4095
```

Note that this materialized view can only be refreshed using the full refresh option as the SQL used to create the materialized view has a PL/SQL function in the `select` clause. We can see the restriction by running the `EXPLAIN_MVIEW` procedure for this materialized view.

```
EXEC DBMS_MVIEW.EXPLAIN_MVIEW ( 'LAND_PARCELS_CITY1' );

Select possible, related_text, msgtxt
From mv_capabilities_table
Where capability_name = 'REFRESH_FAST_AFTER_INSERT';

POSSIBLE
---------
RELATED_TEXT
------------------------------------------------------------------
MSGTXT
------------------------------------------------------------------
N

mv references PL/SQL function that maintains state
```

Streams based replication

Oracle provides two basic technologies for general replication between two or more Oracle databases, namely, advanced replication and streams based replication. The streams based replication provides a more efficient solution than the one based on advanced replication. Rather than synchronously capturing changes in the materialized view logs, which adds overhead to the transaction, streams mine the redo logs created by the updates to the database and create change logs to send to the target database system. This architecture also allows the change to be replicated to more than one system much more quickly. The advanced replication also has some restrictions with SDO_GEOMETRY types that are harder to work around.

In this section, we will describe the streams based replication for the tables with SDO_GEOMETRY columns. Note that even the native streams based replication has restrictions for the tables with SDO_GEOMETRY columns. But procedures can be easily implemented to work around these restrictions, and we will describe these in detail with examples in this section.

The streams can be used to replicate both **data definition language (DDL)** and **data manipulation language (DML)** operations from a source database to one or more target database systems. With streams, replication of a change typically includes three steps, a capture stage, a propagation stage, and an apply stage.

During the capture stage, a capture process records the changes at the source database as one or more **logical change records (LCRs)** and adds them into a queue. If the change was a DML operation, then each LCR encapsulates a row change resulting from the DML operation to a shared table at the source database. If the change was a DDL operation, then an LCR encapsulates the DDL change that was made to a shared database object at a source database. Then a propagation process sends the staged LCR to another queue typically in a database that is separate from the source database where the LCR was captured. The LCR may be propagated to a number of queues before it arrives at a destination database. At the target or destination database, an apply process consumes the changes by applying the LCR to the shared database object. During the apply process, the LCR can be directly applied to the target database or it can be processed before it is applied to the target database. In a streams replication environment, an apply handler can perform customized processing of the LCR and then apply the LCR to the shared database object. If the users want to skip some changes from being applied to the target database, then these skip actions can be encoded as part of the apply handler. The following figure shows the typical flow in a streams based replication system:

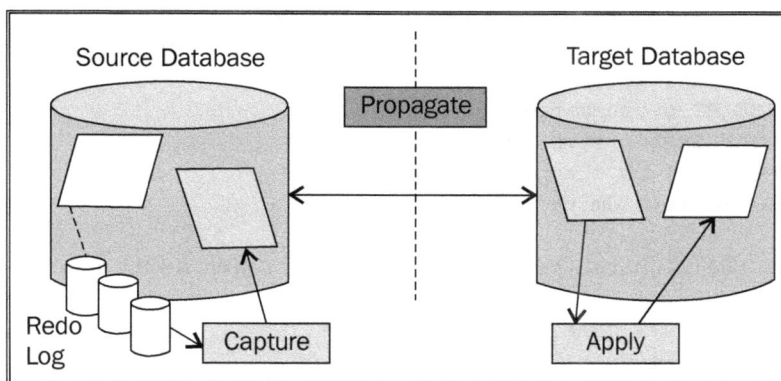

Setting up the database

The streams replication requires the database to be set up with a few required parameters. The database should be GLOBAL_NAME enabled and the archive log should also be enabled. Setting up these parameters are common DBA tasks and are outside the scope of this book. Please refer to the Oracle documentation to understand how to do these tasks. In our example, we do provide the code for doing these operations, but these can vary for each database. We use SOURCE as the global name of the source database and TARGET as the global name of the target database. The domain name for the global name is used as example.oracle.com in the following script, but it should be set according to one's own system domain name. Make sure the TNSNAMEs entries are entered correctly for both the source and target systems; that is, the source and the target systems should be able to tsnping each other using the names SOURCE and TARGET. Once these TNSNAMEs are set up, execute the following steps on the source database. Make sure these TNSNAMEs entries for both the source and target are set up at both the source and target databases. We will use /app/oracle/source/streams/exp_dp as the directory for managing all the replication related objects. Make sure that this directory is created on both the source and the target systems of the databases.

```
-- make sure these are executed on the SOURCE database system
Alter DATABASE rename global_name to source.example.oracle.com;

-- create a separate tablespace  to hold all the streams data
Create TABLESPACE strm_tb DATAFILE
```

```
       '/app/oracle/dbs/source/strm_tb.dbf' SIZE 25M;
-- create a user to manage the streams replication
Create USER strmadmin
IDENTIFIED BY strmadminpw
DEFAULT TABLESPACE strm_tb
QUOTA UNLIMITED ON strm_tb;
GRANT DBA TO strmadmin;

-- grant the required privileges to the streams admin user
Begin
DBMS_STREAMS_AUTH.GRANT_ADMIN_PRIVILEGE(
grantee => 'strmadmin',
grant_privileges => true);
End;
/
Create Or Replace directory STRMS_DP_DIR As '/app/oracle/source/
streams/exp_dp';

CONNECT strmadmin/strmadminpw;
Create DATABASE LINK target CONNECT TO strmadmin IDENTIFIED BY
strmadminpw
Using 'TARGET.EXAMPLE.ORACLE.COM';

-- this should return some rows back from the target database
Select count(*) From dba_tables@target;
```

Next, we do the setup on the TARGET database to create the streams admin user and also create the directory object to manage the replication. Make sure these directory structures exist on the target system before running the following script.

```
-- make sure these are executed on the TARGET database system
Alter DATABASE rename global_name to target.example.oracle.com;
-- create a separate tablespace  to hold all the streams data
Create TABLESPACE strm_tb DATAFILE
   '/app/oracle/dbs/target/strm_tb.dbf' SIZE 25M;
-- create a user to manage the streams replication
Create USER strmadmin
IDENTIFIED BY strmadminpw
DEFAULT TABLESPACE strm_tb
QUOTA UNLIMITED ON strm_tb;
GRANT DBA TO strmadmin;

-- grant the required privileges to the streams admin user
Begin
```

```
DBMS_STREAMS_AUTH.GRANT_ADMIN_PRIVILEGE(
grantee => 'strmadmin',
grant_privileges => true);
End;

Create Or Replace directory STRMS_DP_DIR As '/app/oracle/target/
streams/exp_dp';

CONNECT strmadmin/strmadminpw;
Create DATABASE LINK source CONNECT TO strmadmin IDENTIFIED BY
strmadminpw
Using 'SOURCE.EXAMPLE.ORACLE.COM';

-- this should return some rows back from the source database
Select count(*) From dba_tables@source;
```

Now, connect to the SOURCE database and set up the archive log destination directory for the target. The source database will use these directories to send the logs to the target database, so make sure these directories really exist on the target system. The following set of SQL commands show how to set up the source database in the archive log mode. This is very important as the database logs are used for the streams based replication.

```
-- run these steps on the source database
-- check if the database is enabled for archive log
archive log list;
-- refer to the DBA guide on more information on these steps
Alter SYSTEM set LOG_ARCHIVE_DEST_1='SERVICE=target ARCH OPTIONAL
NOREGISTER TEMPLATE=/app/oracle/target/streams/exp_dp/source_
arch_%t_%s_%r.redo';
Alter SYSTEM set LOG_ARCHIVE_DEST_STATE_1=ENABLE;

shutdown immediate;
-- the next step depends on how the pfile is managed
startup mount pfile=init1.ora
Alter DATABASE archivelog;
Alter DATABASE open;
Alter DATABASE force logging;
-- check to make sure archive log is turned on
archive log list;
Select force_logging Fromv$database;

Alter system switch logfile;
```

Enabling replication

Now, both the databases are configured and are ready for replication to be enabled. Oracle Streams provides several methods for enabling the replication between a source and a target. In our example, we will use the procedure that generates a script that in turn can be used to enable the replication for a given schema. Before we run the script, we should make sure that the directory object on the target is set correctly so that the script can be generated. The following script tries to create a sample file in the directory and if it works as expected, users can proceed to the next stage:

```
-- run this on the source database
SET SERVEROUT ON

Declare
v_buff Varchar2(2000);
fhandle UTL_FILE.FILE_TYPE;
Begin
dbms_output.put_line('WRITE');
fhandle:= UTL_FILE.FOPEN('STRMS_DP_DIR','test.txt','W');
UTL_FILE.put_line(fhandle,' Attempt to write');
UTL_FILE.FCLOSE(fhandle);
End;
```

If the file `test.txt` is created successfully, then run the following procedure on the source database system. In our example, we are using only one source and one target database system. The streams allow a third system to manage the replication processing, but in our example, we just use two systems to keep the example simple.

```
-- connect to the source database
Begin
DBMS_STREAMS_ADM.MAINTAIN_SCHEMAS(
schema_names => 'BOOK',
source_directory_object => 'STRMS_DP_DIR',
destination_directory_object => 'STRMS_DP_DIR',
source_database => 'source',
destination_database => 'target',
perform_actions => false,
script_name => 'instantiation.sql',
script_directory_object => 'STRMS_DP_DIR',
dump_file_name => 'book_rep.dmp',
capture_queue_table => 'rep_capture_queue_table',
capture_queue_name => 'rep_capture_queue',
capture_queue_user => NULL,
apply_queue_table => 'rep_dest_queue_table',
```

```
apply_queue_name => 'rep_dest_queue',
apply_queue_user => NULL,
capture_name => 'capture_book',
propagation_name => 'prop_book',
apply_name => 'apply_book',
log_file => 'export_book.log',
include_ddl => true,
instantiation => DBMS_STREAMS_ADM.INSTANTIATION_SCHEMA);
End;
```

When this procedure completes successfully, it will create a file called
`instantiation.sql` in `/app/oracle/target/streams/exp_dp` directory. Now,
go to that directory and execute the script from the source database system. This
script will ask for the TNS name of the source database, the stream's admin user
and password followed by the TNS name of the target database, and the stream's
admin user and password. After the source and target information, it will ask for the
information for a third system. Since we are only using two systems in our example,
you can just enter any dummy string values for this. After these values are entered,
it will take a while to execute the script. Since this can be error prone if the directory
structures are not set up correctly, it is a good idea to spool the output to a log file for
later inspection in case of errors. So, execute the following steps to run the script on
the source database system:

```
-- go to the directory where the script is generated
-- connect to the source database system
spool instantiation.log
-- this next command is in one line
@instantiation.sql source  strmadmin  strmadminpw target strmadmin
strmadminpw  source  strmadmin strmadminpw
```

The first part of the script will use a data pump to move the data from the source
database to the target database. This simply exports the data from the source and
imports the same into the target database. Once the script completes, you can connect
to the target database as the book user, and the book schema tables from the source
should be present in the target system. Once the script successfully completes, there
should be replication processes setup on both the source and the target databases.
Connect to each database as the STRMADIN user and look for streams-related objects.
The following SQL command can be used to check the capture process on the source
and apply the process on the target we are running:

```
-- connect to the target database
-- the following should return one row
Select APPLY_NAME, QUEUE_NAME, APPLY_DATABASE_LINK, STATUS
From DBA_APPLY
```

```
Where APPLY_NAME = 'APPLY_BOOK';

-- connect to the source database
-- the following should return one row
Select CAPTURE_NAME, QUEUE_NAME, SOURCE_DATABASE, STATUS
From DBA_CAPTURE
Where CAPTURE_NAME = 'CAPTURE_BOOK';
```

If the previous steps confirm that the apply and capture processes are active, then connect to the source database and create a new test table and verify that the table propagates to the target database. Most of the tables in the BOOK schema have the SDO_GEOMETRY column and hence streams will not replicate these tables by default. Note that the initial copy of the tables does work for all the tables in the schema, but any subsequent DML operations on these tables with SDO_GEOMETRY column will not be replicated by default.

Extending streams to support SDO_GEOMETRY

Oracle Streams has an additional procedure that needs to be set up to support tables with SDO_GEOMETRY columns. The stream calls this technology as **Extended Data Type Support** (EDS). This requires an additional EXTENDED_DATATYPE_SUPPRT package supplied by Oracle. More information on how to obtain this package is given in this white paper: http://www.oracle.com/technetwork/database/features/availability/maa-edtsoverview-1-128507.pdf.

In this paper, check the section on using streams for replication of data. With the EDS feature, the tables with SDO_GEOMETRY can be replicated using the streams. The EDS uses triggers and staging tables on the source and target databases to achieve the geometry replication. For each table that has a geometry column, the EDS creates a staging table with columns equivalent to the source table. The staging table is created with columns of basic types so that all column values are supported by the standard streams replication, for example, the SIDEWALKS table with three columns has a staging table that looks as follows:

```
Create Table "BOOK"."L$SIDEWALKS" (
    dmltype Varchar2(1)
  , "O$FID" Number
  , "O$FEATSUBTYPE" Number
  , "O$SYS_NC00004$" Number
  , "O$SYS_NC00005$" Number
  , "O$SYS_NC00006$" Number
```

```
    ,  "O$SYS_NC00007$" Number
    ,  "O$SYS_NC00008$" Number
    ,  "FID" Number
    ,  "FEATSUBTYPE" Number
    ,  "GEOMETRY" RAW(1)
    ,  "SYS_NC00004$" Number
    ,  "SYS_NC00005$" Number
    ,  "SYS_NC00006$" Number
    ,  "SYS_NC00007$" Number
    ,  "SYS_NC00008$" Number
    ,  "SYS_NC00009$" BLOB
    ,  "SYS_NC00010$" BLOB
    )
```

The staging table has a column to specify the type of the operation so that the trigger on the target side can apply the right action to the actual user table. The EDS triggers on the source table convert the SDO_GEOMETRY into a BLOB and store that in the staging table. This staging table is then replicated to the target database. On the target database, the EDS created triggers move the data from the staging table to the actual user table. All of this data movement from and to the staging tables happens under the covers via the EDS created triggers. The following figure shows the general flow of data when the EDS is used to replicate the SDO_GEOMETRY data:

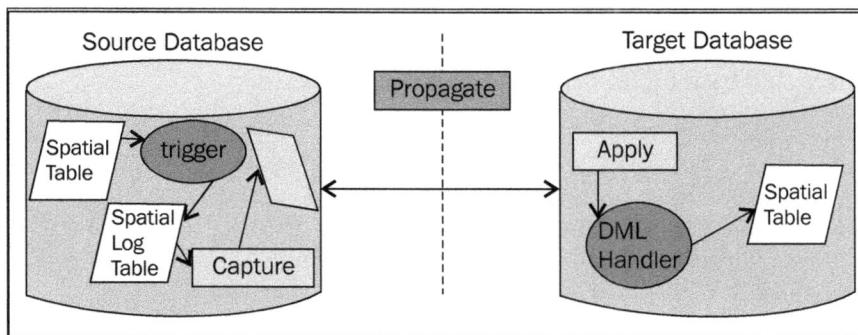

Enabling EDS for geometry tables

Using the EDS provided procedures, we need to generate scripts that are required to register the Oracle Spatial tables with EDS. We first create the script required on the target database. The following procedure call will generate two scripts in the directory we created before for the replication process:

```
-- run this procedure on the TARGET database
-- first the STRMADMIN user needs more privileges for EDS
-- EDS provides the following script in the zip files
@create_adm.sql strmadmin strmadminpw

-- now generate the scripts for the target database
Begin
 extended_datatype_support.set_up_destination_schemas(
    schema_names =>'BOOK', apply_user =>'STRMADMIN', apply_name
=>'APPLY_BOOK',
source_database =>'SOURCE', destination_queue_name=>'REP_DEST_QUEUE',
perform_actions=>FALSE, script_directory_object=>'STRMS_DP_DIR',
script_name_prefix=>'book_eds_1');
End;
```

This generates two scripts, namely, book_eds_1_dest_data_1.sql and book_
eds_1dest_streams_2.sql. Load both these scripts into the target database while connected as the STRMADMIN user. These scripts will create the staging tables and the corresponding triggers on the target database. Now, we need to do the same setup on the source database by first creating the required scripts.

```
-- connect to the SOURCE database
-- first the STRMADMIN user needs more privileges for EDS
-- EDS provides the following script in the zip files
@create_adm.sql strmadmin strmadminpw

-- now generate the scripts for the source database
Begin
   extended_datatype_support.set_up_source_schemas(
schema_names=>'BOOK', capture_name=>'CAPTURE_BOOK',
propagation_name=>'PROP_BOOK', source_database=>'SOURCE',
source_queue_name =>'STRMADMIN.REP_CAPTURE_QUEUE',
destination_queue_name =>'STRMADMIN.REP_DEST_QUEUE@TARGET.REGRESS.
RDBMS.DEV.US.ORACLE.COM',
perform_actions=>FALSE, add_capture_rules=>TRUE,
add_propagation_rules=>TRUE,
script_directory_object=>'STRMS_DP_DIR',
script_name_prefix=>'book_eds_1');
End;
```

This will generate two script files in the directory pointed to by `STRMS_DP_DIR` objects, namely, `book_eds_1_source_data_1.sql` and `book_eds_1_source_streams_2.sql`. Load these two scripts into the source database while connected as the `STRMADMIN` user. When the scripts complete, the system will be set up with all the required staging tables and triggers for the apply process. Before the scripts are loaded into the database, the capture process on the database should be stopped so that the DDL operations done for creating the triggers and staging tables are not propagated to the target database.

```
-- connect as STRMADMIN user on the source database
EXEC dbms_capture_adm.stop_capture('CAPTURE_BOOK');
```

Now, load the two generated scripts into the source database while connected as the `STRMADMIN` user. Then connect to the source database and start the capture process to continue with the replication process.

```
-- connect as STRMADMIN user on the source database
EXEC dbms_capture_adm.start_capture('CAPTURE_BOOK');
```

Now, connect to both the source and target database to make sure the capture and the apply processes are still running. Once these are verified, pick a table with `SDO_GEOMETRY` column and do a DML operation on the source database and verify that the corresponding operation is replicated on the target database.

Physical and logical standby database

Oracle provides two features that support high availability of the database. In this context, availability is the extent to which a database is accessible on demand. The availability of a database is measured by the perception of an end user. A highly available system is accessible to the end users 24/7. A database might become unavailable due to hardware or software problems. The only way to build a highly available system in the face of hardware failure is to have a secondary or an alternate database running on a separate hardware system. A software failure or data corruption can also make the database unavailable. In such cases, a separate system can be used as a failover system such that all the users and applications can transparently use the second system while the primary system is being repaired. In high availability architecture, the main database is called the primary database and the secondary database is called the standby database. The standby databases can also be used to run reports and queries as the data in the standby database is essentially a copy of the primary database. In many organizations, the standby databases are used to offload long running reporting type queries to reduce the load on the primary database. The standby databases offer other advantages such as rolling upgrades and patches as well.

Oracle provides different technologies for building highly available systems. The Oracle automatically transfers data from the main database to a standby database, and in case of failures, it will switch over to the standby database. It provides two ways to create a standby database; a logical standby and a physical standby.

Physical standby

In the physical standby model, two identically configured databases operate on two identically configured hosts. The systems should be identical with respect to all the resources used by the database, including the operating system and the I/O systems. The database schemas and tables should also be the same on both the primary and the standby database. In the conventional mode, the standby database is operated in the standby mode; that means this database cannot be used for running any queries while in this mode. If the primary database fails for some reason, then this standby database can be quickly converted to the primary database to make the database available to the users.

The primary database accepts all the database transactions. The changes are shipped to the standby database by automatically sending the redo logs to the standby database and applying the changes to that instance. With this model, the standby database is always in sync with the primary database, but with a slight delay due to the network latency incurred for copying the change logs to the standby database. The actual setup of the physical standby is outside the scope of this book, as we focus more on the logical standby concepts in this book.

Logical standby

Logical standby is different from physical standby in several ways. The logical standby database structure does not have to match the schema structure of the primary database. It uses the LogMiner technology that transforms the archived redo logs into the DML statements (such as insert/update/delete). These DML statements are shipped to the standby database and the SQL is applied there. One of the main differences is the ability to open the tables on the standby database for updates, and the logical standby database can have additional indexes or materialized views that do not exist on the primary database. This is very useful in situations where the standby database used to run reporting queries to offload the work from the primary database. If additional indexes are required to improve the performance of these reporting queries, these indexes can be created on the standby database without affecting the operations of the primary database.

Another difference of logical standby is that the whole database does not have to be replicated at the standby site. Users can set up skip rules that are used to skip certain tables from being replicated at the standby site. When the DML or DDL operations on these skipped tables are noticed in the redo logs, the SQL corresponding to these tables is skipped at the standby site. With this feature, selective or limited tables can be replicated at the standby site. If only some of the tables in the primary database are required to be highly available, only those tables can be replicated at the standby site; this means that the hardware and other resources allocated for the standby database can be much smaller or different from the primary database.

Setting up a logical standby database

Before setting up the logical standby, the user tables need to be analyzed to make sure all the tables can be supported by the logical standby as it does not support all data types and storage attributes. In this example, we want to use all our tables in the BOOK schema in setting up the logical standby instance. We can check if these tables are supported by the logical standby or not by running the following SQL query:

```
Select table_name, column_name From DBA_LOGSTDBY_UNSUPPORTED  Where
owner='BOOK';


TABLE_NAME                            COLUMN_NAME
-----------------------------------   -----------------------------------
BASE_ADDRESSES                        GEOMETRY
BUILDING_FOOTPRINTS                   GEOMETRY
CITY_FURNITURE                        GEOMETRY
LAND_PARCELS                          GEOMETRY
PLANNING_NEIGHBORHOODS                GEOMETRY
ROAD_CLINES                           GEOMETRY
SIDEWALKS                             GEOMETRY
WATER_AREA                            GEOMETRY
```

As seen from the results, the tables with the SDO_GEOMETRY column are not supported by the logical standby. But with the next database release (12*c*), this restriction is lifted and tables with the SDO_GEOMETRY columns are fully supported with the logical standby. So, the rest of the steps in this chapter only apply to the Oracle 12*c* database release.

The logical standby also requires the rows that are changed on the primary database be uniquely identified so that the changes can be applied on the standby database. Since the physical organization of the logical standby database is different from that of the primary database, the ROWIDs on the standby will be different from the ROWIDs on the primary, for each row in the table. So, the ROWID cannot be used to uniquely identify a row in the table. If there is no primary key, Oracle logs the shortest non-null unique key as part of the Update statement to identify the modified row. In the absence of a Primary Key and non-null unique key, all columns of a limited size are logged as part of the Update statements; that means all columns except those with the types LONG, LONG RAW, LOG, and objects. Since identifying a row using these other columns can be expensive, it is always recommended that the tables participating in the logical standby have a Primary Key or a non-null unique index so that the apply process on the standby can be done efficiently.

The first part of the logical standby setup requires setting up a physical standby database. After the physical standby database is ready, the redo apply process on the standby database is stopped before converting it into a logical standby database. A standby could be running as a physical standby database for some time before turning that into a logical database, that is, any existing physical standby database can be converted to a logical standby database anytime. In this following example, we only describe the main components of setting up the logical standby, as detailed steps for setting up a logical standby are outside the scope of this book:

```
SQL> Alter DATABASE RECOVER MANAGED STANDBY DATABASE CANCEL;
```

On the primary database, a log miner dictionary must be built into the redo data so that the log miner part can correctly interpret the changes it sees in the redo logs. So the following SQL command should be executed from the primary database:

```
-- Build a Dictionary in the Redo Data
SQL> EXECUTE DBMS_LOGSTDBY.BUILD;
```

Now, go to the standby database and alter it to be a logical standby database.

```
-- alter the database to be a logical standby database
SQL> Alter DATABASE RECOVER TO LOGICAL STANDBY <NewDatabaseName>;
```

After this, some initialization parameters need to be adjusted to reflect the new state of the standby database followed by the commands to start the database in the logical standby mode.

```
SQL> SHUTDOWN IMMEDIATE;
SQL> STARTUP MOUNT;
SQL> Alter DATABASE OPEN RESETLOGS;
-- Start applying redo data to the logical standby database:
SQL> Alter DATABASE START LOGICAL STANDBY APPLY;
```

At this time, the standby database is ready to be the logical standby for the primary database. Any DML or DDL operations that happen on the primary will now be shipped to the standby database applied there. The standby database can also be directly accessed at this time to run queries against it. New tables and indexes can be created in this database as if this is a regular database without affecting the primary database.

Skipping rules

The logical standby database can be configured to skip changes to some of the tables in the primary database; for example, in our BOOK schema, we have created a temporary table, LAND_PARCELS_INVALID, to store and rectify the invalid geometry data. There is no reason to replicate this table on the standby database, as this is a temporary table that is only used to invalid data. We can just add rules so that the logical standby will skip the DML against this table.

```
SET SERVEROUTPUT ON
Begin
    DBMS_LOGSTDBY.SKIP(
        stmt => 'DML',
        schema_name => 'BOOK',
        table_name => 'LAND_PARCELS_INVALID',
        proc_name => NULL
    );
End;
```

Working with logical standby

Once the logical standby is set up and the skip rules are in place, we can run some operations to see how they get applied on the standby site. We will create a new table to use in this example, as all the other tables that are already part of the BOOK schema will be automatically initialized at the standby site as part of the initial setup.

Synchronizing the standby and primary databases

When reporting, the queries are run on the standby database. It is important to make sure the standby database is in some logically consistent mode with respect to the primary database; for example, a batch process would update the LAND_PARCELS table with the changes that are coming in from a GIS frontend. After the last change is committed, a program should be started on the standby database to generate a report about the LAND_PARCELS created for the month. But there is no deterministic time at which point these changes will be available at the standby database. It is guaranteed that the changes from the primary will be eventually propagated to the standby database, but there is no predetermined time interval after which the changes will be available at the standby database. This is because the logical standby pushes the changes to the standby database in an asynchronous fashion. So, the DML operations that happen on the primary might not come to the standby database in the same order as they happen on the primary database. But the DDL operations are serialized; this means that if a DML operation happens after a particular DDL operation, then that DML will not be applied on the standby until the DDL is applied first.

Similarly, if a DML operation happened before the DDL operation on the primary database, then that DML is applied to the standby database before the DDL is applied. So, a DDL operation can be used as a synchronizing point on the standby database to check if the changes from the primary database are applied to the standby database. The following procedures show a simple method that can be used on the standby database to synchronize the changes with the primary database. First, we create a procedure that can be used to create a new table on the primary database with a unique name. For this to work, a sequence is first created to generate unique numbers, and those numbers are used to create unique table names. This procedure must be created on the primary database.

```
Create Or Replace procedure sync_logical (link_name Varchar2) As
    cnt number := 0;
    v_role Varchar2(20);
    primary Varchar2(31);
    v_table_name Varchar2(30);

Begin
```

```
primary := '@' || link_name;
execute immediate 'Select database_role From v$database' || primary
     Into v_role;
 If v_role != 'PRIMARY' Then
    raise_application_error(-20502,'ERROR, expected link to
    point to primary');
End If;
 execute immediate 'call sys.sync_logical_ddl' ||primary||'(:1)'
     Using out v_table_name;

 -- wait for table
 while (cnt = 0) loop
   -- See if the table exists
   Select count(1) Into cnt
   From dba_tables
   Where owner = 'BOOK' and table_name = v_table_name;

   If (cnt = 0) Then
     dbms_lock.sleep(10);
   End If;
 End Loop;
 -- break out of the loop when we see the sync table on standby
End;
```

Next, we create the synchronization procedure on the standby database that will check for the existence of the marker table. This procedure takes a database link name as the only parameter. This database link should be a link to the primary database that can be accessed from the standby. So this link should be created before the procedure can be executed from the standby database. It should be created by the system user on the primary database (or to some other DBA user on the primary database).

```
Create or Replace procedure sync_logical (link_name Varchar2) As

    cnt number := 0;
   v_role Varchar2(20);
   primary Varchar2(31);
   v_table_name Varchar2(30);

Begin
```

```
    primary := '@' || link_name;
    execute immediate 'Select database_role From v$database' || primary
         Into v_role;
     If v_role != 'PRIMARY' Then
        raise_application_error(-20502,'ERROR, expected link to point to
primary');
     End If;
      execute immediate 'call sys.sync_logical_ddl' ||primary||'(:1)'
           using out v_table_name;

     -- wait for table
     while (cnt = 0) Loop
       -- See if the table exists
       Select count(1) Into cnt
       From dba_tables
       Where owner = 'BOOK' and table_name = v_table_name;

       If (cnt = 0) Then
         dbms_lock.sleep(10);
       End If;
     End Loop;
     -- break out of the loop when we see the sync table on standby
     End;
```

With these two procedures in place, it will be easy to synchronize the primary and standby databases; for example, every night at 11 PM, the last batch job finishes making the changes to the primary database. Then at 11 PM, the sync procedure can start on the standby database, and look for the sync table to appear. When it sees the sync table on the standby, the reporting applications can start on the standby database. Note that this does not have to be a time-based operation. The batch job on the primary database can send a message to the standby database when the processing is done, and then that can trigger this sync procedure on the standby database.

OLTP and OLAP databases

The database systems are usually designed either as an OLTP database or as an OLAP database. An OLTP database is typically characterized by a large number of short transactions that take no more than a few seconds (or minutes) to complete. The main emphasis of the OLTP systems is on very fast query processing, maintaining data integrity in multi-user environments. The number of transactions per second usually measures the performance of an OLTP database. The OLTP databases generally contain detailed and current data, and the data models are usually designed using the third normal form. In these systems, the data models are designed to have very little data duplication so that the database system can sustain high transaction rates. On the other hand, an OLAP database is typically characterized by a low volume of transactions. The queries are often very complex and involve aggregation of data from multiple tables. For OLAP systems, the response time to queries is an effective measure of performance. These systems are widely used by the applications that typically generate reports using multiple attributes. The OLAP databases usually store historical, aggregated data and there is a fair degree of data duplication among different tables. Since several tables are usually joined to produce results on OLAP databases, it is common to create tables in a denormalized form to avoid the costly JOIN operations at runtime.

In our example schema, the tables are designed to work well in an OLTP database. The schema is designed for use with a typical citywide GIS application. This GIS is used to make changes to land parcel registrations, maintaining road and address information, conditions of the sidewalks, and so on. All of these are simple transactional operations that require short response times to the GIS application users. On the other hand, if someone wants to see a report of all house sales grouped by street and neighborhood, it would require a JOIN of at least three tables. If such operations are common, then it is useful to define new tables that store the results of such a JOIN in a de-normalized table. When an OLTP database is used to derive an OLAP database, users need to identify such queries and figure out which OLTP tables need to be joined to generate tables for the OLAP database.

The database tables are usually two-dimensional entities. In OLAP, these two-dimensional entities are transformed into n-dimensional entities. In the OLAP world, these higher dimension tables are usually referred to as cubes, even though most of these have more than three dimensions. A dimension in OLAP is a perspective used to look at the data; for example, a user might want to look at the data grouped by time, age of the customer, and by item type. Once these dimensions are defined, an OLAP system should be able to find data in any groups of dimensions in any order; for example, users might want to look at sales data by quarter, by customer age group, and by income level. The same data can also be grouped by income group, customer age group, and by quarter. In addition, there are also hierarchies in OLAP that can be used to drill up or drill down different levels of aggregation.

Spatial OLAP (SOLAP)

In spatial applications, the dimensions can include geographic elements. A SOLAP system supports three types of spatial dimensions, namely, geometric spatial dimensions, non-spatial dimensions, and mixed spatial dimensions. The non-geometric dimensions are like the dimensions used in the traditional OLAP and do not involve any spatial predicates. Sometimes, the dimensions look like the spatial dimensions, the previous example using city, county, and state level hierarchies. But these can be done via a named hierarchy without using any spatial or geometric calculations; for example, a text address has this hierarchy information and it can be used in an OLAP query without doing any spatial processing. This type of spatial dimension is the only one supported by the conventional (non-spatial) OLAP tools. In this case, the spatial analysis will be incomplete if certain addresses are incomplete or have names for entities that are not standard; for example, with Washington, DC instead of Washington City, it might not able to identify both as the same spatial entity. The other two spatial dimensions try to minimize this potential problem by using spatial processing to derive the geographic hierarchies. For this concept to work, the geometric spatial dimensions should comprise of geometric shapes for entities at all levels of details; for example, polygon shapes to represent states counties points to represent the cities. The mixed spatial dimensions may have geometric shapes for a subset of levels and named hierarchies for other levels.

In practice, it can be very expensive to compute these spatial relationships at runtime, as spatial predicates tend to be expensive compared to their traditional relational predicates. For this reason, it is a common practice to use the spatial shapes to derive the named hierarchies and build the OLAP cubes that only have non-spatial dimensions. The spatial information can be used in building the cubes, but the cubes themselves will only have traditional dimensions.

Let's look at examples of OLAP cubes containing geographic dimensions. We will show how to derive these geographic dimensions using the Oracle Spatial operations and how to use them in building the OLAP cubes. For this example, we will use two new tables instead of the existing tables in our schema to better demonstrate the idea of SOLAP, CUSTOMER_DATA and PURCHASE_DATA.

```
Create Table CUSTOMER_DATA (
name Varchar2(200),
customer_id Number,
address Varchar2(400),
gender Char(1),
age_group Number,
phone_number Varchar2(20),
location SDO_GEOMETRY,
  Constraint c_d_pk Primary Key (customer_id));

Create Table PURCHASE_DATA (
Customer_id Number,
txn_id Number,
txn_date DATE,
price Number,
item_type Varchar2(100),
item_id Number,
  Constraint p_d_pk Primary Key (txn_id));
```

The CUSTOMER_DATA table has the typical information collected by a merchant from their customers. The customer does not supply the location column but it is derived using a geocoding process using the address column. The PURCHASE_DATA table is used to record the typical information associated with the purchases. With these two tables, the merchant is interested in finding how customers in different geographies spend money on the different types of items in different months of the year. All of these types of questions can be answered by joining these two tables. But this operation would get very expensive, if there are millions of customers and millions of purchase transactions every year. In addition, there could be hundreds of millions of purchase transactions accumulated over the years. In such a scenario, an OLAP database should be used to answer the aggregation queries. Since these queries do not require the latest and up-to-the-last minute information from these two tables, the merchant can run a batch job every month or every week to convert the data from the OLTP database to an OLAP database. While doing this conversion, they can also use additional spatial geographic hierarchy data to convert the database into a SOLAP system. The geographic hierarchy is stored in a reference table called COUNTRY_HIERARCHY.

```
Create Table country_hierarchy (
area_id Number,
name Varchar2(200),
admin_level Number,
parent_area_id Number,
geometry SDO_GEOMETRY,
  Constraint c_h_pk Primary Key (area_id) );
```

This has polygon geometry for each administrative area along with its level in the hierarchy, for example, the USA will have three levels; country is level 1, state is level 2, and county is level 3. With this hierarchy in place, we can map the customer information into this geographic hierarchy. Note that this hierarchy does not have to be based on fixed administrative units. One can easily define the artificial boundaries based on some region or distance-based grouping, for example, one can generate polygons based on drive time analysis and use them as geometries in this hierarchy instead of counties. We also add another column called PARENT_AREA_ID to this table to explicitly store the hierarchical information. Note that the values for the PARENT_ AREA_ID can be easily derived using the following script:

```
Create Table temp_hierarchy
as Select a.area_id, b.area_id parent_area_id
From country_hierarchy a, country_hierarchy b
Where b.admin_level = 1 and a.admin_level = 2
  and sdo_relate(a.geometry, b.geometry, 'mask=INSIDE+COVEREDBY') =
'TRUE';

Insert Into temp_hierarchy
Select a.area_id, b.area_id parent_area_id,
From country_hierarchy a, country_hierarchy b
Where b.admin_level = 2 and a.admin_level = 3
  and sdo_relate(a.geometry, b.geometry, 'mask=INSIDE+COVEREDBY') =
'TRUE';

-- update the hierarchy table with the parent_area_id values
Update country_hierarchy a set parent_area_id =
(Select parent_area_id From temp_hierarchy b Where
 b.area_id = a.area_id);
```

Once the geographic hierarchy is defined and the corresponding polygon geometry is populated, we can then use these in defining the SOLAP tables.

We define a new table to store the denormalized data from these three tables in our SOLAP database named CUSTOMER_HISTORY_DATA. The following SQL scripts show how to construct this table from our three base tables. Assume that the COUNTRY_ HIERARCHY table has a spatial index already built on it.

```
-- create the table to store all the data in a de-normalized form
Create Table customer_history_data (
Customer_id Number,
txn_id Number,
txn_date DATE,
location SDO_GEOMETRY,
level1_area_id Number,
level2_area_id Number,
level3_area_id Number,
item_type  Varchar2(100),
price Number,
gender CHAR(1),
age_group Number);

-- create a temporary table to get the parent_id at the lowest
-- admin level for each customer
-- make sure the customer_data table has a spatial_index on LOCATION
Create Table temp_customer_data_admin
as Select a.*, b.area_id
From customer_data a, country_hierarchy b
Where b.admin_level = 3
  and sdo_inside(a.location, b.geometry) = 'TRUE';

-- now populate the main table for OLAP database
Insert Into customer_history_data
Select a.customer_id, d.txn_id, d.txn_date, a.location,
       a.area_id, b.parent_area_id, c.parent_area_id,
       d.item_type, d.price, a.gender, a.age_group
From temp_customer_data_admin a,
     country_hierarchy b,
     country_hierarchy c,
     purchase_data d
Where a.customer_id = d.customer_id
  and a.area_id = b.area_id
  and b.parent_area_id = c.area_id;
```

Once the CUSTOMER_HISTOR_DATA table is created, then the standard OLAP type queries can be executed against it. Since we have materialized all the geometric hierarchy relationships, regular relational queries can be used to efficiently determine the different aggregation values at different geographic hierarchy levels. Since the PARENT_ID values are also populated, it will be easy to drill up or drill down the hierarchy, for example, a mapping dashboard can show a report of all sales to people aged between 30 and 40 for the state of California. From there, we can drill down easily to show the same data aggregated at the county level for all the California counties. Since the PARENT-CHILD relationship is explicitly stored in this table, a simple relational query can be used to identify all the child records corresponding to the state of California.

Summary

Replicating or maintaining the different versions or copies of the spatial data is often left to the GIS application. However, there are several easy ways to implement features that can be used to manage this replication at the database level. In this chapter, we discussed several techniques for replicating data at the table, schema, and database level. We showed examples for streams and materialized view based replication. We explained the logical and physical standby features in the context of replication. We also covered the concepts of OLAP and Spatial OLAP and described how a geometric hierarchy can be used in SOLAP to enrich the queries that can be executed in the traditional OLAP. The examples demonstrated how the spatial hierarchies can be used without executing any expensive spatial operations at runtime.

In the next chapter, we'll describe the concept of Oracle partitioning that can be used to achieve manageability and high performance while dealing with very large tables. Since the spatial applications tend to generate large volumes of data, this feature can be very useful for large scale spatial databases.

5
Partitioning of Data Using Spatial Keys

Spatial applications tend to generate large volumes of data, especially as the scale of observation of the world's surface extends to large parts of the Earth's surface. With the increasing data, database models have to adapt to deal with large volumes of spatial data that are not seen in traditional GIS applications. GIS applications expect all the related data in one feature layer, even if the feature layer contains millions of features. Oracle database supports a feature called partitioning that can break large tables at the physical storage level to smaller units while keeping the table as one object at the logical level. In this chapter, we cover the following five topics that are useful for managing large volumes of spatial data:

- Introduction to partitioning
- Time-based partitioning
- Spatial key based partitioning
- Implementing space curves based partitioning
- High performance loading of the spatial data

Introduction to partitioning

Oracle partitioning, an extra cost option of Oracle Database Enterprise Edition, enhances the manageability, performance, and availability of a wide variety of applications. While such technology and cost is beyond the normal practitioner, we will introduce many concepts that apply to the efficient organization of large volumes of data via partitioning. Partitioning allows tables and indexes to be subdivided into smaller pieces, enabling these database objects to be managed and accessed at a finer level of granularity.

Oracle partitioning support is one of the very useful features for managing tables with very large volumes of data. When tables are large, creating indexes, collecting statistics, and other data management tasks take a long time. If the full table scans are used in some execution plans, the queries will take a very long time to execute if the table is large (hundreds of gigabytes in size). Operations such as table recovery and index maintenance will take a long time if large tables are not partitioned, for example, if the data is corrupted over a few rows in the table, only the partition containing the corrupted rows has to be recovered. Since the partition is much smaller than the whole table, this operation will take significantly less time than recovering the whole table. Similarly, the backup operations will take much less time if a partition of the table needs to be backed up instead of the whole table. When the table is partitioned, SQL queries can run much faster when partition pruning happens at query time. Partition pruning is very important in reducing the response times for queries on the large partitioned tables. In partition pruning, the optimizer analyses the From and Where clauses of the SQL statement to eliminate the unneeded partitions from query execution. This elimination, or pruning, enables the database to perform table or index access operations on only the subset of the table data that is relevant for the query.

Partitioning methods

Oracle provides three basic methods for partitioning a table, namely, range, hash, and list. When a table is partitioned, one of the columns or a set of columns is designated as the partition key. Range partitioning maps table rows in the partitions based on ranges of values of the partition key established for the table. The range partition is the most common type of partitioning used and it is very often used with date ranges. Hash partitioning maps table rows to partitions based on a hashing algorithm. The hashing algorithm strives to evenly distribute the table rows among all the partitions, giving each partition approximately the same number of rows. Hash partitioning is the main alternative to range partitioning. This is especially true when the data to be partitioned does not have an obvious range partitioning key. List partitioning enables users to explicitly control how different rows are mapped to different partitions. The main advantage of this partitioning method is that a user can completely control how different rows are mapped to different partitions, and the rows that are not related in a natural way can be mapped to the same partition.

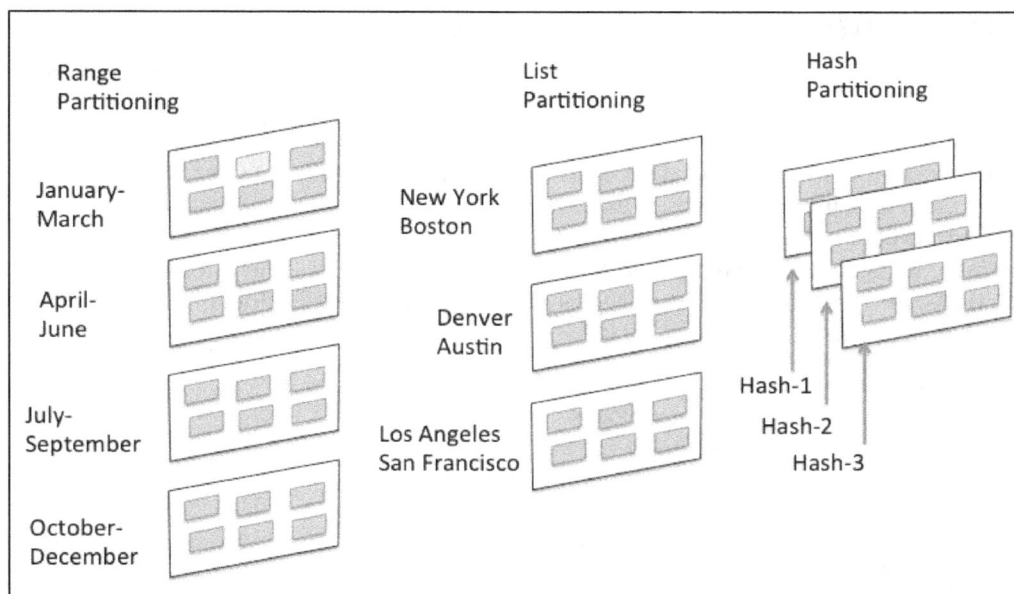

In addition to these basic partitioning methods, Oracle supports composite partitioning, which is a combination of the basic partitioning methods. The composite partitioning can be used to partition the tables at different levels; that is, one method can be chosen to partition the table at the first level, then a second partitioning method can be used to partition each of the first level partitions.

Partitioning of spatial tables

Spatial tables can be partitioned using any of the previously described methods. However, spatial indexing imposes some restrictions on the type of partitioning used for the spatial tables. When a table is partitioned, the index on the table can be created as global or local. A global index treats the table as a non-partitioned table and creates one large index to cover all the rows of the table. A local index is partitioned the same way as the table and each partition of the table will have one index.

Oracle Spatial only supports the range-based partitioning approach for spatial indexes; that is, an Oracle Spatial table can be partitioned using any of the methods, but if a partitioned (or local) spatial index is required, then only the range-based partitioning must be used. Due to this restriction, the rest of the discussion in this chapter will focus on the range-based partitioning.

Partitioning spatial indexes

A range-partitioned table must have a column that can be used as the partition key column. If the table has natural partition keys such as date ranges or timestamps, then columns are well suited for partitioning the data; for example, data can be partitioned by quarter if new data is added to the table every quarter. For spatial tables, such columns can be used as partition keys. Let's take the base_addresses table and create a partitioned table by adding a new column to store a date value. The table is then range-partitioned based on the date values. The initial table has four partitions, and we show the required SQL commands to add additional partitions to the table as required:

```
Create Table base_addresses_part
(FID                 Number NOT NULL,
 ADDRESS             Varchar2(100),
 ADDR_NUM            Number,
 ADDR_N_SFX          Varchar2(10),
 ST_NAME             Varchar2(60),
 ST_TYPE             Varchar2(6),
 LONGITUDE           Number,
 LATITUDE            Number,
 ZIPCODE             Varchar2(5),
 IN_DATE             DATE,
 GEOMETRY            MDSYS.SDO_GEOMETRY)
partition by range (IN_DATE) (
partition p0 Values less than ('01-APR-2012'),
partition p1 Values less than ('01-JUL-2012'),
partition p2 Values less than ('01-OCT-2012'),
partition p3 Values less than ('01-JAN-2013'));


-- move the data from the original table to this new table
-- see the usage of DBMS_RANDOM to create random date values for
-- exisiting data

Insert Into base_addresses_part
Select FID, ADDRESS, ADDR_NUM, ADDR_N_SFX, ST_NAME, ST_TYPE,
LONGITUDE,
       LATITUDE, ZIPCODE,
       TO_DATE(TRUNC(DBMS_RANDOM.VALUE(2455928,2455928+364)),'J'),
       GEOMETRY
  From base_addresses ;
```

```
Select count(*) From base_addresses_part partition(p0);
Select count(*) From base_addresses_part partition(p1);
Select count(*) From base_addresses_part partition(p2);
Select count(*) From base_addresses_part partition(p3);
```

The last four SQL commands show the number of rows in each partition of the table. Since we used a random procedure to generate the date values for the rows in the table, the rows are nearly uniformly distributed among all partitions.

Creating spatial local indexes

One of the advantages of partitioned spatial indexes is that they are easier to create and maintain even on very large tables. One useful trick is to create the indexes for each partition in parallel so that the index creation takes much less time than for an index on a full table.

```
Insert Into USER_SDO_GEOM_METADATA Values ('BASE_ADDRESSES_PART',
'GEOMETRY',
SDO_DIM_ARRAY(SDO_DIM_ELEMENT('X', 5900000, 6100000, .05),
SDO_DIM_ELEMENT('Y', 2000000, 2200000, .05)),  2872);

Create Index BASE_ADDRESSES_PART_SIDX on BASE_ADDRESSES_PART(GEOMETRY)
indextype is mdsys.spatial_index
LOCAL UNUSABLE;

Alter Index BASE_ADDRESSES_PART_SIDX REBUILD PARTITION p0;
Alter Index BASE_ADDRESSES_PART_SIDX REBUILD PARTITION p1;
Alter Index BASE_ADDRESSES_PART_SIDX REBUILD PARTITION p2;
Alter Index BASE_ADDRESSES_PART_SIDX REBUILD PARTITION p3;
```

Note that the index is created with the keyword LOCAL to specify that we want the index to be partitioned the same way the table is partitioned. The index is also created as UNUSBALE so that the Create Index command does not really create the index. It just creates the required metadata, but does not create the actual index. The following Alter Index statements create the index for each partition. Ideally, these commands should be issued in different sessions so that each partition-based index could be created in parallel. This can also be done via a DBMS_SCHEDULE job for each partition instead if creating several sessions from the command line. Note that if there is some problem with one of the index partitions, users only need to rebuild just that partition's index. When the date nears the end of the year, the table needs a new partition to store the data for the next quarter. The users can alter the table and add a new partition to contain the data for the next quarter as shown in the following SQL example:

```
Alter Table BASE_ADDRESSES_PART ADD PARTITION p4
Values LESS THAN ('01-APR-2013');

Alter Index BASE_ADDRESSES_PART_SIDX REBUILD PARTITION p4;
```

The other index maintenance and manageability operations supported for partitioned indexes will be covered in the next few sections.

Spatial partitioning of tables

Tables with geometry can also be partitioned with a spatial key that is, the partitioning is done such that the rows that are close to each other in space will likely be placed in the same partition. There are several methods for using the spatial key to partition the data, and these methods vary depending on the type (point, line, or polygon) of spatial data. One of the main advantages of using a spatial partitioning key is the spatial pruning that comes into play during the query execution. In this section, we explain this concept and show how spatial pruning can help reduce query runtimes. First we look at tables with point data and explain the spatial partitioning of tables.

Single column key

Range-based partitioning requires a scalar (number, character, date, and so on) value as the partitioning key. From the geometry value, we can use either X or Y as this scalar value as the partitioning key. If the table has non-point geometry data, we need to map the POLYGON or LINE geometry in the spatial column to a point so that the X or Y values of the point can be used for partitioning. We will look at different methods to map non-point geometry to a point for the purpose of partitioning in later sections. We use the BASE_ADDRESSES table in this example as it already has a point geometry column.

Before a table is partitioned using either X or Y values, the data should be analyzed first to figure out what values are good for partition ranges. The distribution of the data along the X and Y axis will determine whether the partitioning should be done on the X or Y axis; for example, data for the US is better suited for partitioning along the X axis (as the extent along the X axis is bigger than the extent along the Y axis), while the data for Italy is better suited for partitioning along the Y axis. When a table is partitioned, it is very important to try to keep the number of rows in each partition approximately equal. If one partition has many more rows than other partitions, the work on that larger partition will dominate the work on other partitions leading to unbalanced query execution times. And even the index maintenance workload will vary significantly between different partitions. So, it is very important to create partitions that have nearly the same number of rows in each partition.

Note that this is not an issue to be considered if a natural partition boundary such as DATE is used as the partition key. The following SQL example shows how to analyze the data to find appropriate partition boundaries for the BASE_ADDRESSES table. We use the SQL NTILE function that can take a range of scalar values and divide them into buckets with equal numbers of rows. For this example, we want to create five partitions based on the X values of the GEOMETRY column:

```
Select MIN(x), MAX(x), bucket, COUNT(*)
From (
    Select NTILE (5) OVER (ORDER BY X) bucket, x
    From
    (Select a.geometry.sdo_point.x  X From BASE_ADDRESSES a)
    )
Group By bucket
Order By bucket;
```

MIN(X)	MAX(X)	BUCKET	COUNT(*)
5979545.4	5991274.18	1	41374
5991274.45	5999120.69	2	41374
5999120.69	6003758.99	3	41374
6003759.18	6008657.77	4	41374
6008657.93	6022316.76	5	41374

Once we know these partition boundaries, we can create our partition table with these range values as follows:

```
Drop Table base_addresses_part;
Create Table base_addresses_part
(FID  Number NOT NULL,
 ADDRESS                                    Varchar2(100),
 ADDR_NUM                                   Number,
 ADDR_N_SFX                                 Varchar2(10),
 ST_NAME                                    Varchar2(60),
 ST_TYPE                                    Varchar2(6),
 LONGITUDE                                  Number,
 LATITUDE                                   Number,
 ZIPCODE                                    Varchar2(5),
 GEOMETRY                                   MDSYS.SDO_
GEOMETRY)
partition by range (geometry.sdo_point.x) (
partition p0 Values less than (5991275),
partition p1 Values less than (5999121),
partition p2 Values less than (6003759),
```

```
partition p3 Values less than (6008658),
partition p4 Values less than (6022317));
```

```
Insert Into BASE_ADDRESSES_PART
Select FID, ADDRESS, ADDR_NUM, ADDR_N_SFX, ST_NAME, ST_TYPE,
LONGITUDE, LATITUDE, ZIPCODE, GEOMETRY
From base_addresses ;
commit;
```

Now, if the number of rows in each partition is checked, they will be approximately equal in number. Once this table is created, the rest of the steps for creating the index in parallel are the same as the ones we used for the DATE based partitioning in the previous example.

Spatial partition pruning

Once the spatial index is created, we can look at the root MBR of the index to see how the spatial index is partitioned in space. Since the geometry values in each partition of the table are spatially partitioned, the root MBR for each index partition will only cover the space occupied by the rows in the corresponding table partition. We can see the values of the root MBR with the following SQL example, and when we plot these MBRs, we can see that they horizontally partition the space:

```
Select SDO_ROOT_MBR MBR From USER_SDO_INDEX_METADATA
Where SDO_INDEX_NAME = 'BASE_ADDRESSES_PART_SIDX';
```

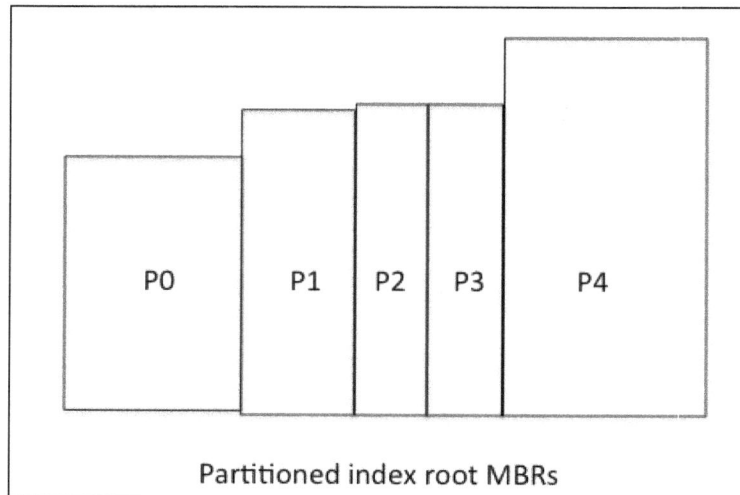

Partitioned index root MBRs

There are three main concepts we should see from this example. First, notice that there are five root MBRs for this index. Since there are five partitions in the base table, there is one logical index for each partition, and hence five root MBRs corresponding to this index. Second, all the Y values of the MBRs are approximately the same. Since we did not control how rows are placed in each partition based in the Y values of the geometry, these values are randomly distributed over all the partitions. Third, we see that the X values of the root MBRs nicely align with the partition boundaries of the base table. If the data in the base table is non-point geometry data, then this alignment with the base table partitions will not be as clean as it is for the point geometry data. We will see these examples in the later sections.

When a spatial query is executed, the query gets executed for each partition of the base table if no partition key is specified in the query. Since we based our partition on the X values of the geometry, there is no natural key that can be specified at query runtime. So, the optimizer has no way to do any partition pruning based on the predicates in the query. But internally, spatial can do the partition pruning since the root MBR for each partition gives the extent of the data in that partition. We use the following SQL example to explain this concept:

```
Select * From BASE_ADDRESSES_PART
Where SDO_FILTER(GEOMETRY,
SDO_GEOMETRY(2003, 2872,NULL,
SDO_ELEM_INFO_ARRAY(1, 1003, 3),
SDO_ORDINATE_ARRAY(5990121, 21001199, 5999758.99, 2112488)))
='TRUE';
```

The query MBR spans the root MBRs of the second and third partitions. When the spatial operator SDO_FILTER is executed for each partition, the first step will be to compare the root MBR of the partition with the MBR of the query geometry. If the geometries do not overlap, the whole partition is skipped immediately. For this query, only the second and third partition's indexes are scanned to find the result set for the given query. When these tables are very large, this spatial pruning can result in substantial savings in runtime for the complex spatial queries.

If the partitioning is done using both the X and Y values, the spatial pruning would have been much better since the current schema did not eliminate any partitions based on the Y values of the query MBR. In the next section, we'll look at multi-column range-based indexing using spatial partitioning.

Multi-column partition key

We can use the X and Y values of the point geometry as the partition keys. Before we define the partition ranges, we first do the analysis on the rows in the table to figure out what ranges make sense for the given data set. For this example, we will divide the X range into three buckets and the Y range into three buckets for each of the ranges of X. So, we will have a total of nine partitions in our example:

```
-- first find out the ranges for X
Select MIN(x), MAX(x), bucket, COUNT(*)
From (
  Select NTILE (3) OVER (ORDER BY X) Bucket, X From
  (Select a.geometry.sdo_point.x  X From base_addresses a)
    )
Group By bucket
Order By bucket;
```

MIN(X)	MAX(X)	BUCKET	COUNT(*)
5979545.4	5996750.66	1	68957
5996750.73	6005322.31	2	68957
6005322.34	6022316.76	3	68956

```
-- next find the ranges for Y for each of the buckets of X
Select MIN(x), MAX(x), MIN(y), MAX(y), bucketX, bucketY, COUNT(*) From
(
  Select NTILE(3) OVER (PARTITION BY BUCKETX ORDER BY Y) BUCKETY,
BUCKETX, X, Y
  From (
    Select NTILE (3) OVER (ORDER BY X) BucketX, X, Y From
    (Select a.geometry.sdo_point.x  X, a.geometry.sdo_point.y  Y
    From BASE_ADDRESSES a)
    ))
Group By bucketX, buckety
Order By bucketx, buckety;
```

MIN(X)	MAX(X)	MIN(Y)	MAX(Y)	BUCKETX	BUCKETY	COUNT(*)
5981379.77	5996749.85	2086252.26	2099543.77	1	1	22986
5980817.17	5996750.66	2099544.8	2106613.54	1	2	22986
5979545.4	5996747.06	2106613.54	2118768.33	1	3	22985
5996751.77	6005322.19	2086114.17	2099523.7	2	1	22986

5996753.51	6005322.31	2099523.7	2109210.68	2	2	22986
5996750.73	6005322.22	2109210.68	2122488.91	2	3	22985
6005322.34	6021807.09	2085905.8	2097583.03	3	1	22986
6005322.5	6018751.71	2097584.85	2109613.03	3	2	22985
6005322.4	6022316.76	2109613.03	2130454.19	3	3	22985

With these X and Y ranges, we can create a partitioned table that has roughly the same number of rows in each partition. The multi-column range partition in Oracle is a little tricky to use as the partition keys are not suitable for continuous key values, such as X and Y values in our example. Single key partitioning is suitable for continuous key values, but multi-key partitioning is applied slightly differently on the key columns. This can be explained easily with a simple example:

```
Create Table test_part (x number, y number)
partition by range (x,y)
( partition p0 Values less than (2,1),
  partition p1 Values less than (2,2),
  partition p2 Values less than (3,1),
  partition p3 Values less than (3,2));


Insert Into test_part Values (1,1);
Insert Into test_part Values (1,2);

Select * From test_part partition (p0);
```

The table is created with two columns, x and y, and the partition keys are defined as (x,y). The first partition has the range of values for which x is less than 2 and y is less than 1. The second partition has the range of values for which x is less than 2 and y is less than 2. With this schema, one would expect the two input rows to go into partition p1, but in fact both the rows go into partition p0. The reason is that when the row values are compared to the partition keys, the x values are less than 2, that is, the upper limit for p0, and the rows are immediately placed in partition p0, that is, if the condition on the first key of the range partition is satisfied with a less than predicate, no further checking is done. But if the first key comparison is true for equal to predicate, then the second key is checked. This means we need to normalize the values of x in order to get the correct distribution of rows in each partition for our example. The following SQL example shows how this can be done using a Case statement and two additional columns to explicitly store the x and y values in the partitioned table:

```
Drop Table base_addresses_part;
Create Table base_addresses_part
(FID   Number NOT NULL,
 ADDRESS                                    Varchar2(100),
```

```
            ADDR_NUM                                Number,
            ADDR_N_SFX                              Varchar2(10),
            ST_NAME                                 Varchar2(60),
            ST_TYPE                                 Varchar2(6),
            LONGITUDE                               Number,
            LATITUDE                                Number,
            ZIPCODE                                 Varchar2(5),
            GEOMETRY                                MDSYS.SDO_
GEOMETRY,
 X Number,
 Y Number)
partition by range (X, Y) (
partition p0 Values less than (1, 2099544),
partition p1 Values less than (1, 2106614),
partition p2 Values less than (1, 2118769),
partition p3 Values less than (2, 2099524),
partition p4 Values less than (2, 2109211),
partition p5 Values less than (2, 2122489),
partition p6 Values less than (3, 2097584),
partition p7 Values less than (3, 2109614),
partition p8 Values less than (3, 2130455));


            Insert Into BASE_ADDRESSES_PART
            Select FID, ADDRESS, ADDR_NUM, ADDR_N_SFX, ST_NAME, ST_TYPE,
            LONGITUDE, LATITUDE, ZIPCODE,  GEOMETRY,
            Case When (a.geometry.sdo_point.x BETWEEN 0 AND 5996750) Then 1
                 When (a.geometry.sdo_point.x BETWEEN 5996750 AND 6005322) Then 2
                 When (a.geometry.sdo_point.x BETWEEN 6005322 AND 6022317) Then 3
                 Else 4
            End ,
            a.geometry.sdo_point.y
            From BASE_ADDRESSES a;
```

Now, if we look at the rows in each partition, the number of rows will come out to be approximately equal in each partition. The rest of the steps for creating the index are the same as the ones in the previous examples. Now that the space is partitioned along X and Y directions, we will get much better spatial partition pruning for the spatial queries.

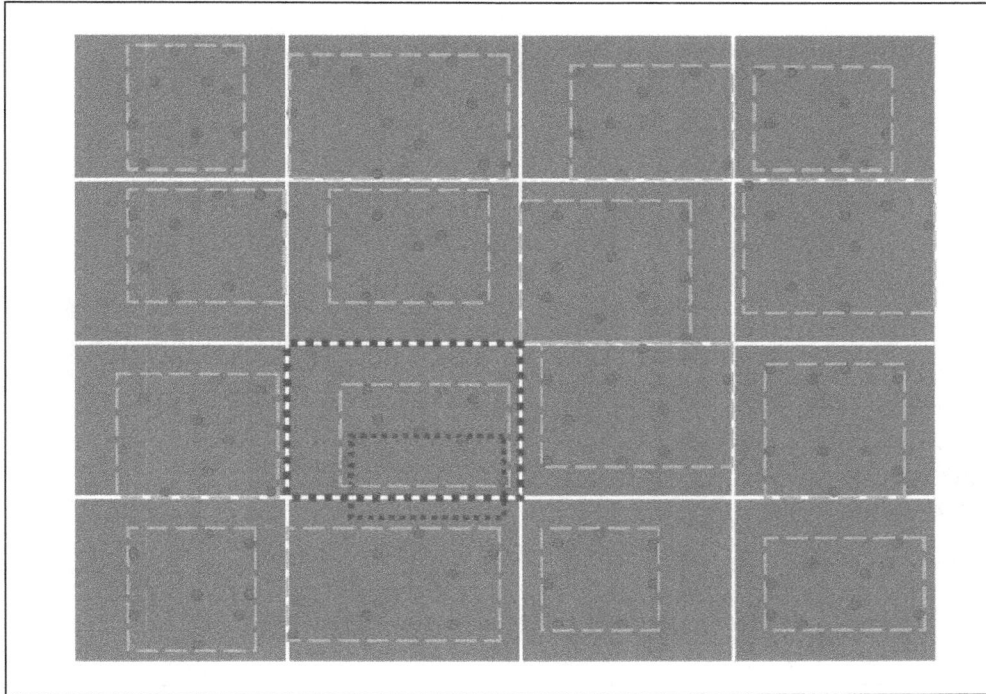

Suppose the table has 16 partitions as shown in the previous figure. If the query geometry's MBR spans two of these partitions, then spatial pruning can easily eliminate 14 of these partitions from query consideration.

Combining spatial and non-spatial partitioning

Often it will be useful to combine non-spatial and spatial partitioning keys to define the partitioning method for a table. This is useful when a key like a DATE is always present in the Where clause of most of the SQL queries against the candidate table. In such cases, the optimizer will provide the first level pruning of the partitions. But inside each of the high level partitions if the number of rows is very large, the spatial pruning will help performance at the second level. We can extend the previous example by adding a YEAR column to the table and using that as the partition key in addition to the x and y columns.

The following figure shows the different levels of pruning that happen for spatial queries with other partition keys in the Where clause:

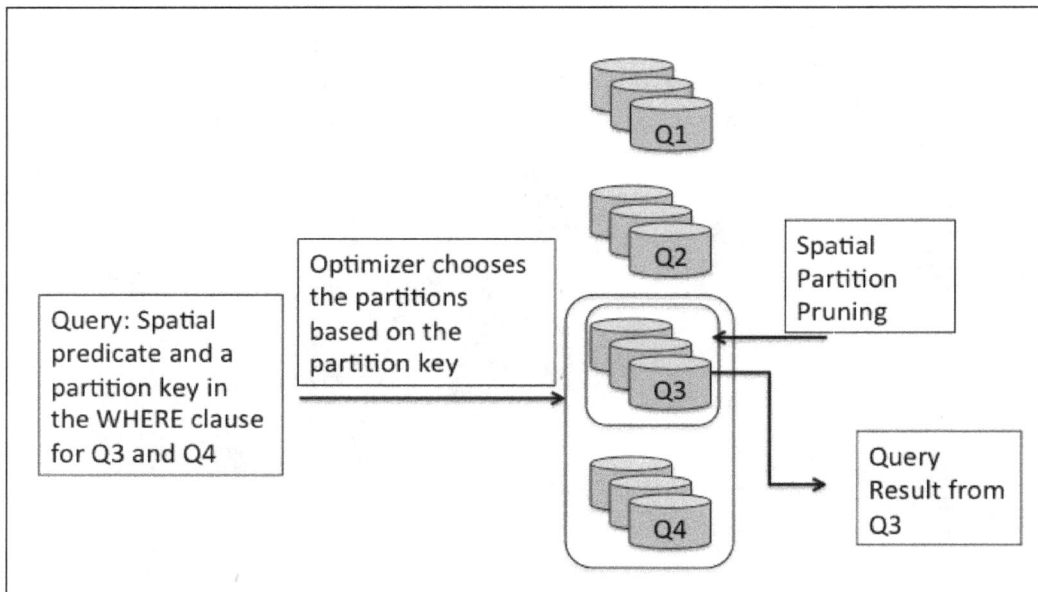

The actual SQL for this example is left as an exercise to the reader.

Space curve based partitioning

In this section, we will describe an example to show how to use a space curve value as the partition key to achieve partitioning that is similar to X and Y-based partitioning. This approach has some advantages and disadvantages when compared to multi-key range-based partitioning. We can use a single partition key instead of the multi-column partition key. If a new partition is to be added by splitting any of the existing partitions, multi-key partitions are harder to manage if only one partition needs to be split based on the first value of the partition key. On the other hand, the multi-key partitioning provides an easy-to-use method for coming up with initial partitions that have the same number of rows in each partition.

For this example, we will use the `land_parcels` table to show the concept of spatial-based partitioning for tables with non-point geometry data. Before we create the partitioned table, we will first revisit the space curve concepts. We learned about the Morton code method in *Chapter 2, Importing and Exporting Spatial Data*, but that method is not directly very useful for use with partitioning concepts. It can be used to create the initial partitions, but splitting the partitions after the initial table creation and assigning the rows to the new partitions gets harder with our Morton code method. In the following SQL example, we will use an Oracle Spatial-supplied method for generating the space curves. Oracle Spatial uses something that is very similar to the Morton codes, but the ordering of the codes is based on a z-curve. Technically, it works just like the Morton code method, so we will not go into a detailed explanation about how these codes are generated.

The Oracle Spatial-supplied function is called `HHENCODE_BYLEVEL`, and it is in the `MDSYS.MD` package. Normal users do not have execute privileges on this package, so users will not be able to use it without a DBA granting them the required privileges on the `MD` package. In the 12*c* release of Oracle DB, there is a public function under the `SDO_UTIL` package that can be used for this purpose, but in the current example, we will stick to this `MD` packaged function as this will work with releases prior to 12*c*. The `HHENCODE_BYLEVEL` function takes as input x and y values along with the lower and upper bound and a level value. The level defines how many tiles are generated for the given input space. Level 1 will have four tiles, level 2 will have 16 tiles, and so on. Initially, we want to create the table with four partitions, so we need to use the level 1 tiles of the `HHENCODE` function. We first create the following function that is easier to use than the Oracle-supplied function for the purpose of using it in our partitioning example:

```
Create or Replace function hh_code(geom SDO_GEOMETRY,
                        min_x number, max_x number,
                        min_y number, max_y number, lvl number)
Return varchar2 as
g SDO_GEOMETRY;
hhcode Varchar2(40);
begin
  -- generate a point for line/polygon geometry
  g := sdo_geom.sdo_pointonsurface(geom, 0.05);0.05);
  hhcode := MD.HHENCODE_bylevel(g.sdo_point.x, min_x, max_x, lvl,
                g.sdo_point.y, min_y, max_y, lvl);
  Return  substr(hhcode, 1, 2*lvl);
end;
```

This function generates a code of type `Varchar2`. Since we want to generate the initial table with four partitions, we need to know the values that can be used as the partition ranges. We create a simple table with rows and generate the `HHCODE` that can be used as the range values for the partition key.

```
Create Table hhcode_tab (x number, y number);
Insert Into hhcode_tab Values (0,0);
Insert Into hhcode_tab Values (0,1);
Insert Into hhcode_tab Values (1,0);
Insert Into hhcode_tab Values (1,1);

Select hhcode From (
   Select hh_code(SDO_GEOMETRY(2001, null,
     sdo_point_type(x,y,null),null,null), 0,1, 0,1, 1)  hhcode
   From hhcode_tab )
Order By hhcode;

HHCODE
-------------------------------------------------------------
00
40
80
C0
```

This shows that `00`, `40`, `80`, and `c0` are the initial range values for the partition key. Note that whatever the space occupied by the data is the first level codes are always going to be these four values. Based on this, we proceed to create the partition table that is based on the `land_parcels` table.

```
Create Table land_parcels_part
(FID   Number NOT NULL,
 MAPBLKLOT                                    Varchar2(9),
 BLKLOT                                       Varchar2(9),
 BLOCK_NUM                                    Varchar2(5),
 LOT_NUM                                      Varchar2(4),
 FROM_ST                                      Varchar2(5),
 TO_ST                                        Varchar2(5),
 STREET                                       Varchar2(20),
 ST_TYPE                                      Varchar2(4),
 ODD_EVEN                                     Varchar2(1),
 part_id                                      Varchar2(40),
 GEOM                                         MDSYS.SDO_
GEOMETRY)
partition by range (part_id) (
partition p0 Values less than ('40'),
```

```
partition p1 Values less than ('80'),
partition p2 Values less than ('C0'),
partition p3 Values less than ('FF'));


Insert Into LAND_PARCELS_PART
Select FID, MAPBLKLOT, BLKLOT, BLOCK_NUM, LOT_NUM, FROM_ST,
      TO_ST, STREET, ST_TYPE, ODD_EVEN,
    hh_code(geometry, 5970000, 6030000, 2080000, 2140000, 1),
geometry
From land_parcels;

-- check the number of rows in each partition
Select COUNT(*), part_id
From LAND_PARCELS_PART
Group By PART_ID;

  COUNT(*) PART_ID
---------- -----------
     56874 00
     18607 40
     59671 80
     19050 C0
```

As seen from the previous SQL results, the number of rows in each partition varies by a large factor. In such cases, we can pick a partition with many rows and split it into smaller partitions so that the number of rows in each partition is comparable. Note that with the multi-column partition key example, we are able to balance the number of rows in each partition at the initial partition creation time. But even in that case, if new rows are added over a period of time, the number of rows in each partition will become unbalanced. So, in this section, we describe the steps required to split existing partitions so that the rows from a partition with too many rows can be distributed over smaller partitions. Since our partitioning scheme in this example is based on the HHENCODE function, any time we want to split a partition, we just use a code that is one level higher than the current code used for the partition key. But first, we will create the metadata and the spatial indexes so that the example will be complete:

```
Insert Into USER_SDO_GEOM_METADATA Values ('LAND_PARCELS_PART',
'GEOM',
SDO_DIM_ARRAY(SDO_DIM_ELEMENT('X', 5900000, 6100000, .05), SDO_DIM_
ELEMENT('Y', 2000000, 2200000, .05)),  2872);

Create Index LAND_PARCELS_PART_SIDX on LAND_PARCELS_PART(GEOM)
INDEXTYPE IS mdsys.spatial_index
```

```
LOCAL UNUSABLE;

Alter Index LAND_PARCELS_PART_SIDX REBUILD PARTITION p0;
Alter Index LAND_PARCELS_PART_SIDX REBUILD PARTITION p1;
Alter Index LAND_PARCELS_PART_SIDX REBUILD PARTITION p2;
Alter Index LAND_PARCELS_PART_SIDX REBUILD PARTITION p3;
```

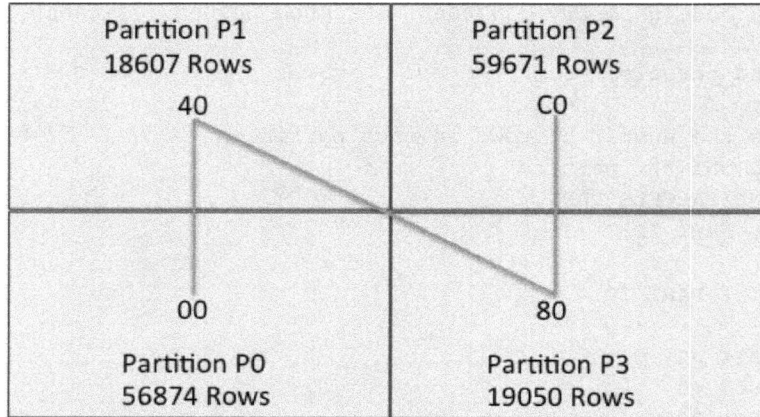

Before we further process these partitions, let's first take a look at how the HHENCODE function splits the input space. The previous figure shows the ordering of the partitions in the input space. Since we are doing a first level split, the four partitions are ordered as shown in the figure.

Now let's split partition **P0** since it has many more rows than **P1** and **P3**. The same logic can be used to split partition P2 as well, but we will stick to **P0** in this example. First, we need to compute the new part_id for those rows that are currently in partition **P0**. Then, we split the partition into four new partitions and the rows from partition **P0** will be automatically distributed over the newly created partitions.

```
-- update the row with a Level 2 HHCODE
Update land_parcels_part
set part_id = hh_code(geom, 5970000, 6030000, 2080000, 2140000, 2)
Where part_id < '40';
```

Now let's look at the part_id values for the new partitions so that we know what range values are required for the new partitions we want to create.

```
Select count(*), part_id From land_parcels_part
Group by part_id;

COUNT(*)    PART_ID
----------  -----------------------
       188  0002
      9415  1002
     14970  2002
     32301  3002
     18607  40
     59671  80
     19050  C0
```

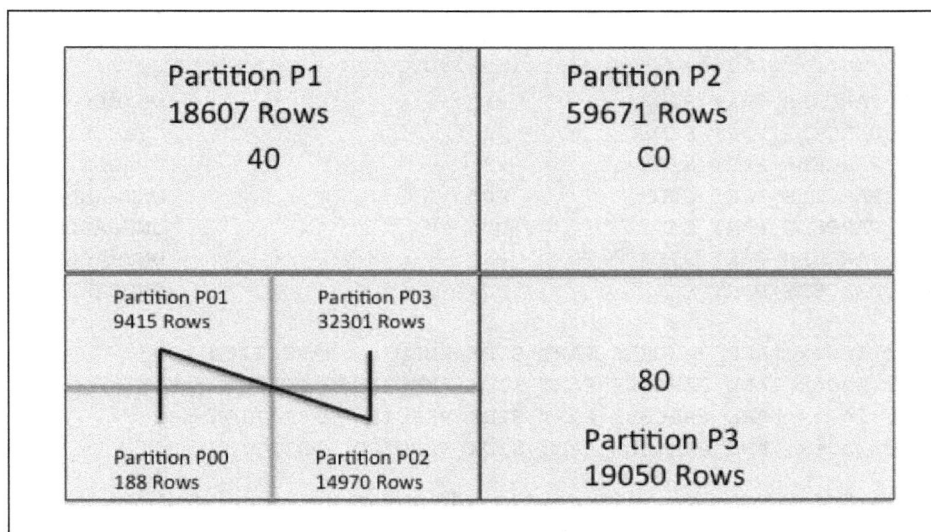

The previous figure shows the distribution of the new partitions in the input space. Only partition P0 will be split, so all the data in these new partitions will still have partition key values less than the partition key values for the other partitions. Note that the key length is now more than the key length for the other partitions, but since these keys are sorted in a lexicographic order, the new key values are still ordered before the keys of the other partitions. Using these new partition range values, we now proceed to split the current partition P0 into four new partitions:

```
Alter Table land_parcels_part split partition p0
at ('1002') Into (partition p00, partition p0);
```

```
Alter Table land_parcels_part split partition p0
at ('2002') Into (partition p01, partition p0);

Alter Table land_parcels_part split partition p0
at ('3002') Into (partition p02, partition p03);
```

At this time, the spatial index on the old partition **P0** is disabled and the spatial index corresponding to the four new partitions is also disabled. We need to rebuild the index for these partitions so that the table is operational again.

```
Select index_name, partition_name, status
From user_ind_partitions
Where index_name = 'LAND_PARCELS_PART_SIDX';
```

INDEX_NAME	PARTITION_NAME	STATUS
LAND_PARCELS_PART_SIDX	P1	USABLE
LAND_PARCELS_PART_SIDX	P2	USABLE
LAND_PARCELS_PART_SIDX	P3	USABLE
LAND_PARCELS_PART_SIDX	P00	UNUSABLE
LAND_PARCELS_PART_SIDX	P01	UNUSABLE
LAND_PARCELS_PART_SIDX	P02	UNUSABLE
LAND_PARCELS_PART_SIDX	P03	UNUSABLE

```
Alter Index LAND_PARCELS_PART_SIDX REBUILD PARTITION p00;
Alter Index LAND_PARCELS_PART_SIDX REBUILD PARTITION p01;
Alter Index LAND_PARCELS_PART_SIDX REBUILD PARTITION p02;
Alter Index LAND_PARCELS_PART_SIDX REBUILD PARTITION p03;
```

The same splitting process can be used to split any of the other partitions. The way these HHCODE values are generated ensures that creating a new part_id value using a higher-level value with the HHCODE function will keep the rows in the same partition until the explicit splitting of the partition is invoked. This way, the users have flexibility to update the part_id value in place and look at the distribution of the rows with the new part_id values. If required, these values can be updated again with a higher-level HHCODE value until a satisfactory distribution of rows is achieved.

Spatial partitioning versus non-spatial partitioning

One of the main advantages of partitioning based on spatial keys is the spatial pruning during query execution. When deciding on a partitioning key for spatial tables, it is important to choose the right partitioning strategy. Let's consider a spatial table that gets new data every quarter. Let's also assume that the queries are always constrained by specific time ranges. In such cases, it is always better to choose the non-spatial partitioning strategy and choose the time column as the partition key. Since the queries always specify the time range, the Oracle server will do the partition pruning based on the time range specified. Now let's consider another spatial table that has no natural partition key like the table in the previous example. Let's also assume that the queries do not have any specific pattern based on the non-spatial columns of the table. In such cases, it is always better to use a spatial key for partitioning, as spatial partition pruning will come into play for most of the queries. In practice, the query patterns will be somewhere in the middle of these two extremes. So, it is important to understand the query patterns before deciding on a specific partitioning strategy.

Parallel queries and partitioning

One of the benefits of partitioning is the ability to parallelize spatial queries. If the query window spans multiple partitions (as is likely to happen with DATE based partitioning), then each partition can be queried in parallel. But the default behavior of partitioned spatial queries is not suitable for queries that span a large number of partitions on systems that have a large number of processors available for the queries. When a spatial query is issued against a partitioned spatial table and no partition key is used in the Where clause, then the spatial internally rewrites the query so that only those partitions that can potentially interact with the given window geometry are used. This is done by doing the spatial pruning query internally to find the list of partitions that can potentially interact with the window geometry, and this list of partitions is added to the Where clause. So, the user query is transformed to a query with additional predicates in the Where clause to explicitly constrain the partitions for the query. If the query window is going to touch many partitions, this might add overhead that is more expensive than doing the whole query in parallel without the rewrite. There is no easy way to determine the right behavior for a generic table. So, it will be useful to analyze the queries and the execution plans to see what is actually happening with the queries. The following SQL example shows the default execution plan for queries on partitioned queries:

```
set pages 30000
SQL> explain plan for
Select /*+ parallel (BASE_ADDRESS_PART, 4) */ FID
From BASE_ADDRESSES_PART
Where SDO_FILTER(GEOMETRY,
SDO_GEOMETRY(2003, 2872,NULL, SDO_ELEM_INFO_ARRAY(1, 1003, 3),
SDO_ORDINATE_ARRAY(5990121, 21001199, 5999758.99, 2112488))) ='TRUE';

Select ID, OPERATION, NAME From Table(dbms_xplan.display);


PLAN_TABLE_OUTPUT
-------------------------------------------------------------------
Plan hash value: 117716759


-------------------------------------------------------------------
```

Id	Operation	Name
0	SELECT STATEMENT	
1	NESTED LOOPS	
2	VIEW	VW_NSO_1
3	HASH UNIQUE	
4	COLLECTION ITERATOR PICKLER FETCH	SDO_PQRY
5	TABLE ACCESS BY USER ROWID	BASE_ADDRESSES_PART

The important item to notice in the execution plan is the keyword SDO_PQRY. When this function appears in the plan, one can be certain that spatial is rewriting the query to add the partition list to the Where clause. If the system has many CPUs available for queries, it might be useful to disable this rewrite operation. This can be done via an event that can be set at the session level or the DB level.

```
Alter session set events '54669 trace name context forever';

explain plan for
Select /*+ parallel (BASE_ADDRESS_PART, 4) */ FID
From BASE_ADDRESSES_PART
Where SDO_FILTER(GEOMETRY,
SDO_GEOMETRY(2003, 2872,NULL, SDO_ELEM_INFO_ARRAY(1, 1003, 3),
SDO_ORDINATE_ARRAY(5990121, 21001199, 5999758.99, 2112488))) ='TRUE';
```

```
Select ID, OPERATION, NAME From Table(dbms_xplan.display);

| Id  | Operation                           | Name             |

|  0  | SELECT STATEMENT                    |                  |

|  1  |  PARTITION RANGE ALL                |                  |
|  2  |   TABLE ACCESS BY LOCAL INDEX ROWID|BASE_ADDRESSES_PART  |
|* 3  |    DOMAIN INDEX                     |BASE_ADDRESSES_PART_SIDX|


Predicate Information (identified by operation id):
---------------------------------------------------

   3 - access("MDSYS"."SDO_FILTER"("GEOMETRY","MDSYS"."SDO_
GEOMETRY"(2003,2872,NULL,"SDO_ELEM_INFO_ARRAY"(1,1003,3),"SDO_
ORDINATE_ARRAY"(5990121,21001199,5999758.99,2112488)))='TRUE')
```

As seen in the previous example, the execution plan is slightly different and the partition list explicitly lists that all four partitions are used in the query (the PSTART and PSTOP keywords in the previous plan). If there are at least four CPUs available for the query, then each partition is queried in parallel.

High performance loading

Next, we show a very useful trick to add large volumes of new data to an existing partitioned table already indexed. We want to add this data in such a way as to avoid disruption to queries against the table, and at the same time, reduce the cost of index maintenance on the partitioned table. There are two cases we can consider. In the first case, the new data that is coming in does not have to be available for queries as soon as the data is acquired. This can be supported with a staging table to collect the incoming data. In the second case, the incoming data should be available for queries as soon as the data is acquired. In this case, we cannot use a staging table and the data has to be directly inserted into the base table.

Loading with a staging table

In this case, the data is coming in at a very fast rate or is being moved from an OLTP database to a data warehouse. In both cases, we can use a staging table to collect the incoming data so that the index on the base table does not have to be maintained during this massive insert operation.

To illustrate this example, we will first create a table to contain the data that is currently in the partition p4 of the base_addresses_part table. We will then truncate the partition p4 and show how it can be added back to the partitioned table with an Alter Table command with minimum disruption to the base_addresses_part table:

```
-- create the table to hold the partition p4 data
Create Table base_addresses_part_20
as Select * From base_addresses_part partition(p4);

-- truncate partition p4 rows from the table
Alter Table base_addresses_part truncate partition (p4);

-- metadata for the new table
Insert Into USER_SDO_GEOM_METADATA Values ('BAE_ADDRESSES_PART_20',
  'GEOM',
SDO_DIM_ARRAY(SDO_DIM_ELEMENT('X', 5900000, 6100000, .05),
  SDO_DIM_ELEMENT('Y', 2000000, 2200000, .05)),  2872);

-- create an index on the smaller table
Create Index BASE_ADDRESSES_PART_20_SIDX on BASE_ADDRESSES_
  PART_20(GEOM)
indextype is mdsys.spatial_index;

-- Now add this data back to the big table
Alter Table BASE_ADDRESSES_PART EXCHANGE PARTITION P4
WITH Table BASE_ADDRESSES_PART_20 INCLUDING INDEXES
  WITHOUT VALIDATION;
```

The main points in this example are the creation of the spatial index on the smaller table that has data for partition p4 and the Alter Table command that swaps the data from the smaller table into the larger table. In some applications, a new set of data that is in very large batches comes in periodically. In such cases, it would be very expensive to insert this data one row at a time into the large table that is already spatially indexed. If this data is coming in periodically, then it is better to collect it in a separate table until the batch of data is complete and then create the spatial index on it. Add a new partition to the base table to hold this new data and use the Alter Table EXCHANGE PARTITION command to move the data and index into the large table. This Alter Table command takes less than a second to execute and the user queries against the larger table are not affected by this operation.

Loading without a staging table

For this scenario, the data coming in has to be available for users immediately after it is inserted into the base table. In such cases, the spatial index has to be updated for each row inserted into the table. This can lead to a lot of contention on the index nodes and can cause the load operation to slow down. There are some recommended best practices to improve the performance of the ingest operation:

- If possible, do not commit after each row is inserted. The spatial has commit call back operations that are activated on a commit operation. During the commit call back operation, the index metadata is read and the rows are inserted into the index. Until that time, the incoming rows are placed in an internal staging table, and these rows are also available for queries while waiting for the commit operation. So, if the commit operation is done after a batch of rows is inserted, then the overhead of updating the index is reduced. Typical batch sizes are in the range of 1000 to 5000 rows. If the batch gets too large, each commit operation can take too long and can start affecting queries and other commit operations.

- If the client programs inserting into the spatial tables use connection pooling, make sure the number of connections is minimized. Note that the commit operations on the spatial index table are serialized. So, the more connections there are, the more contention there is when it comes to committing the transactions on different connections. So, try to minimize the number of connections that are used to update the same spatial table.

- If multiple tables are being updated, it is better to dedicate some connections to update specific tables to minimize the number of connections updating the same table.

- If the spatial table is partitioned, try to dedicate each connection to a specific partition of the table. Since each partition's index is maintained independently, the contention on the index will be minimized.

Summary

In this chapter, we learned about the concept of Oracle partitioning and how it can be useful for managing tables with large numbers of rows. We explained how Oracle Spatial supports the partitioning concept for the spatial index. The different types of partitioning models are also covered that are suitable for different types of applications. We showed examples of performance gains that can be obtained during index creation and query time using partitioning. The question of when to use spatial partitioning and non-spatial partitioning is also addressed in this chapter. In addition, we also described the best practices for ingesting large volumes of spatial data with the help of partitioning.

In the next chapter, we'll learn how to use Oracle Spatial capabilities to solve real world problems. We'll see how to access and work with different properties and attributes of the SDO_GEOMETRY object and describe several PL/SQL functions and procedures to extend the functionality provided by Oracle Spatial.

6
Implementing New Functions

To create new functions that use and extend those offered by Oracle's Spatial and Locator products, the SDO_GEOMETRY object, its attributes and its structure must be thoroughly understood. Therefore this chapter will start by building some functions that will help you understand, access, and process an SDO_GEOMETRY object's attributes. In building these functions, those Oracle SDO_GEOMETRY methods and SDO_GEOM and SDO_UTIL package functions that relate to the processing of the SDO_GEOMETRY attributes and structure, will be introduced and used. This chapter will present information that will help the reader understand how to:

- Expose or create additional properties for the SDO_GEOMETRY object
- Manipulate the SDO_ORDINATE_ARRAY in SQL and PL/SQL
- Create functions to expose SDO_ELEM_INFO_ARRAY properties
- Use SDO_ELEM_INFO_ARRAY to process SDO_ORDINATE_ARRAY correctly
- Organize functions via object types and packaging
- Sort geometries

Later chapters will build on the knowledge gained in this chapter when creating functions that solve specific real-world problems.

Background to programming SDO_GEOMETRY

A spatial data type such as SDO_GEOMETRY defines a storage structure and processing functionality that is exposed through standardized SQL functionality.

If the type's storage structure is not accessible outside of the functions provided with the product, the ability to extend the core functionality of the type becomes dependent on the richness of the supplied functions. If that functionality is limited, nothing can be done because insufficient knowledge exists for accessing the underlying data, unless the vendor provides additional tools for accessing the data through low-level languages, such as C.

In Oracle's case, the SDO_GEOMETRY storage structure uses a SQL object type system that is composed of records/types, numbers, and arrays. This, coupled with powerful spatial and non-spatial functionality (for example, SQL99 analytics and pipelining), which is supplied with the database, creates a rich extensibility platform that allows for new functions to be constructed (using PL/SQL, a precursor to ANSI's SQL/PSM language - Java will be covered in another chapter). This can be used to solve all sorts of business problems.

This chapter gives a taste for the sorts of additional functions one can build to extend the basic offerings associated with SDO_GEOMETRY.

Exposing additional SDO_GEOMETRY properties

The standard SDO_GEOMETRY object has a number of constructors and some public methods. The SDO_GEOMETRY object, at 11*g*R2 (the number of constructors and methods that exist have increased since SDO_GEOMETRY object's inception back in Oracle 8*i*) looks like this (presented slightly differently from how it appears in the Oracle database itself):

```
Create Or Replace Type SDO_GEOMETRY As Object (
   Sdo_Gtype      Number,
   Sdo_Srid       Number,
   Sdo_Point      Sdo_Point_Type,
   Sdo_Elem_Info  SDO_ELEM_INFO_ARRAY,
   Sdo_Ordinates  SDO_ELEM_INFO_ARRAY,
   -- Additional Constructors To The Default
   Constructor Function SDO_GEOMETRY(Wkt In Clob,
                                     Srid In Integer Default Null)
        Return Self As Result,
   Constructor Function SDO_GEOMETRY(Wkt  In varchar2,
                                     Srid In Integer Default Null)
        Return Self As Result,
```

```
Constructor Function SDO_GEOMETRY(Wkb In Blob,
                                  Srid In Integer Default Null)
       Return Self As Result,
-- Member Functions Or Methods
Member Function Get_Gtype   Return Number    Deterministic,
Member Function Get_Dims    Return Number    Deterministic,
Member Function Get_Lrs_Dim Return Number    Deterministic,
Member Function Get_Wkb     Return Blob      Deterministic,
Member Function Get_Wkt     Return Clob      Deterministic,
Member Function St_Coorddim Return Smallint  Deterministic,
Member Function St_Isvalid  Return Integer   Deterministic
);
```

However, when creating new algorithms or uses for SDO_GEOMETRY data, we are often faced with the need to create new functions that extend the basic methods on offer. Given that we cannot, effectively, change the SDO_GEOMETRY type, we need to find efficient, effective, and supportable ways of doing what is needed.

The sections that follow show some ways to achieve what is required.

Permissions

Firstly, in order to be able to create new PL/SQL procedures, packages, and types, the relevant permissions must exist for the schema that will hold the objects. A **Database Administrator (DBA)**, (if appropriate), needs to ensure that the schema user has the following permissions (expressed as GRANT statements):

```
GRANT Create PROCEDURE To <user>;
GRANT Create Type      To <user>;
```

Examining an SDO_GEOMETRY's dimensionality

The OGC SFA 1.2/SQLMM standard requires a spatial object to have an inspection property called ST_Dimension (not to be confused with ST_CoordDim). The standard defines this as follows:

> *The dimension of an* ST_Geometry *value is less than or equal to the coordinate dimension.*

This is not a useful definition for implementation. A more useful one is:

Within a traditional mathematical framework based on Euclidean geometry (Abbott, E.A., 1884, Flatland: A romance of many dimensions: OxFord, Blackwell. 100 pp.), a single point has a topological dimension of zero. Similarly, a line or curve connecting two points is one-dimensional, a plane or surface is two-dimensional, and a volume is three-dimensional (Jones, R. R., Wawrzyniec, T. F., Holliman, N. S., McCaffrey, K. J. W., Imber, J. & Holdsworth, R. E. 2008. Describing the dimensionality of geospatial data in the earth sciences – recommendations For nomenclature. Geosphere. 4 354-359).

In short:

- A (MULTI)POINT has a dimension of 0
- A (MULTI)LINESTRING has a dimension of 1
- A (MULTI)POLYGON has a dimension of 2
- A solid/volume has a dimension a 3

> The collection (SDO_GTYPE = DL04) and SOLID objects (SDO_GTYPE = 3L08/3L09) are not supported by the ST_GEOMETRY standard.

Why might such a function be useful? Other than compliance with the relevant standards, such a function may simplify PL/SQL code (when testing for points, lines, or polygons), or improve semantic expression, if the dimension concept is more familiar to programmers (than individual gtype analysis). Another reason for the function is that GET_GTYPE itself does not provide any dimensional data for collections (yet, for an application a clear use-case may exist for processing them).

The following two "snippets" of PL/SQL show how ST_Dimension may be useful.

Use of Get_GType to select point types, or points and lines	```If (p_geom.get_gtype() In (1,5)) Then Return 0; End If; If (p_geom.get_gtype() In (1,2,5,6)) Then Return 0; End If;```
Use of ST_Dimension to select point types, or points and lines	```If (ST_Dimension(p_geom) = 0) Then Return 0; End If; If (ST_Dimension(p_geom) In (0,1)) Then Return 0; End If;```

Oracle's SDO_GEOMETRY type does not have a method for returning the dimensionality of an SDO_GEOMETRY object, but it can be created by one of two methods. The first builds a function from a "first principles" analysis of the SDO_GTYPE attribute, and the second takes advantage of the fact that Oracle's ST_GEOMETRY object type (first introduced with 10gR1) does have such a method.

First principles

A function built from first principles, but modified for solids (collections (p_geom.get_gtype() = 4) will be implemented in the section "Understanding and using SDO_ELEM_INFO) is as follows:

```
Create Or Replace
Function ST_Dimension(p_geom in SDO_GEOMETRY)
  Return Integer Deterministic
AuthId Current_User
As
    v_gtype pls_integer := 0;
Begin
    If ( p_geom Is null or p_geom.get_gtype() Is null ) Then
       Return NULL;
    End If;
    v_gtype := MOD(a.geom.sdo_gtype,1000);
       If ( v_gtype in (1,5) ) Then Return 0;
    ElsIf ( v_gtype in (2,6) ) Then Return 1;
    ElsIf ( v_gtype in (3,7) ) Then Return 2;
    ElsIf ( v_gtype in (8,9) ) Then Return 3;
    ElsIf ( v_gtype = 4 )      Then Return -4; -- Deliberate
    Else                            Return -1;
    End If;
End ST_Dimension;
```

Reusing the existing ST_GEOMETRY ST_Dimension method

A second method, that uses the ST_Dimension method of the ST_GEOMETRY (included from Oracle 10gR1 onwards) object type, is as follows:

```
Create Or Replace
Function ST_Dimension(p_geom in SDO_GEOMETRY)
  Return Integer Deterministic
AuthId Current_User
As
Begin
```

```
      If ( p_geom Is null or p_geom.get_gtype() Is null ) Then
         Return NULL;
      End If;
      Return MdSys.ST_Geometry(p_geom).ST_Dimension();
   End ST_Dimension;
```

The following SQL tests the "first principles" function against a suitable sample of
geometry types. Its results are compared against some SDO_GEOMETRY methods and
attributes, and direct use of the ST_GEOMETRY ST_DIMENSION function.

```
With geoms As (
   Select SDO_GEOMETRY(2001,NULL,
                       SDO_POINT_TYPE(1,2,NULL),NULL,NULL) As geom
     From dual Union All
   Select SDO_GEOMETRY(2002,NULL,NULL,
                       SDO_ELEM_INFO_ARRAY(1,2,1),
                       SDO_ORDINATE_ARRAY(1,1,2,1)) As geom
     From dual Union All
   Select SDO_GEOMETRY(2003,NULL,NULL,
                       SDO_ELEM_INFO_ARRAY(1,1003,3),
                       SDO_ORDINATE_ARRAY(3,3,4,4)) As geom
     From dual Union All
   Select SDO_GEOMETRY(2004,null,null,
                       SDO_ELEM_INFO_ARRAY(1,1,1, 3,2,1, 7,1003,3),
                       SDO_ORDINATE_ARRAY(1,2, 1,1,2,1, 3,3,4,4))
             As geom
     From dual Union All
   Select SDO_GEOMETRY(2005,NULL,NULL,
                       SDO_ELEM_INFO_ARRAY(1,1,2),
                       SDO_ORDINATE_ARRAY(1,2,3,4)) As geom
     From dual Union All
   Select SDO_GEOMETRY(2006,NULL,NULL,
                       SDO_ELEM_INFO_ARRAY(1,2,1,5,2,1),
                       SDO_ORDINATE_ARRAY(1,2,3,4,5,6,7,8)) As geom
     From dual Union All
   Select SDO_GEOMETRY(2007,NULL,NULL,
                       SDO_ELEM_INFO_ARRAY(1,1003,3,5,1003,3),
                       SDO_ORDINATE_ARRAY(1,1,2,2,5,5,6,6)) As geom
     From dual Union All
   Select SDO_GEOMETRY(3008,NULL,NULL,
                       SDO_ELEM_INFO_ARRAY(1,1007,3),
                       SDO_ORDINATE_ARRAY(1,1,1,3,3,3)) As geom
     From dual
)
Select a.geom.sdo_gtype     As sdo_gtype,
       a.geom.ST_CoordDim() As ST_coordDim,
       a.geom.Get_Dims()    As get_dims,
```

```
        a.geom.get_gtype()    As get_gtype,
        ST_Dimension(a.geom) As ST_Dimension,
        Case When a.geom.sdo_gtype = 3008
             Then -1 -- avoids crashing ST_Dimension
             Else mdsys.ST_Geometry(a.geom).ST_Dimension()
          End As ST_GeomDim
   From geoms a;
-- Results
--
SDO_GTYPE ST_COORDDIM GET_DIMS GET_GTYPE ST_DIMENSION ST_GEOMDIM
--------- ----------- -------- --------- ------------ ----------
     2001           2        2         1            0          0
     2002           2        2         2            1          1
     2003           2        2         3            2          2
     2004           2        2         4           -4         -1
     2005           2        2         5            0          0
     2006           2        2         6            1          1
     2007           2        2         7            2          2
     3008           3        3         8            3         -1

8 rows Selected
```

In the results, note how ST_GEOMETRY's ST_DIMENSION function does not report the dimensionality of a collection (this is correct behavior as per the standard) or a solid (solids not defined by implemented standard — crashes if attempted to apply to a solid), whereas the "first principles" function handles solids.

Understanding and using SDO_ ORDINATES

When creating new spatial processing functions, sometimes only the SDO_GEOMETRY's array of ordinates (SDO_ORDINATES) need be processed. Two functions will be created that use two different methods for processing the ordinates. These functions are drawn from real-life situations that will be described.

Two things need to be understood about the SDO_ORDINATES attribute of the SDO_GEOMETRY object.

Firstly, the name of the SDO_GEOMETRY attribute that stores ordinate data is called SDO_ORDINATES, which is different from the underlying data type, which is called SDO_ORDINATE_ARRAY. This is the same as in a database table; an attribute may be called GID, but its data type Integer SDO_ORDINATE_ARRAY is defined as follows:

```
Create Or Replace  SDO_ORDINATE_ARRAY
    As VARRAY (1048576) Of Number;
```

> A VARRAY is a variable array.
>
> The 1048576 ordinate limit can be removed at 11gR2 via execution of the sdoupggeom.sql script. See "A.3 Increasing the Size of Ordinate Arrays to Support Very Large Geometries" of Oracle® Spatial Developer's Guide, 11g Release 2 (11.2), Part Number E11830-04.

Secondly, the structure of the geometry coordinate data stored within this array is as follows:

SDO_GTYPE	SDO_ORDINATE_ARRAY
200X	XYXYXYXY.....
300X	XYZXYZXYZ....
330X	XYMXYMXYM...
430X	XYMZXYMXXYMZ....

Note that this is not an array of another structure, for example, Oracle did not define an array of a coordinate object as follows:

```
Create Or Replace Type SDO_COORDINATE Is Object (
   x  Number,
   y  Number,
   z  Number,
   w  Number
);

Create Or Replace Type SDO_COORDINATE_ARRAY
   As VARRAY (1048576) Of SDO_COORDINATE;
```

While more "readable", such a structure would create problems as the need to modify the definition of the coordinate had to change over time (for example, support for the introduction of solids with 11g added new attributes to MDSYS. VERTEX_TYPE). All the issues that apply to modifying SDO_GEOMETRY apply here.

Oracle therefore chose a flexible structure that stores the individual ordinates of a set of coordinates as a flattened list of ordinate values. To clarify how the flattening occurs, look at the following list of 2D coordinates For a linestring:

```
{10,10}, {10,14}, {6,10}, {14,10}
```

The required SDO_ORDINATE_ARRAY constructor shows how the ordinates are flattened and organized:

```
SDO_ORDINATE_ARRAY(10,10, 10,14, 6,10, 14,10)
```

When processing an SDO_GEOMETRY object's SDO_ORDINATES its "flattened" nature must always be kept in mind.

Correct "unpacking" of the SDO_ORDINATES in PL/SQL requires knowledge of whether there are 2, 3, or 4 ordinates in a coordinate. This is determined by the dimension element of the SDO_GEOMETRY object's SDO_GTYPE value, which can be accessed via the SDO_GEOMETRY method Get_Dims. A PL/SQL anonymous blocks shows this unpacking.

```
Set ServerOutput On Size Unlimited
Declare
  v_geom SDO_GEOMETRY :=
              SDO_GEOMETRY(2002,NULL,NULL,
                 SDO_ELEM_INFO_ARRAY(1,2,1),
                 SDO_ORDINATE_ARRAY(10,10,10,14,6,10,14,10));
Begin
  For i In v_geom.sdo_ordinates.first..
            v_geom.sdo_ordinates.last Loop
     If (mod(i,v_geom.get_dims())=1) Then
       dbms_output.put('('||v_geom.sdo_ordinates(i)||',');
     Else
       dbms_output.put_line(v_geom.sdo_ordinates(i)||')');
     End If;
  End Loop;
End;
/
anonymous block completed
(10,10)
(10,14)
(6,10)
(14,10)
```

Another way of processing the ordinates is as follows:

```
Set ServerOutput On Size Unlimited
Declare
  v_geom        SDO_GEOMETRY :=
                 SDO_GEOMETRY(2002,NULL,NULL,
                    SDO_ELEM_INFO_ARRAY(1,2,1),
                    SDO_ORDINATE_ARRAY(10,10,10,14,6,10,14,10));
  v_num_coords pls_integer;
  v_ord        pls_integer;
Begin
  v_num_coords := v_geom.sdo_ordinates.count /
                    v_geom.get_dims();
  For i In 1..v_num_coords Loop
     v_ord := ((i*v_geom.get_dims()) - 1);
```

```
        dbms_output.put_line('('||v_geom.sdo_ordinates(v_ord)||
                             ','||v_geom.sdo_ordinates(v_ord+1)
                        ||')');
   End Loop;
End;
/
anonymous block completed
(10,10)
(10,14)
(6,10)
(14,10)
```

Finally, there is one other approach that is based on the SDO_UTIL.GetVertices function (See "32 SDO_UTIL Package (Utility)" in Oracle® Spatial Developer's Guide, 11*g* Release 2 (11.2), Part Number E11830-04.):

```
SDO_UTIL.GetVertices(geometry in SDO_GEOMETRY)
Return VERTEX_SET_TYPE;
```

- Description: Returns the coordinates of the vertices of the input geometry.

- Parameter: Geometry for which to return the coordinates of the vertices.

- Usage notes: This function returns an object of MDSYS.VERTEX_SET_TYPE, which consists of a table of objects of MDSYS.VERTEX_TYPE. Oracle Spatial defines the type MDSYS.VERTEX_SET_TYPE as:

```
Create Type vertex_set_type As Table Of mdsys.vertex_type;
```

- Oracle Spatial defines the object type VERTEX_TYPE as:

```
Create Type mdsys.vertex_type As Object
    (x    NUMBER,
     y    NUMBER,
     z    NUMBER,
     w    NUMBER,
     [possibly followed by additional 11g+ attributes]
     id   NUMBER);
```

We can use this function to transform the SDO_ORDINATES into a more easily understood form as the following code shows:

```
Set ServerOutput On Size Unlimited
Declare
   v_geom SDO_GEOMETRY :=
              SDO_GEOMETRY(2002,NULL,NULL,
                  SDO_ELEM_INFO_ARRAY(1,2,1),
                  SDO_ORDINATE_ARRAY(10,10,10,14,6,10,14,10));
```

```
  v_vertices mdsys.vertex_set_type;
Begin
  v_vertices := SDO_UTIL.GetVertices(v_geom);
  For i In 1..v_vertices.count Loop
    dbms_output.put_line('('||v_vertices(i).x||
                         ','||v_vertices(i).y||')');
  End Loop;
End;
/
anonymous block completed
(10,10)
(10,14)
(6, 10)
(14,10)
```

The choice of which method to use can vary depending on specific needs. However, one thing can be said; while the use of SDO_UTIL.GetVertices renders the processing a little more comprehensible, it comes at a cost of greater memory use and slower execution speed.

We will now use two of these array processing methods to implement two functions:

ST_RoundOrdinates	A function which will round the ordinate numbers to improve readability and decrease storage costs.
ST_FixOrdinates	A function that replaces NULL values in the SDO_ORDINATES array with a supplied value.

Rounding ordinate values

Often Oracle Locator/Spatial users want to know how to round the individual ordinates of the coordinates of an SDO_GEOMETRY object. The most common reason for this is because users understand coordinate data to have a stated precision (For example, "the data is accurate to 1 cm"), and expect SDO_ORDINATES data to represent that precision when visualized.

In addition, the rounding of ordinates (removing superfluous decimal digits) reduces storage and potentially increases query performance. The effect of this is covered in another chapter.

Background

Oracle imposes no precision limit on the NUMBER that records an ordinate of a coordinate except, which is imposed by the NUMBER data type itself.

Yet a statement of precision is required when processing spatial data, for example, to determine if two adjacent coordinates are the same when checking the validity of the description of a spatial object, some sort of precision model is required. When, for example, are the following two coordinates equal, at the centimeter, millimeter, or some other decimal place?

```
515343.140982008,5216871.7926608
515343.141052428,5216871.7931635
```

The spatial metadata associated with a Table's SDO_GEOMETRY column describes a tolerance (SDO_TOLERANCE in an SDO_DIM_ELEMENT of an SDO_DIM_ARRAY array accessed by an XXXX_SDO_GEOM_METADATA object). However, Oracle does not automatically enforce this tolerance value by rounding the ordinate values of an SDO_GEOMETRY column to a specific precision as **Data Manipulation Language (DML)** actions (Insert or Update) occur. The main reason For this is that the metadata SDO_TOLERANCE value is defined as:

> *[the] distance that two points can be apart and still be considered the same (see "1 Spatial Concepts, 1.5.5 Tolerance " in Oracle® Spatial Developer's Guide, 11g Release 2 (11.2), Part Number E11830-04).*

Put another way, this is a distance between two coordinates. It is not the same as the precision of a single ordinate's decimal number value. Thus, applying a tolerance to a stored ordinate would be incorrect, because the tolerance is a distance and not a statement of ordinate precision. Therefore, it is better that ordinates are left alone, unless a user application or business requirement mandates a change via a more appropriate method.

Situations where applying precision may be necessary

Here are a few situations where the application of precision to an ordinate value may be justified.

- When loading data from an external source, such as an ESRI shapefile, there is a mismatch between the way ordinate values are stored in that source (a shapefile stores ordinates as in double precision format) and Oracle's NUMBER based storage.

  ESRI shapefiles' ordinate storage is double precision. The following SDO_GEOMETRY object, loaded from a shapefile, shows the mismatch between storage types. (The 28355 SRID is For "GDA94 / MGA zone 55" and the data is from 1:25,000 scale data, that is, +/- 15m on the ground.)

```
SDO_GEOMETRY(2003,28355,NULL,
            SDO_ELEM_INFO_ARRAY(1,1003,1),
            SDO_ORDINATE_ARRAY(515343.140982008,5216871.7926608,
                        515358.672727787,5216871.7926608,
                        515344.832,5216879.5,
                        515343.140982008,5216871.7926608))
```

- When processing two SDO_GEOMETRY objects with differing ordinate precision against each other (for example, Sdo_Difference and Sdo_Union).

  Take a polygon and linestring geometries and compute the difference between the two. The polygon has lower precision than the linestring. Also, note that we gave a tolerance value of 0.05 (5cm) to the function for its processing.

```
Select SDO_GEOM.Sdo_Difference(
           SDO_GEOMETRY(2002,28355,NULL,
             SDO_ELEM_INFO_ARRAY(1,2,1),
             SDO_ORDINATE_ARRAY(515350.171,5216870.41,
                                 515343.435,5216876.411,
                                 515352.113,5216878.676
           )),
           SDO_GEOMETRY(2003,28355,NULL,
             SDO_ELEM_INFO_ARRAY(1,1003,1),
             SDO_ORDINATE_ARRAY(515343.1,5216871.8,
                                 515358.7,5216871.8,
                                 515344.8,5216879.5,
                                 515343.1,5216871.8)),
           0.05) as geom
   From dual;
-- Results
--
GEOM
-----------------------------------------------------------
SDO_GEOMETRY(2006,28355,NULL,
    SDO_ELEM_INFO_ARRAY(1,2,1,7,2,1,11,2,1),
    SDO_ORDINATE_ARRAY(515344.159777912,5216876.60017054,
                        515343.435,5216876.411,
                        515344.005752361,5216875.9025254,
                        515348.610753375,5216871.8,
                        515350.171,5216870.41,
                        515348.153198659,5216877.64247269,
                        515352.113,5216878.676))
```

- The projection or transformation of an SDO_GEOMETRY object from one coordinate system to another will change the number of decimal digits in the resulting ordinates.

  Let's assume we have a longitude/latitude value (generated via a click on a map) and we want the Google Mercator Map coordinates for that point.

```
Select SDO_CS.TransForm(
        SDO_GEOMETRY(2002,8311,NULL,
                SDO_ELEM_INFO_ARRAY(1,2,1),
                SDO_ORDINATE_ARRAY(147.123,-32.456,
                                    147.672,-33.739)),
            3785) As geom
   From dual;
```

```
-- Results
--
GEOM
--------------------------------------------------------
SDO_GEOMETRY(2002,3785,NULL,
     SDO_ELEM_INFO_ARRAY(1,2,1),
     SDO_ORDINATE_ARRAY(16377657.4439788,-3800381.82007675,
                     16438771.8444243,-3970070.49100647))
```

Being that the input data was only specified to 0.001 of a degree, an output—in meters—specified to eight decimal places seems somewhat excessive!

What is a suitable precision of an ordinate?

What constitutes a suitable precision value depends on the original "sensor" that recorded the original value:

1. Manually surveyed (theodolite and surveyor) may be both accurate and precise, recording observations down to 1 mm.

2. High precision differential GPS may record ordinate values down to 1 cm.

3. Cheap hand-held GPS may record a specific coordinate value to a few meters or tens of meters of accuracy.

4. Data scanned from old paper/Mylar maps may be only accurate to ±10 to ±20 meters!

5. Satellite data is variable in accuracy and precision.

Implementing an ordinate rounding function

The following function will process any SDO_GEOMETRY object, and round all its ordinates to the supplied number of decimal places. (A fuller version of this function is available that allows a different number of decimal places to be set for each ordinate of the geometry rather than just one.)

```
Create Or Replace
Function ST_RoundOrdinates(p_geom       In SDO_GEOMETRY,
                           p_dec_places In Number Default 3)
  Return SDO_GEOMETRY Deterministic
AuthId Current_User
Is
  v_geom       SDO_GEOMETRY := p_geom; -- Copy so it can be edited
  v_dec_places pls_integer  := NVL(p_dec_places,3);
Begin
  If ( p_geom Is null ) Then
     Return null;
  End If;
```

```
    -- Process possibly independent sdo_point object
    If ( v_geom.Sdo_Point Is Not Null ) Then
      v_geom.sdo_point.X := Round(v_geom.sdo_point.x,v_dec_places);
      v_geom.Sdo_Point.Y := Round(v_geom.Sdo_Point.Y,v_dec_places);
      If ( p_geom.get_dims() > 2 ) Then
        v_geom.sdo_point.z := Round(v_geom.sdo_point.z,
                                    v_dec_places);
      End If;
    End If;
    -- Now let's round the ordinates
    If ( p_geom.sdo_ordinates Is not null ) Then
      <<while_vertex_to_process>>
      For v_i In 1..v_geom.sdo_ordinates.COUNT Loop
        v_geom.sdo_ordinates(v_i):= Round(p_geom.sdo_ordinates(v_i),
                                          v_dec_places);
      End Loop while_vertex_to_process;
    End If;
    Return v_geom;
End ST_RoundOrdinates;
```

This function can be tested as follows:

```
With geoms As (
Select SDO_GEOMETRY(3001,NULL,
                  SDO_POINT_TYPE(1.1111,2.2222,3.3333),
                  NULL,NULL) As geom
   From dual Union All
Select SDO_GEOMETRY(3001,NULL,NULL,
                  SDO_ELEM_INFO_ARRAY(1,2,1),
                  SDO_ORDINATE_ARRAY(1.1111,2.2222,3.3333))
         As geom
   From dual Union All
Select SDO_GEOMETRY(3002,NULL,
                  SDO_POINT_TYPE(1.1111,2.2222,3.3333),
                  SDO_ELEM_INFO_ARRAY(1,2,1),
                  SDO_ORDINATE_ARRAY(1.1111,2.2222,999.499,
                                     3.3333,4.4444,9999.999))
         As geom
   From dual Union All
Select SDO_GEOMETRY(3003,null,
                  SDO_POINT_TYPE(10.1111,10.2222,3.3333),
                  SDO_ELEM_INFO_ARRAY(1,1003,1),
                  SDO_ORDINATE_ARRAY(5.1111,10.2222,50.3333,
                                     20.1111,10.2222,50.8333,
                                     20.1111,20.2222,51.3333,
```

```
                                    5.1111,20.2222,50.8333,
                                    5.1111,10.2222,50.3333))
          As geom
     From dual
   )
   Select ST_RoundOrdinates(a.geom,1) As geom
     From geoms a;


   -- Results
   --
   GEOM
   ----------------------------------------------------------------
   SDO_GEOMETRY(3001,NULL,SDO_POINT_TYPE(1.1,2.2,3.3),NULL,NULL)
   SDO_GEOMETRY(3001,NULL,NULL,
                SDO_ELEM_INFO_ARRAY(1,2,1),
                SDO_ORDINATE_ARRAY(1.1,2.2,3.3))
   SDO_GEOMETRY(3002,NULL,SDO_POINT_TYPE(1.1,2.2,3.3),
                SDO_ELEM_INFO_ARRAY(1,2,1),
                SDO_ORDINATE_ARRAY(1.1,2.2,999.5,3.3,4.4,10000))
   SDO_GEOMETRY(3003,NULL,SDO_POINT_TYPE(10.1,10.2,3.3),
                SDO_ELEM_INFO_ARRAY(1,1003,1),
                SDO_ORDINATE_ARRAY(5.1,10.2,50.3,20.1,10.2,50.8,
                        20.1,20.2,51.3,5.1,20.2,50.8,5.1,
                        10.2,50.3))
```

Swapping ordinates

When working with spatial data, there is often a need to swap the ordinates of a coordinate, for example, while we humans speak about "Latitude/Longitude", all spatial database types store the ordinate data as Longitude/Latitude. So, a common problem is the construction of an SDO_GEOMETRY object using geographic ordinates and ordering them:

```
Latitude,Longitude,Latitude,Longitude,Latitude,Longitude....
```

When in fact the order should be:

```
Longitude,Latitude,Longitude,Latitude,Longitude,Latitude....
```

The re-ordering of the ordinates can be done using a function that operates only on SDO_ORDINATE_ARRAY. The following function enables the reordering of the X and Y ordinates:

```
Create Or Replace
Function ST_SwapOrdinates(p_geom in SDO_GEOMETRY)
  Return SDO_GEOMETRY
Authid Current_User
Is
  v_geom      SDO_GEOMETRY := p_geom;
  v_dim       pls_integer;
  v_temp_ord  Number;
  v_start_ord pls_integer;
Begin
  If ( p_geom Is null ) Then
    Return NULL;
  End If;
  -- Get geometry properties needed for processing
  v_dim := p_geom.Get_Dims();
  -- SDO_POINT Handled separately to SDO_Ordinates
  If ( v_geom.sdo_point Is not null ) Then
    v_temp_ord          := p_geom.sdo_point.x;
    v_geom.sdo_point.x := p_geom.sdo_point.y;
    v_geom.sdo_point.y := v_temp_ord;
  End If;
  -- Process the geometry's ordinate array
  If ( p_geom.sdo_ordinates Is not null ) Then
    For i In 1 .. (p_geom.sdo_ordinates.COUNT/v_dim) Loop
      v_start_ord := (i-1)*v_dim;
      v_temp_ord := p_geom.sdo_ordinates(v_start_ord+1); -- save X
      v_geom.sdo_ordinates(v_start_ord + 1) :=
              p_geom.sdo_ordinates(v_start_ord + 2);
      v_geom.sdo_ordinates(v_start_ord + 2) := v_temp_ord;
    End Loop;
  End If;
  Return V_Geom;
End ST_SwapOrdinates;
```

Testing this function with SDO_POINT objects produces the following code:

```
With geoms As (
Select SDO_GEOMETRY(2001,8307,
                 SDO_POINT_TYPE(-32.23,147.11,NULL),NULL,NULL)
          As geom
  From dual union all
Select SDO_GEOMETRY(3001,8307,
                 SDO_POINT_TYPE(-32.23,147.11,999),NULL,NULL)
          As geom
```

```
      From dual
  )
  Select a.geom.sdo_point as point,
         ST_SwapOrdinates(a.geom).sdo_point As swappedPoint
    From geoms a;

  --Results
  --a
  POINT                                 SWAPPEDPOINT
  ------------------------------------- -------------------------------
  SDO_POINT_TYPE(-32.23,147.11,NULL) SDO_POINT_TYPE(147.11,-32.23,NULL)
  ---
  SDO_POINT_TYPE(-32.23,147.11,999)  SDO_POINT_TYPE(147.11,-32.23,999)
```

Testing with lines shows the following code:

```
  With geoms As (
  Select SDO_GEOMETRY(2002,8307,SDO_POINT_TYPE(-32.75,147.8,30),
                    SDO_ELEM_INFO_ARRAY(1,2,1),
                    SDO_ORDINATE_ARRAY(-32.23,147.11,-32.4,148.1))
             As geom
    From dual union all
  Select SDO_GEOMETRY(3302,null,null,SDO_ELEM_INFO_ARRAY(1,2,1),
                    SDO_ORDINATE_ARRAY(-32.23,147.11,0,
                                       -32.4,148.1,10))
             As geom
    From dual union all
  Select SDO_GEOMETRY(4302,8307,null,SDO_ELEM_INFO_ARRAY(1,2,1),
           SDO_ORDINATE_ARRAY(-32.23,147.11,0,-456,
                                       -32.4,148.1,10,NULL)) As geom
    From dual
  )
  Select a.geom.sdo_ordinates as ords,
         ST_SwapOrdinates(a.geom).sdo_ordinates As swappedOrds
    From geoms a;

  -- Results
  --
  ORDS
  -----------------------------------------------------------
  SWAPPEDORDS
  -----------------------------------------------------------
  SDO_ORDINATE_ARRAY(-32.23,147.11,-32.4,148.1)
  SDO_ORDINATE_ARRAY(147.11,-32.23,148.1,-32.4)
  SDO_ORDINATE_ARRAY(-32.23,147.11,0,-32.4,148.1,10)
  SDO_ORDINATE_ARRAY(147.11,-32.23,0,148.1,-32.4,10)
```

```
SDO_ORDINATE_ARRAY(-32.23,147.11,0,-456,-32.4,148.1,10,null)
SDO_ORDINATE_ARRAY(147.11,-32.23,0,-456,148.1,-32.4,10,null)
```

A fuller version of ST_SwapOrdinates that allows for the reordering of any combination of the XYZM ordinates is supplied with the book.

Understanding and using SDO_ELEM_INFO

The new functionality created so far did not require looking at the logical structure of an SDO_GEOMETRY object, as described by the SDO_ELEM_INFO attribute. Understanding and being able to process the contents of this VARRAY is important, when developing or implementing more complex functionality.

The SDO_ELEM_INFO (see "2.2.4 SDO_ELEM_INFO", in the Oracle® Spatial Developer's Guide 11*g* Release 2 (11.2)) attribute is organized as a set of elements, each having three values or members. Each member of a triplet has a name: SDO_OFFSET (actually SDO_STARTING_OFFSET), SDO_ETYPE, and SDO_INTERPRETATION. These members are described in the "Section 2.2.4 SDO_ELEM_INFO", in the Oracle® Spatial Developer's Guide 11*g* Release 2 (11.2) (edited) by the following Table:

SDO_OFFSET	Indicates the offset within the SDO_ORDINATES array where the first ordinate For this element is stored.	
SDO_ETYPE	Indicates the type of the element. values 1, 2, 1003, and 2003 are considered simple elements and values 4, 1005, 2005, 1006, and 2006 are considered compound elements).	
SDO_INTERPRETATION	This describes how the SDO_ETYPE should be interpreted:	
	1	Can mean the element's vertices are connected by straight line segments.
	2	The element's vertices describe a connected sequence of circular arcs.
	3	The element is an optimized rectangle.
	4	The element is a circle (a Form of circular arc).
	n>1	If geometry is point cluster, the number of points; if a compound element the number of subelements.

Unpacking SDO_ELEM_INFO in SQL and PL/SQL

The SDO_ELEM_INFO array can be "unpacked" and presented in its triplet Form in SQL by pivoting on the triplet number.

```
Select Trunc((rownum-1)/3,0)+1 As triplet,
       Sum(Decode(Mod(rownum,3),1,e.column_value,null)) As offset,
       Sum(Decode(Mod(rownum,3),2,e.column_value,null)) As etype,
       Sum(Decode(Mod(rownum,3),0,e.column_value,null)) As interp
   From Table(Select SDO_GEOMETRY(2002,NULL,NULL,
                     SDO_ELEM_INFO_ARRAY(1,4,2,1,2,1,3,2,2),
                     SDO_ORDINATE_ARRAY(52,60,
                                        52.7,67,52.5,67,50,65)
                   ).sdo_elem_info
             From dual) e
 Group By Trunc((rownum - 1) / 3,0)
 Order By Trunc((rownum - 1) / 3,0),2;

-- Results
--
TRIPLET OFFSET ETYPE INTERP
------- ------ ----- ------
      1      1     4      2
      2      1     2      1
      3      3     2      2
```

The Table function takes the SDO_ELEM_INFO array, and returns each value as a single column value on its own row. The Trunc ((rownum-1)/3,0) equation generates a triplet number (starting from 0), which is used to pivot the offset, etype, and interpretation values for that triplet into a single line as separate attributes. In this Form human interpretation is easier.

All the functions which will be built that need to process the SDO_ELEM_INFO attribute do so, but in a different way to SQL. In PL/SQL it is faster to iterate over the VARRAY as follows:

```
Set ServerOutput On Size Unlimited
Declare
  v_triplets       pls_integer;
  v_offset         pls_integer;
  v_etype          pls_integer;
  v_interpretation pls_integer;
  v_geom           SDO_GEOMETRY :=
    SDO_GEOMETRY(2002,NULL,NULL,
```

```
                     SDO_ELEM_INFO_ARRAY(1,4,2,1,2,1,3,2,2),
                     SDO_ORDINATE_ARRAY(52,60,52.7,67,52.5,67,50,65));
Begin
  v_triplets := ( ( v_geom.sdo_elem_info.COUNT / 3 ) - 1 );
  <<triplet_Extraction>>
  For v_i In 0 .. v_triplets Loop
      v_offset           := v_geom.sdo_elem_info(v_i * 3 + 1);
      v_etype            := v_geom.sdo_elem_info(v_i * 3 + 2);
      v_interpretation := v_geom.sdo_elem_info(v_i * 3 + 3);
      dbms_output.put_line('Triplet('||v_i||')=(' ||
                          v_offset||','||
                          v_etype||','||
                          v_interpretation||')');
  End Loop triplet_Extraction;
End;
/
Show errors

-- Results
--
Triplet(0)=(1,4,2)
Triplet(1)=(1,2,1)
Triplet(2)=(3,2,2)

No Errors.
```

First the number of triplets is calculated (a valid SDO_ELEM_INFO structure always has a total number of values that is a multiple of three), then a loop is constructed that iterates over each triplet, and finally each value in a triplet being processed is extracted from the SDO_ELEM_INFO array by calculating its offset and accessing its value.

Dimensionality of compound objects

The "first principles" ST_Dimension function implemented earlier in this chapter did not support compound (DL04) objects. To handle these compound objects the original function could be modified to examine the object's SDO_ELEM_INFO array, and return the maximum dimension that exists within it. Or a new function could be written that checks to see. If a specific dimensioned object exists. This function, called ST_hasDimension, classifies each GTYPE element according to its dimensionality and compares it to the supplied value; returning true (1), if there is a match. Where the GTYPE element indicates a compound object, the code then iterates over the SDO_ELEM_INFO array inspecting each triplet's ETYPE member, classifying it as a point, line, or polygon, and returning true (1), if there is a match.

```
Create Or Replace
Function ST_hasDimension(p_geom in SDO_GEOMETRY,
                         p_dim  in Integer default 2)
  Return Integer Deterministic
AuthId Current_User
As
   v_etype pls_integer := 0;
   v_gtype pls_integer := -1;
Begin
   If ( p_geom Is null or p_geom.get_gtype() Is null ) Then
      Return 0;
   End If;
   v_gtype := MOD(p_geom.sdo_gtype,100);   -- DLTT
   If (  ( v_gtype in (1,5) and p_dim = 0 )
      Or ( v_gtype in (2,6) and p_dim = 1 )
      Or ( v_gtype in (3,7) and p_dim = 2 )
      Or ( v_gtype in (8,9) and p_dim = 3 ) ) Then
         Return 1;
   End If;
   If ( v_gtype = 4 ) Then
     <<element_Extraction>>
     For v_i In 0..( (p_geom.sdo_elem_info.COUNT / 3) - 1) Loop
        v_etype := p_geom.sdo_elem_info(v_i * 3 + 2);
        -- check etype value until find required dimension p_dim
        If ( (v_etype = 1                      and p_dim=0)
          Or (v_etype in (2,4)                 and p_dim=1)
          Or (v_etype in (1003,2003,1005,2005) and p_dim=2)
          Or (v_etype in (1007,1006,2006)      and p_dim=3) ) Then
           Return 1;
        End If;
     End Loop element_Extraction;
     Return 0;
   Else
      Return 0;
   End If;
End ST_hasDimension;

-- Results
--
SDO_GTYPE GET_DIMS GET_GTYPE REQUIRED_DIM ST_HASDIMENSION
--------- -------- --------- ------------ ---------------
     2001        2         1            0               1
     2002        2         2            1               1
     2003        2         3            2               1
```

2004	2	4	0	1
2004	2	4	1	1
2004	2	4	2	1
2004	2	4	3	0
2005	2	5	0	1
2006	2	6	1	1
2007	2	7	2	1
3008	3	8	3	1

11 rows Selected

Detecting if circular arcs exist

Oracle's support for circular arcs in an SDO_GEOMETRY object provide a really powerful element for describing humanly constructed objects, such as cul-de-sacs, road alignments, and forest inventory plots. When constructing algorithms to process such objects (An ESRI shapefile cannot contain linestrings or polygons defined using circular arcs), it is necessary to know if the object contains circular curves in order to choose how to process an element.

What constitutes a geometry having circular arc elements? By examining its SDO_ELEM_INFO attribute to detect, if any element is described by a triplet that contains a circular arc etype or interpretation value.

Determining if an SDO_GEOMETRY contains circular arcs is done by coding a new function is ST_hasCircularArcs, which will return 1, if the description of the linestring or polygon contains one or more circular arcs and 0 if not.

```
Create Or Replace
Function ST_hasCircularArcs(p_geom in SDO_GEOMETRY)
  Return Integer Deterministic
Authid Current_User
Is
  v_etype          pls_integer;
  v_interpretation pls_integer;
  v_elements       pls_integer;
  v_elem_info      mdsys.SDO_ELEM_INFO_ARRAY;
Begin
  If ( p_geom Is null ) Then
    Return null;
  End If;
  If ( ST_Dimension(p_geom) = 0 ) Then -- Doesn't apply to points
    Return 0;
  End If;
  v_elements  := ((p_geom.sdo_elem_info.COUNT/3)-1);
```

```
    v_elem_info := p_geom.sdo_elem_info;
    <<element_Extraction>>
    For v_i IN 0 .. v_elements Loop
      v_etype          := v_i * 3 + 2;
      v_interpretation := v_etype + 1;
      If (      ( v_elem_info(v_etype)            = 2 And
                  v_elem_info(v_interpretation)   = 2 )
           Or ( v_elem_info(v_etype)            In (4,1005,2005) )
           Or ( v_elem_info(v_etype)            In (1003,2003) And
                  v_elem_info(v_interpretation) In (2,4) ) ) Then
          Return 1;
      End If;
    End Loop element_Extraction;
    Return 0;
  End ST_hasCircularArcs;
```

This function works by extracting each combination of SDO_ETYPE and SDO_
INTERPRETATION (c.f., interpretation table in section 2.2.4 of the Oracle Spatial
documentation), and detecting those that relate to circular arcs. A geometry has
circular arc segments where the SDO_ETYPE/SDO_INTERPRETATION pairs, described
in the following table, occur:

SDO_ETYPE	SDO_INTERPRETATION	Meaning
2	2	Linestring made up of a connected sequence of circular arcs
4	n > 1	Compound linestring with some vertices connected by straight line segments and some by circular arcs
1005 or 2005	n > 1	Compound polygon with some vertices connected by straight line segments and some by circular arcs
1003 or 2003	2	Polygon made up of a connected sequence of circular arcs that closes on itself
1003 or 2003	4	Circle type described by three distinct non-collinear points, all on the circumference of the circle

We can test this using the following SQL:

```
With geoms As (
Select 'Plain Point' As gtype,
       SDO_GEOMETRY(3001,8307,SDO_POINT_TYPE(-32.2,147.1,3),
                    NULL,NULL) As geom
  From dual Union All
Select 'Plain line' As gtype,
       SDO_GEOMETRY(2002,8307,NULL,
           SDO_ELEM_INFO_ARRAY(1,2,1),
           SDO_ORDINATE_ARRAY(-32.5,147.6,-33.0,147.9)
       ) As geom
  From dual Union All
Select 'Compound_line' As gtype,
       SDO_GEOMETRY(2002,NULL,NULL,
           SDO_ELEM_INFO_ARRAY(1,4,2, 1,2,1, 3,2,2),
           SDO_ORDINATE_ARRAY(0,0,70,70,50,70,50,50)
       ) As geom
 From dual Union All
Select 'Arc polygon' As gtype,
       SDO_GEOMETRY(2003,null,null,
           SDO_ELEM_INFO_ARRAY(1,1003,2),
           SDO_ORDINATE_ARRAY(15,115,20,118,15,120,10,118,15,115)
         ) As geom
  From dual Union All
Select 'Compound polygon' As gtype,
       SDO_GEOMETRY(2003,NULL,NULL,
           SDO_ELEM_INFO_ARRAY(1,1005,2, 1,2,1, 5,2,2),
           SDO_ORDINATE_ARRAY(6,10,10,1,14,10,10,14,6,10)
       ) As geom
  From dual Union All
Select 'Compound polygon in Collection' As gtype,
       SDO_GEOMETRY(2004,NULL,NULL,
           SDO_ELEM_INFO_ARRAY(1,1,1,3,1005,2,3,2,1,7,2,2),
           SDO_ORDINATE_ARRAY(0,0,6,10,10,1,14,10,10,14,6,10)
       ) As geom
  From dual)
Select gtype, ST_hasCircularArcs(a.geom) as hasCircularArc
  From geoms a;

-- Results
--
GTYPE                            HASCIRCULARARC
-------------------------------  --------------
Plain Point                                   0
Plain line                                    0
Compound_line                                 1
Arc polygon                                   1
Compound polygon                              1
Compound polygon in Collection                1

  6 rows Selected
```

Testing optimized rectangles

A corollary to this function is the one that tests for the existence of optimized rectangles.

SDO_ETYPE	SDO_INTERPRETATION	**Meaning**
1003, 2003	3	Rectangle type (sometimes called optimized rectangle). A bounding rectangle such that only two points, the lower-left and the upper-right, are required to describe it.

This test is relatively simple and is encapsulated in the following function, ST_hasOptimizedRectangles:

```
Create Or Replace
Function ST_hasOptimizedRectangles(p_geom In SDO_GEOMETRY)
  Return Integer deterministic
Is
  v_etype           pls_integer;
  v_interpretation  pls_integer;
  v_elements        pls_integer := 0;
  v_rectangle_count Number := 0;
Begin
  If ( p_geom Is null ) Then
    Return null;
  End If;
  If ( ST_Dimension(p_geom) in (0,1) ) Then -- Not Points or Lines
    Return 0;
  End If;
  v_elements := ((p_geom.sdo_elem_info.COUNT/3)-1);
  <<element_Extraction>>
  For v_i IN 0 .. v_elements Loop
    v_etype          := p_geom.sdo_elem_info(v_i * 3 + 2);
    v_interpretation := p_geom.sdo_elem_info(v_i * 3 + 3);
    If ( ( v_etype in (1003,2003) And
           v_interpretation = 3 ) ) Then
       v_rectangle_count := v_rectangle_count + 1;
    End If;
  End Loop element_Extraction;
  Return v_rectangle_count;
End ST_hasOptimizedRectangles;
```

This function can be tested as follows:

```
With geoms As (
Select SDO_GEOMETRY(2002,null,null,
                    SDO_ELEM_INFO_ARRAY(1,2,1),
                    SDO_ORDINATE_ARRAY(1,1,10,10)) As geom
  From dual Union all
Select SDO_GEOMETRY(2003,null,null,
                    SDO_ELEM_INFO_ARRAY(1,1003,3),
                    SDO_ORDINATE_ARRAY(1,1,10,10)) As geom
  From dual
)
Select ST_hasOptimizedRectangles(a.geom) as hasRects
  From geoms a;

-- Results
--
HASRECTS
--------
       0
       1
```

Counting the rings of a polygon

Before continuing to investigate the uses of the SDO_ELEM_INFO attribute, another investigative function that will allow us to count the number of polygon rings within a polygon geometry needs to be constructed. Oracle's current (11*g*R2) function set contains a function called SDO_UTIL.GetNumElem that can count the number of elements (whole SDO_GEOMETRY elements), but it does not contain a function that can count the rings in a polygon (or the subelements of a linestring, for example, circular arcs and linestrings). Counting the number of rings can be done by examining the SDO_ELEM_INFO array as follows:

```
Create Or Replace
Function ST_NumRings(p_geom       In SDO_GEOMETRYSDO_GEOMETRY,
                     p_ring_type in Integer default 0
                     /* 0 = ALL; 1 = OUTER; 2 = INNER */ )
  Return Number Deterministic
AuthId Current_User
Is
  c_i_ring_type Constant pls_integer   := -20120;
  c_s_ring_type Constant Varchar2(100) :=
     'p_ring_type must be one of 0(ALL),1(OUTER),2(INNER) only.';
  v_elements     pls_integer := 0;
```

```
        v_ring_count   pls_integer := 0;
        v_etype        pls_integer;
    Begin
      If ( p_geom Is null ) Then
        Return null;
      End If;
      If ( p_ring_type Is null Or p_ring_type not in (0,1,2) ) Then
        raise_application_error(c_i_ring_type,c_s_ring_type);
      End If;
      If ( ST_hasDimension(p_geom,2)=0 ) Then -- not a polygon
        Return 0;
      End If;
      v_elements := ((p_geom.sdo_elem_info.COUNT/3)-1);
      <<element_Extraction>>
      For v_i In 0 .. v_elements Loop
          v_etype := p_geom.sdo_elem_info(v_i * 3 + 2);
          If ( (v_etype In (1003,1005,
                            2003,2005) And 0 = p_ring_type)
            Or (v_etype In (1003,1005) And 1 = p_ring_type)
            Or (v_etype In (2003,2005) And 2 = p_ring_type)) Then
             v_ring_count := v_ring_count + 1;
          End If;
      End Loop element_Extraction;
      Return v_ring_count;
    End ST_NumRings;
```

This function can be used as follows:

```
    With geom As (
    Select row_number() Over (Order By 1) As rin, geom
      From (
      Select SDO_GEOMETRY(2003,NULL,NULL,
                        SDO_ELEM_INFO_ARRAY(1,1003,3),
                        SDO_ORDINATE_ARRAY(50,135,60,140)
            ) As geom
        From dual Union All
      Select SDO_GEOMETRY(2003,NULL,NULL,
                    SDO_ELEM_INFO_ARRAY(1,1003,3, 5,2003,3),
                    SDO_ORDINATE_ARRAY(50,135,60,140,51,136,59,139)
            ) As geom
        From dual Union All
      Select SDO_GEOMETRY(2003,NULL,NULL,
                    SDO_ELEM_INFO_ARRAY(1,1003,3,5,2003,3,9,2003,3),
                    SDO_ORDINATE_ARRAY(50,135,60,140,51,136,
                                       59,139,52,137,53,138)
               ) As geom
        From dual Union All
```

```
      Select SDO_GEOMETRY(2003,NULL,NULL,
                 SDO_ELEM_INFO_ARRAY(1,1003,3, 5,1003,3),
                 SDO_ORDINATE_ARRAY(50,135,60,140,61,136,69,139)
             ) As geom
        From dual Union All
      Select SDO_GEOMETRY(2003,NULL,NULL,
               SDO_ELEM_INFO_ARRAY(1,1005,2, 1,2,1, 5,2,2, 11,2003,3),
               SDO_ORDINATE_ARRAY(6,10,10,1,14,10,10,14,6,10,8,9,12,12)
             ) As geom
        From dual
    )
 )
Select a.rin, a.geom.sdo_elem_info,
       Case b.ringType
            When 0 Then 'ALL'
            When 1 Then 'OUTER'
            Else 'INNER'
        End As rType,
       ST_NumRings(a.geom,b.ringType) As rCnt
  From geom a,
       /* Three Ring Types 0, 1 and 2 */
       (Select LEVEL-1 as ringType
          From DUAL Connect By Level <= 3) b
 Order By a.rin, b.ringType;

-- Results
--
RIN GEOM.SDO_ELEM_INFO                                     RTYPE RCNT
--- ------------------------------------------------------ ----- ----
  1 SDO_ELEM_INFO_ARRAY(1,1003,3)                          ALL      1
  1 SDO_ELEM_INFO_ARRAY(1,1003,3)                          OUTER    1
  1 SDO_ELEM_INFO_ARRAY(1,1003,3)                          INNER    0
  2 SDO_ELEM_INFO_ARRAY(1,1003,3,5,2003,3)                 ALL      2
  2 SDO_ELEM_INFO_ARRAY(1,1003,3,5,2003,3)                 OUTER    1
  2 SDO_ELEM_INFO_ARRAY(1,1003,3,5,2003,3)                 INNER    1
  3 SDO_ELEM_INFO_ARRAY(1,1003,3,5,2003,3,9,2003,3)        ALL      3
  3 SDO_ELEM_INFO_ARRAY(1,1003,3,5,2003,3,9,2003,3)        OUTER    1
  3 SDO_ELEM_INFO_ARRAY(1,1003,3,5,2003,3,9,2003,3)        INNER    2
  4 SDO_ELEM_INFO_ARRAY(1,1003,3,5,1003,3)                 ALL      2
  4 SDO_ELEM_INFO_ARRAY(1,1003,3,5,1003,3)                 OUTER    2
  4 SDO_ELEM_INFO_ARRAY(1,1003,3,5,1003,3)                 INNER    0
  5 SDO_ELEM_INFO_ARRAY(1,1005,2,1,2,1,5,2,2,11,2003,3) ALL      2
  5 SDO_ELEM_INFO_ARRAY(1,1005,2,1,2,1,5,2,2,11,2003,3) OUTER    1
  5 SDO_ELEM_INFO_ARRAY(1,1005,2,1,2,1,5,2,2,11,2003,3) INNER    1

 15 rows Selected
```

Examining the operation of the ST_NumRings function

The ST_NumRings function is fairly simple; it iterates over the SDO_ELEM_INFO array, and examines the ETYPE value in the OFFSET, ETYPE, and INTERPRETATION triplet looking For those values that describe the rings (boundaries) of a polygon. These values are: 1003, 1005 (1003/5 values describe the outer boundaries of a polygon.), 2003, 2005 (2003/5 values describe the inner boundaries, holes, of a polygon) an appropriate count is computed.

Counting compound subelements

SDO_GEOMETRY linestrings and polygon boundaries can be described using any combination of vertex-connected linestrings or a sequence of circular arcs. These are known as compound elements. The 4, 1005, and 2005 SDO_ETYPES (see the preceding table) are just such elements.

```
Create Or Replace
Function ST_NumSubElems(p_geom    in SDO_GEOMETRY,
                        p_subArcs in Integer default 0)
  Return Number Deterministic
AuthId Current_User
Is
  c_i_sub_arcs      Constant pls_integer   := -20121;
  c_s_sub_arcs      Constant Varchar2(100) :=
                    'p_subArcs must be 0 (No) Or 1 (Yes).';
  v_elements        pls_integer := 0;
  v_sub_elem_count  pls_integer := 0;
  v_nCoords         pls_integer := 0;
  v_compound_count  pls_integer := 0;
  v_offset          pls_integer := 0; -- sdo_elem_info triplet
  v_etype           pls_integer := 0;
  v_interpretation  pls_integer := 0;
Begin
  If ( p_geom Is null ) Then
     Return null;
  End If;
  If ( ST_Dimension(p_geom) = 0 ) Then -- Not points
     Return 0;
  End If;
  If ( p_subArcs Is null or p_subArcs not in (0,1) ) Then
     raise_application_error(c_i_sub_arcs,c_s_sub_arcs,true);
  End If;
  v_elements := ((p_geom.sdo_elem_info.COUNT / 3) - 1);
```

```
   v_compound_count := 0;
   <<element_Extraction>>
   For v_i In 0 .. v_elements Loop
       v_offset         := p_geom.sdo_elem_info(v_i * 3 + 1);
       v_etype          := p_geom.sdo_elem_info(v_i * 3 + 2);
       v_interpretation := p_geom.sdo_elem_info(v_i * 3 + 3);
       If ( v_etype in (4,1005,2005) ) Then
           -- Compound elements with sub-elements follow
           -- v_interpretation gives number of sub-elements
           -- Worry only about compound circular arc sub-elements
           v_sub_elem_count := v_sub_elem_count + v_interpretation;
           -- Record when processing a compound element's sub-elemnt
           v_compound_count := v_interpretation;
       ElsIf ( v_interpretation = 2 ) Then
           If (p_subArcs = 0) Then
               -- Only count If compound sub-element
               If ( v_compound_count > 0 ) Then
                 v_sub_elem_count := v_sub_elem_count + 1;
                 v_compound_count := v_compound_count - 1;
               End If;
           Else
               -- Any circular arc which Is compound needs counting.
               If ( v_i = v_elements /* last triplet */ ) Then
                   v_nCoords := (p_geom.sdo_ordinates.count-v_offset) /
                               p_geom.get_dims();
               Else
                   v_nCoords := (p_geom.sdo_elem_info((v_i+1)*3+1)
                               - v_offset) / p_geom.get_dims();
               End If;
               v_sub_elem_count := v_sub_elem_count +
                               ((v_nCoords-1)/2);
               -- Have already counted the sub-element of compound elm
               If ( v_compound_count > 0 ) Then
                   v_sub_elem_count := v_sub_elem_count - 1;
               End If;
           End If;
       End If;
   End Loop element_Extraction;
   Return v_sub_elem_count;
End ST_NumSubElems;
```

Testing this with a collection of linestrings and polygons results in the following code:

```
With cGeoms As (
Select SDO_GEOMETRY(2002,NULL,NULL,
          SDO_ELEM_INFO_ARRAY(1,4,2,1,2,1,7,2,2),
          SDO_ORDINATE_ARRAY(10,78,10,75,20,75,20,78,15,80,10,78)
       ) As geom
  From dual Union All
Select SDO_GEOMETRY(2002,NULL,NULL,
          SDO_ELEM_INFO_ARRAY(1,4,2,1,2,1,3,2,2),
          SDO_ORDINATE_ARRAY(52,60,52.7,67,52.5,67,50,65)
       ) As geom
  From dual Union All
Select SDO_GEOMETRY(2002,NULL,NULL,
          SDO_ELEM_INFO_ARRAY(1,4,2,1,2,1,3,2,2),
          SDO_ORDINATE_ARRAY(52,60,52.7,67,52.5,67,50,65,
                             50,63,51,62,51,61,52,60)
       ) As geom
  From dual Union All
Select SDO_GEOMETRY(2002,NULL,NULL,
          SDO_ELEM_INFO_ARRAY(1,2,2),
          SDO_ORDINATE_ARRAY(15,115,20,118,15,120,10,118,15,115)
       ) As geom
  From dual Union All
Select SDO_GEOMETRY(2003,NULL,NULL,
          SDO_ELEM_INFO_ARRAY(1,1003,2),
          SDO_ORDINATE_ARRAY(15,115,20,118,15,120,10,118,15,115)
       ) As geom
  From dual Union All
Select SDO_GEOMETRY(2003,NULL,NULL,
          SDO_ELEM_INFO_ARRAY(1,1005,2,1,2,1,5,2,2),
          SDO_ORDINATE_ARRAY(6,10,10,1,14,10,10,14,6,10)
       ) As geom
  From dual Union All
Select SDO_GEOMETRY(2003,NULL,NULL,
          SDO_ELEM_INFO_ARRAY(1,1005,5,1,2,1,3,2,2,7,2,1,
                              9,2,2,13,2,1),
          SDO_ORDINATE_ARRAY(118.5,711.3,230.4,754.8,219.3,782.3,
                             207.6,809.5,97.5,762.0,99.0,758.3,
                             100.7,754.6,109.8,733.1,118.5,711.3)
       ) As geom
  From dual Union All
Select SDO_GEOMETRY(2003,NULL,NULL,
          SDO_ELEM_INFO_ARRAY( 1,1005,2, 1,2,1, 5,2,2,
                              11,2005,2,11,2,1,15,2,2),
          SDO_ORDINATE_ARRAY( 6,10,10,1,14,10,10,14,6,10,
                             13,10,10,2,7,10,10,13,13,10)
       ) As geom
```

```
       From dual Union All
    Select SDO_GEOMETRY(2007,NULL,NULL,
              SDO_ELEM_INFO_ARRAY(1,1005,2, 1,2,1, 5,2,2,
                                  11,2005,2,11,2,1,15,2,2,
                                  21,1005,2,21,2,1,25,2,2),
              SDO_ORDINATE_ARRAY(6,10,10,1,14,10,10,14,6,10,
                                 13,10,10,2,7,10,10,13,13,10,
                                 106,110,110,101,114,110,
                                 110,114,106,110)
          ) As geom
       From dual
    )
    Select ST_NumSubElems(a.geom,1) As subElems,
           SDO_UTIL.getNumVertices(a.geom) As coords,
           a.geom.sdo_elem_info
       From cGeoms a;

    -- Results
    --
    SUBELEMS COORDS GEOM.SDO_ELEM_INFO
    -------- ------ ---------------------------------------------------
          2      6 SDO_ELEM_INFO_ARRAY(1,4,2,1,2,1,7,2,2)
          2      4 SDO_ELEM_INFO_ARRAY(1,4,2,1,2,1,3,2,2)
          4      8 SDO_ELEM_INFO_ARRAY(1,4,2,1,2,1,3,2,2)
          2      5 SDO_ELEM_INFO_ARRAY(1,2,2)
          2      5 SDO_ELEM_INFO_ARRAY(1,1003,2)
          2      5 SDO_ELEM_INFO_ARRAY(1,1005,2,1,2,1,5,2,2)
          5      9 SDO_ELEM_INFO_ARRAY(1,1005,5,
                                       1,2,1,3,2,2,7,2,1,9,2,2,13,2,1)
          4     10 SDO_ELEM_INFO_ARRAY(1,1005,2,1,2,1,5,2,2,
                                       11,2005,2,11,2,1,15,2,2)
          6     15 SDO_ELEM_INFO_ARRAY(1,1005,2,1,2,1,5,2,2,
                                       11,2005,2,11,2,1,15,2,2,
                                       21,1005,2,21,2,1,25,2,2)

    9 rows Selected
```

The ST_NumSubElems function is slightly more involved than counting individual geometries or rings; this is because an accurate count cannot be computed by inspection of SDO_ETYPE, that is, ST_NumRings, alone. To count the subelements one must look inside compound elements to count the elements that describe them. For example, a compound linestring (4, 1005, 1005) might be made up of a single vertex-connected linestring and a single circular arc (total = 2). This can be done by counting the subelement count in the SDO_INTERPRETATION field. However, where a circular arc element exists, one must also consider that a circular arc element might be made up of a set of connected circular arcs. Counting the adjacent circular arcs can only be done by examination of the SDO_ORDINATE_ARRAY counts in SDO_OFFSET.

Extracting and filtering SDO_GEOMETRY elements

As we have seen, often the processing of an object cannot occur unless its SDO_ELEM_INFO attribute informs an algorithm's decision making. This section will explore the extraction of various elements and subelements from different geometry objects by processing the SDO_ELEM_INFO array.

Introducing the Oracle SDO_UTIL.Extract function

The Oracle API from 10gR1 onwards provides the SDO_UTIL.Extract element extraction function. The function is as follows:

```
SDO_UTIL.Extract(
    geometry In SDO_GEOMETRY,
    element  In NUMBER,
    ring     In NUMBER DEFAULT 0
) Return SDO_GEOMETRY;
```

- Geometry: [Two dimensional] geometry from which to extract the [Return] geometry.

- Element: Number of the element in the geometry: 1 for the first [...], 2 for the second [...], and so on. Geometries with SDO_GTYPE values [...] ending in 1, 2, or 3 have one element; [...] ending in 4, 5, 6, or 7 can have more than one element. For example, a [2007] multipolygon [...] might contain three elements (polygons).

- Ring: [...] This parameter is valid only for specifying a subelement of a polygon with one or more holes or of a point cluster: For a polygon with holes, its first subelement is its exterior ring, its second subelement is its first interior ring, [...], and so on. [...] For a point cluster, its first subelement is the first point in the point cluster; its second subelement is the second point in the point cluster, and so on. The default is 0, which causes the entire element to be extracted.

This function allows the extraction of top-level elements (points, lines, or polygons including the rings of those polygons), but not the subelements of a compound element as in the circular arcs that describe a polygon ring or a line. The sections that follow will examine the extraction of elements at all levels.

Why would one want to do this? To answer this, four different scenarios will be examined:

1. Extracting geometries after intersection.
2. Filtering rings of a polygon by area.
3. Extracting rings as separate geometries.
4. Extracting all geometry subelements as separate geometries.

Scenario 1 – Extracting geometries after intersection

This section will show you how to extract a specific geometry type object (point, multi-line, or multi-polygon) from a compound object. A compound object can be created in many ways, but a common one is as a result of executing an intersection between two polygons.

Let's assume that an application needs to discover the area of intersection for the earlier two polygons. However, when two polygons are intersected (as in the following SQL), the result may not just be a single polygon, but a heterogeneous geometry collection.

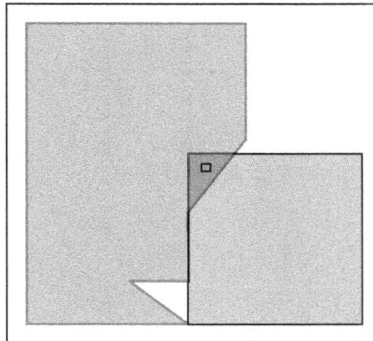

```
Select SDO_GEOM.Sdo_Intersection(
        SDO_GEOMETRY(2003,NULL,NULL,
          SDO_ELEM_INFO_ARRAY(1,1003,1,11,2003,3),
          SDO_ORDINATE_ARRAY(100,0,400,0,400,300,100,300,100,0,
                             125,270,140,280)),
        SDO_GEOMETRY(2003,NULL,NULL,
          SDO_ELEM_INFO_ARRAY(1,1003,1),
          SDO_ORDINATE_ARRAY(-175,0,100,0,0,75,100,75,100,200,
                             200,325,200,525,-175,525,-175,0)),
        0.05) As iGeom
  From dual;

-- Results
--
IGEOM
-------------------------------------------------------------------
SDO_GEOMETRY(
        2004,NULL,NULL,
        SDO_ELEM_INFO_ARRAY(1,2,1, 5,1,1, 7,1003,1, 15,2003,1),
        SDO_ORDINATE_ARRAY(
        100,75, 100,200,
        100,0,
        100,200, 180,300, 100,300, 100,200,
        125,280, 140,280, 140,270, 125,270, 125,280
      ))

1 rows selected
```

The intersection geometries are shown in the following figure:

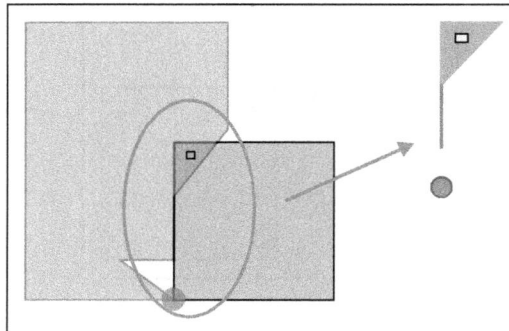

The geometry object returned from the intersection of our two polygons (ST_ Dimension of 2) contains a single POINT (ST_Dimension of 0), a single LINE (ST_Dimension of 1), and a single POLYGON with a ring (ST_Dimension of 2). Other intersections could return other combinations of objects of the same or different dimensions. But the returned collection cannot include an object whose dimension is greater than the maximum of two input objects' dimensions, for example, intersecting two linestrings (ST_Dimension of 1) could result in one or more points (0) or linestrings (1) but not polygons (2).

Implementing the ST_Extract function

Oracle's SDO_UTIL.Extract function can be used to extract whole elements from a geometry collection, that is, whole points, lines, or polygons. It also allows for the extraction of the subelements of a multipoint (one or more individual points), or polygon (one or more individual rings) as SDO_GEOMETRY objects.

In this first scenario, only the higher-level point, line, or polygon elements need to be extracted. Because it is common when intersecting two polygons to be interested only in the resulting polygons, the user will be given the option of extracting all objects or only point, line, or polygon objects via a parameter. The following function, ST_Extract, shows how to implement this processing:

```
Create Or Replace
Function ST_Extract(p_geom     In SDO_GEOMETRY,
                    p_geomType In varchar2 Default 'POLYGON')
  Return SDO_GEOMETRY Deterministic
AuthId Current_User
Is
  c_i_Extract_type Constant pls_integer   := -20120;
  c_s_Extract_type Constant Varchar2(100) :=
  'p_geomType must be one of: LINE,LINESTRING,POINT,POLY,POLYGON';
  v_elements        pls_integer;
  v_egtype          pls_integer;
  v_geom            SDO_GEOMETRY;
  v_egeom           SDO_GEOMETRY; -- Extracted geometry
  v_etype           varchar(5);
Begin
  If ( p_geom Is null Or p_geom.Get_Gtype() <> 4 ) Then
     Return p_geom;
  End If;
  If ( UPPER(p_geomType) Not In
     ('LINE','LINESTRING','POINT','POLY','POLYGON') ) Then
    raise_application_error(c_i_Extract_type,
                            c_s_Extract_type,true);
```

```
         End If;
      v_etype := Replace(
                    Replace(UPPER(p_geomType),'POLYGON','POLY'),
                    'LINESTRING','LINE');
      v_elements := SDO_UTIL.GetNumElem(p_geom);
      For v_element IN 1..v_elements Loop
        -- Extract element with all sub-elements
        v_egeom := SDO_UTIL.Extract(p_geom,v_element,0);
        v_egtype := v_egeom.get_gtype();
      If (  (v_etype = 'POINT' AND v_egtype = 1)
         Or (v_etype = 'LINE'  AND v_egtype = 2)
         Or (v_etype = 'POLY'  AND v_egtype = 3) ) Then
        If ( v_geom Is null ) Then
           v_geom := v_egeom;
        Else
           v_geom := case when v_egtype = 2
                          Then SDO_UTIL.Concat_Lines(v_geom,
                                                     v_egeom)
                          Else SDO_UTIL.Append(v_geom,v_egeom)
                     End;
        End If;
      End If;
    End Loop;
    Return(v_geom);
End ST_Extract;
```

This function can be used to extract the three geometry types from the intersection via many different methods (PL/SQL units or interactive SQL). But the basic use can be seen in the following example SQL:

```
With geoms As (
  Select SDO_GEOMETRY(2003,NULL,NULL,
            SDO_ELEM_INFO_ARRAY(1,1003,1,11,2003,3),
            SDO_ORDINATE_ARRAY(100,0,400,0,400,300,100,300,100,0,
                               125,270,140,280)
         ) As g1,
         SDO_GEOMETRY(2003,NULL,NULL,
            SDO_ELEM_INFO_ARRAY(1,1003,1),
            SDO_ORDINATE_ARRAY(-175,0,100,0,0,75,100,75,100,200,
                               200,325,200,525,-175,525,-175,0)
         ) As g2
    From DUAL
)
Select ST_Extract(b.igeom,e.gType) As geom
  From (Select SDO_GEOM.Sdo_Intersection(a.g1,a.g2,0.05) As iGEOM
```

```
        From geoms a
    ) b,
    /* Generate each possible object type to Extract */
    (Select DECODE(LEVEL,1,'POINT',2,'LINE',3,'POLYGON')
            As gType
        From DUAL
    Connect By Level < 4) e;


-- Results
--
GEOM
--------------------------------------------------------------------
SDO_GEOMETRY(2001,NULL,NULL,
        SDO_ELEM_INFO_ARRAY(1,1,1),
        SDO_ORDINATE_ARRAY(100,0))
SDO_GEOMETRY(2002,NULL,NULL,
        SDO_ELEM_INFO_ARRAY(1,2,1),
        SDO_ORDINATE_ARRAY(100,75,100,200))
SDO_GEOMETRY(2003,NULL,NULL,
        SDO_ELEM_INFO_ARRAY(1,1003,1,9,2003,1),
        SDO_ORDINATE_ARRAY(100,200,180,300,100,300,100,200,
                125,280,140,280,140,270,125,270,125,280))
```

"What is the area of intersection?" can now be answered:

```
Select SDO_GEOM.Sdo_Area(
        ST_Extract(
          SDO_GEOM.Sdo_Intersection (
          SDO_GEOMETRY(2003,NULL,NULL,
           SDO_ELEM_INFO_ARRAY(1,1003,1),
           SDO_ORDINATE_ARRAY(100,0,400,0,400,300,100,300,100,0)),
          SDO_GEOMETRY(2003,NULL,NULL,
           SDO_ELEM_INFO_ARRAY(1,1003,1),
           SDO_ORDINATE_ARRAY(-175,0,100,0,0,75,100,75,100,200,
                          200,325,200,525,-175,525,-175,0)),
          0.005),
          'POLYGON'),
          0.005) As area_of_intersection
    From DUAL;


-- Results
--
AREA_OF_INTERSECTION
-------------------
            4000
```

Scenario 2 – filtering rings of a polygon by area

The need to be able to filter the rings of a multi-polygon geometry by area, for example, removing inner rings less than a certain area in size, is a requirement for some users. This section will show how this can be done with a new function called ST_FilterRings that relies on the same understanding of the SDO_UTIL.Extract function as implemented in ST_Extract. To do this, ST_FilterRings will not only extract each polygon in an SDO_GEOMETRY object, but also extract each and every ring subelement from those polygons where those rings are greater than a declared size.

Implementing the ST_FilterRings function

The input to ST_FilterRings could be a single or multi-polygon geometry. For each, it will extract every ring (outer and inner), compute each ring's area, discard those rings whose area is less than that required, and then construct and return a single (possibly multi) polygon geometry from the resultant rings.

```
Create Or Replace
Function ST_FilterRings(p_geom      in SDO_GEOMETRY,
                        p_area      in number,
                        p_tolerance in number default 0.005,
                        p_unit      in varchar2 default null)
 Return SDO_GEOMETRY Deterministic
AuthId Current_User
Is
   v_num_elems pls_integer;
   v_num_rings pls_integer;
   v_ring      SDO_GEOMETRY;
   v_geom      SDO_GEOMETRY;
   v_unit      Varchar2(50) :=
      case when (p_geom Is null
              or p_geom.sdo_srid Is null
              or p_unit Is null)
           Then NULL
           Else case when SUBSTR(UPPER(p_unit),1,5) like 'UNIT=%'
                     Then p_unit
                     Else 'unit='||p_unit
                End
      End;
Begin
  If (p_geom Is null or ST_hasDimension(p_geom,2)=0) Then
     Return Null;
  End If;
  v_num_elems := SDO_UTIL.GetNumElem(p_geom); -- Count 1003 geoms
  <<process_all_elements>>
  For v_elem_no IN 1..v_num_elems Loop
```

```
      v_num_rings:= ST_NumRings(SDO_UTIL.Extract(p_geom,
                                                 v_elem_no,0),
                           0);
    <<process_all_rings>>
    For v_ring_no IN 1..v_num_rings Loop
      v_ring := SDO_UTIL.Extract(p_geom,v_elem_no,v_ring_no);
      If (v_ring Is not null) Then
        If (SDO_GEOM.Sdo_Area(v_ring,p_tolerance,v_unit)
              > p_area) Then
          If (v_ring_no = 1) Then -- outer ring
            v_geom := case when (v_geom Is null)
                           then v_ring
                           Else SDO_UTIL.Append(v_geom,v_ring)
                      End;
          Else -- inner ring
            v_geom := SDO_GEOM.Sdo_Difference(v_geom,v_ring,
                                              p_tolerance);
          End If;
        End If;
      End If;
    End Loop process_all_rings;
  End Loop process_all_elements;
  Return v_geom;
End ST_FilterRings;
```

The following example shows how it can be used:

```
With testGeoms As (
Select SDO_GEOMETRY(2003,NULL,NULL,
        SDO_ELEM_INFO_ARRAY(1,1003,3),
        SDO_ORDINATE_ARRAY(0,0,1,1)) As geom
  From dual union all
Select SDO_GEOMETRY(2003,NULL,NULL,
        SDO_ELEM_INFO_ARRAY(1,1003,1,11,2003,1,21,2003,1),
        SDO_ORDINATE_ARRAY(0,0,20,0,20,20,0,20,0,0,
                           10,10,10,11,11,11,11,10,10,10,
                           5,5,5,7,7,7,7,5,5,5.0)) As geom
  From dual union all
Select SDO_GEOMETRY(2007,NULL,NULL,
        SDO_ELEM_INFO_ARRAY(1,1003,1,11,2003,1,21,2003,1,
                            31,1003,1),
        SDO_ORDINATE_ARRAY(0,0,20,0,20,20,0,20,0,0,
                           10,10,10,11,11,11,11,10,10,10,
                           5,5,5,7,7,7,7,5,5,5,
                           40,40,50,40,50,50,40,50,40,40.0))
  From dual union all
Select SDO_GEOMETRY(2003,NULL,NULL,
        SDO_ELEM_INFO_ARRAY(1,1005,2,1,2,1,5,2,2),
        SDO_ORDINATE_ARRAY(6,10,10,1,14,10,10,14,6,10.0))
  From dual union all
```

```
Select SDO_GEOMETRY(2007,NULL,NULL,
        SDO_ELEM_INFO_ARRAY(1,1005,2,1,2,1,5,2,2,
                            11,2005,2,11,2,1,15,2,2,
                            21,1005,2,21,2,1,25,2,2),
        SDO_ORDINATE_ARRAY(6,10,10,1,14,10,10,14,6,10,
                           13,10,10,2,7,10,10,13,13,10,
                           106,110,110,101,114,110,110,114,106,110.0))
    From dual union all
Select SDO_GEOMETRY(2007,NULL,NULL,
        SDO_ELEM_INFO_ARRAY(1,1003,3,   5,2003,3,  9,2003,3,
                            13,1003,1,
                            23,2003,3,  27,2003,3),
        SDO_ORDINATE_ARRAY(-10,-10,-1,-1,-9,-9,-8,-8,-5,-5,-3,-3,
                           0,0,20,0,20,20,0,20,0,0,
                           10,10,11,11,  15,15,18,19.0))
    From dual
)
Select b.geom.get_gtype()  As geomType,
        rings_beFore,
        ST_NumRings(b.geom) As rings_after,
        coords_beFore,   case when b.geom Is not null
                            Then SDO_UTIL.getNumVertices(b.geom)
                            Else null
                         End As coords_after
    From (Select ST_NumRings(a.geom) As rings_beFore,
            SDO_UTIL.getNumVertices(a.geom) As coords_beFore,
            ST_FilterRings(a.geom,2.0,0.005) As geom
        From testGeoms a
    ) b;

-- Results
--
GEOMTYPE RINGS_BEFORE RINGS_AFTER COORDS_BEFORE COORDS_AFTER
-------- ------------ ----------- ------------- ------------
 (NULL)            1      (NULL)             2       (NULL)
      3            3           2            15           10
      7            4           3            20           15
      3            1           1             5            5
      7            3           3            15           15
      7            6           4            15           20

  6 rows Selected
```

Note how the `ST_FilterRings` function removed a whole polygon (the first), because that polygon's outer ring's area was less than the supplied value.

Before leaving this function, it is important to note that the SDO_GEOM.Sdo_Area function has an optional last parameter:

> **Unit (Unit of Measure)**: This is a name/value pair string made up of a unit= keyword and an SDO_UNIT value from the SDO_AREA_UNITS Table (for example, 'unit=SQ_KM'). If this parameter is not specified, the unit of measurement associated with the data is assumed. For geodetic data, the default unit of measurement is square meters.

The ST_FilterRings function supplied with the book implements handling for this parameter.

Scenario 3 – Extracting rings as separate geometries

The need to be able to filter the rings of a polygon geometry by area is useful, especially where the result is to be a single geometry. But what if one wanted to extract all the rings of a multi-polygon geometry and return them as a collection of individual polygon SDO_GEOMETRY objects?

This section shows how to extract all of the rings of a (multi) polygon, returning them as separate geometries. Again, the SDO_UTIL.Extract function's ability to extract the rings of a polygon geometry as separate SDO_GEOMETRY objects will be used. To return a collection (or "array") of rings, will require looking at the mechanisms to do this.

Normally, functions (such as ST_FilterRings) return a single object, so how can a function return a collection of objects? Oracle has two different methods for doing this that are of interest to us.

The first is by having the function create, and assign a set of SDO_GEOMETRY objects to a single collection or array variable (either a VARRAY or NESTED Table (See "Chapter 5 PL/SQL Collections and Records", in Oracle® Database PL/SQL Language Reference 11*g* Release 2 (11.2) Part Number E25519-05)), and then have the function return that variable at the end of processing via a single Return statement.

The second method implements a concept known as pipelining. A pipelined function allows each SDO_GEOMETRY object to be pushed out of the function as soon as it is created; no return array needs to be created as in the first method, thereby reducing the need For PGA memory. In addition, pipelined functions can be used in parallel queries improving performance. Because of this, pipelining is a method that can significantly lower the overhead (Reducing function processing over-head is a characteristic of good corporate citizenship especially where the database is shared by lots of applications and people.) of a function that returns multiple rows.

This section will demonstrate the use of the pipelining concept through the construction of a pipelined function called ST_ExtractRings. A non-pipelined version of ST_Extract_Rings function is provided in the code available with this book.

Implementing the ST_ExtractRings function

Because the ST_ExtractRings function will be extracting and returning one or more rings, it will therefore need a collection data type that can be used to return each extracted ring to the user.

The existing MDSYS.SDO_GEOMETRY_ARRAY data type is a possible return collection type.

```
Create Or Replace Type SDO_GEOMETRY_ARRAY
As VARRAY(10485760) Of SDO_GEOMETRY
```

However, two issues affect its use. The first, a relatively minor one, is that there is a fixed upper limit in the VARRAY data type of 10,485,760 objects, which may be an issue for those users managing very large geometric datasets.

The second issue is that any SDO_GEOMETRY returned in the pipeline will be interpreted as a collection of individual columns of data as shown in the following code:

```
Select t.*
  From Table(SDO_GEOMETRY_ARRAY(
             SDO_GEOMETRY(2001,NULL,
               SDO_POINT_TYPE(1,1,NULL),NULL,NULL),
             SDO_GEOMETRY(2002,NULL,NULL,
               SDO_ELEM_INFO_ARRAY(1,2,1),
               SDO_ORDINATE_ARRAY(1,1,2,2)),
             SDO_GEOMETRY(2003,NULL,NULL,
               SDO_ELEM_INFO_ARRAY(1,1003,3),
               SDO_ORDINATE_ARRAY(1,1,10,10)))) t;
-- Results
--
SDO_GTYPE SDO_SRID SDO_POINT
--------- -------- ------------------------
SDO_ELEM_INFO                   SDO_ORDINATES
--------------------------- ----------------------------
     2001 NULL     SDO_POINT_TYPE(1,1,NULL)
NULL                            NULL
     2002 NULL     NULL
SDO_ELEM_INFO_ARRAY(1,2,1)      SDO_ORDINATE_ARRAY(1,1,2,2)
     2003 NULL     NULL
SDO_ELEM_INFO_ARRAY(1,1003,3) SDO_ORDINATE_ARRAY(1,1,10,10)
```

SQL can reconstruct a single `SDO_GEOMETRY` object from its attributes, but only by using a standard `SDO_GEOMETRY` constructor as follows:

```
Select SDO_GEOMETRY(t.sdo_gtype,t.sdo_srid,t.sdo_point,
                    t.sdo_elem_info,t.sdo_ordinates) as geom
  From Table(SDO_GEOMETRY_ARRAY(
             SDO_GEOMETRY(2001,NULL,
                SDO_POINT_TYPE(1,1,NULL),NULL,NULL),
             SDO_GEOMETRY(2002,NULL,NULL,
                SDO_ELEM_INFO_ARRAY(1,2,1),
                SDO_ORDINATE_ARRAY(1,1,2,2)),
             SDO_GEOMETRY(2003,NULL,NULL,
                SDO_ELEM_INFO_ARRAY(1,1003,3),
                SDO_ORDINATE_ARRAY(1,1,10,10))))) t;
GEOM
-----------------------------------------------------------------
SDO_GEOMETRY(2001,NULL,SDO_POINT_TYPE(1,1,NULL),NULL,NULL)
SDO_GEOMETRY(2001,NULL,NULL,SDO_ELEM_INFO_ARRAY(1,2,1),
            SDO_ORDINATE_ARRAY(1,1,2,2))
SDO_GEOMETRY(2003,NULL,NULL,SDO_ELEM_INFO_ARRAY(1,1003,3),
            SDO_ORDINATE_ARRAY(1,1,10,10))
```

To avoid the need to reconstruct the `SDO_GEOMETRY` object after the elements are returned from the `Table` function, two new data types are declared:

```
Create Or Replace Type SDO_GEOMETRY_Obj
As Object (geom SDO_GEOMETRY);

Grant Execute On SDO_GEOMETRY_obj To Public;

Create Or Replace Type sdo_geometries
As Table of SDO_GEOMETRY_Obj;

Grant Execute On sdo_geometries To Public;
```

So that now an `SDO_GEOMETRY` object is correctly returned:

```
Select t.*
  From Table(SDO_GEOMETRIES(
             SDO_GEOMETRY_OBJ(SDO_GEOMETRY(2001,NULL,
                SDO_POINT_TYPE(1,1,NULL),NULL,NULL)),
             SDO_GEOMETRY_OBJ(SDO_GEOMETRY(2002,NULL,NULL,
                SDO_ELEM_INFO_ARRAY(1,2,1),
                SDO_ORDINATE_ARRAY(1,1,2,2))),
             SDO_GEOMETRY_OBJ(SDO_GEOMETRY(2003,NULL,NULL,
                SDO_ELEM_INFO_ARRAY(1,1003,3),
                SDO_ORDINATE_ARRAY(1,1,10,10))))) t;
```

```
GEOM
------------------------------------------------------------------
SDO_GEOMETRY(2001,NULL,SDO_POINT_TYPE(1,1,NULL),NULL,NULL)
SDO_GEOMETRY(2002,NULL,NULL,SDO_ELEM_INFO_ARRAY(1,2,1),
            SDO_ORDINATE_ARRAY(1,1,2,2))
SDO_GEOMETRY(2003,NULL,NULL,SDO_ELEM_INFO_ARRAY(1,1003,3),
            SDO_ORDINATE_ARRAY(1,1,10,10))
```

For ST_ExtractRings to be pipelined it must include the PIPELINED clause in the return clause part of its function definition. In addition, it must call PIPE ROW as each ring is extracted to return the ring.

```
Create Or Replace
Function ST_ExtractRings(p_geom in SDO_GEOMETRY)
  Return sdo_geometries Pipelined
Is
   v_num_rings pls_integer;
   v_ring      SDO_GEOMETRY;
Begin
  If (p_geom Is null Or ST_hasDimension(p_geom,2)=0) Then
     Return;
  End If;
  <<process_all_elements>>
  For v_elem_no In 1..SDO_UTIL.GETNUMELEM(p_geom) Loop
     v_num_rings := ST_NumRings(SDO_UTIL.Extract(p_geom,
                                             v_elem_no,0),
                           0/*ALL*/);
     If ( v_num_rings = 1 ) Then
       PIPE ROW (SDO_GEOMETRY_OBJ(
                      SDO_UTIL.Extract(p_geom,v_elem_no,1))
              );
     Else
       <<process_all_rings>>
       For v_ring_no In 1..v_num_rings Loop
          v_ring := SDO_UTIL.Extract(p_geom,
                                       v_elem_no,v_ring_no);
          If ( v_ring Is not null ) Then
             PIPE ROW(SDO_GEOMETRY_OBJ(v_ring));
          End If;
       End Loop process_all_rings;
     End If;
  End Loop process_all_elements;
  Return;
End ST_ExtractRings;
```

In this function, you will notice that, other than the creation of a new v_ring SDO_GEOMETRY object, no other memory is needed. As each ring geometry is extracted it can be pushed straight into the pipeline (PIPE ROW).

Executing a pipelined function requires the use of the `Table` keyword within an SQL statement. The `Table` function treats the resulting rows from the function like a regular table or row source. If the row source has one or more attributes the table function allows access to the row source's column data. The `Table` function implements what is called collection unnesting.

```
WITH testGeoms As (
Select row_number() over (order by 1) As rin, geom
   From (Select SDO_GEOMETRY(2003,NULL,NULL,
                    SDO_ELEM_INFO_ARRAY(1,1003,3),
                    SDO_ORDINATE_ARRAY(-10.0,-10.0, -1.0,-1.0))
                As geom
     From dual UNION ALL
   Select SDO_GEOMETRY(2007,null,null,
                SDO_ELEM_INFO_ARRAY(1,1003,3, 5,2003,3, 9,2003,3,
                                    13,1003,1,23,2003,3,27,2003,3),
                SDO_ORDINATE_ARRAY(-10,-10,-1,-1,
                                   -9,-9,-8,-8,
                                   -5,-5,-3,-3,
                                    0,0,20,0,20,20,0,20,0,0,
                                   10,10,11,11,
                                   15,15,18,19))
                As geom
     From dual)
)
Select a.rin, t.*
 From testGeoms a,
      Table(ST_ExtractRings(a.geom)) t;

-- Results
--
RIN GEOM
--- -----------------------------------------------------------------
  1 SDO_GEOMETRY(2003,NULL,NULL,SDO_ELEM_INFO_ARRAY(1,1003,3),
             SDO_ORDINATE_ARRAY(-10,-10,-1,-1))
  2 SDO_GEOMETRY(2003,NULL,NULL,SDO_ELEM_INFO_ARRAY(1,1003,3),
             SDO_ORDINATE_ARRAY(-10,-10,-1,-1))
  2 SDO_GEOMETRY(2003,NULL,NULL,SDO_ELEM_INFO_ARRAY(1,1003,3),
             SDO_ORDINATE_ARRAY(-9,-9,-8,-8))
  2 SDO_GEOMETRY(2003,NULL,NULL,SDO_ELEM_INFO_ARRAY(1,1003,3),
             SDO_ORDINATE_ARRAY(-5,-5,-3,-3))
  2 SDO_GEOMETRY(2003,NULL,NULL,SDO_ELEM_INFO_ARRAY(1,1003,1),
             SDO_ORDINATE_ARRAY(0,0,20,0,20,20,0,20,0,0))
  2 SDO_GEOMETRY(2003,NULL,NULL,SDO_ELEM_INFO_ARRAY(1,1003,3),
             SDO_ORDINATE_ARRAY(10,10,11,11))
  2 SDO_GEOMETRY(2003,NULL,NULL,SDO_ELEM_INFO_ARRAY(1,1003,3),
             SDO_ORDINATE_ARRAY(15,15,18,19))

 7 rows Selected
```

The function processes a geometry object as follows. Firstly it determines how many elements are in the geometry (1 is a single polygon, =>1 if it is a multi-polygon) ("Collection Unnesting: Examples" under "Select" in the Oracle® Database SQL Language Reference 11*g* Release 2 (11.2) Part Number E26088-02.) and iterates over each element. Then it determines how many rings exist within that extracted geometry via the ST_NumRings function. Each ring is then extracted via the SDO_UTIL.Extract function. As soon as each ring is extracted, the PIPE ROW call is used to push each extracted ring out of the function. Because all the data is being pushed out using PIPE ROW calls, there is nothing to return at the end of the function, hence the empty Return call.

Scenario 4 – Extracting all geometry subelements as separate geometries

The SDO_UTIL.Extract function is very useful for extracting the elements and subelements of a SDO_GEOMETRY object, but notice what happens when it is used on a compound polygon object with multiple rings:

> And in "Table 2-2 values and Semantics in SDO_ELEM_INFO" in Oracle® Spatial Developer's Guide 11*g* Release 2(11.2) Part Number E11830-04.

```
With geom As (
/* Extract top level polygons */
Select row_number() over (order by 1) As rin,
       ST_NumRings(b.gElem) As rings, gelem
  From (Select SDO_UTIL.Extract(a.geom,LEVEL,0) As gElem
         From (Select SDO_GEOMETRY(2007,NULL,NULL,
                      SDO_ELEM_INFO_ARRAY(1,1005,2, 1,2,1, 5,2,2,
                                          11,2005,2,11,2,1,15,2,2,
                                          21,1005,2,21,2,1,25,2,2),
                      SDO_ORDINATE_ARRAY(
                            6,10,10, 1,14,10,10,14, 6,10,
                            13,10,10, 2, 7,10,10,13,13,10,
                            21,25,25,16,29,25,25,29,21,25))
                      As geom
                From dual
             ) a
         Connect By Level <= SDO_UTIL.getNumElem(a.geom)
       ) b
)
/* Extract rings of each polygon as polygon */
Select 'Ring ' || t.column_value || ' out of ' || c.rings As ring,
       SDO_UTIL.Extract(c.gElem,1,t.column_value) As gElem
  From geom c,
       /* Generate ring number for all rings in each polygon */
       Table(CAST(MULTISET(Select level
                           From dual
```

```
                   Connect By Level <= c.rings)
               As mdsys.SDO_NUMBER_ARRAY)) t
   Where t.column_value <= c.rings
   Order By c.rin, t.column_value;

-- Results
--
RING            GELEM
--------------  -------------------------------------------------------
Ring 1 out of 2 SDO_GEOMETRY(2003,NULL,NULL,
                   SDO_ELEM_INFO_ARRAY(1,1005,2,1,2,1,5,2,2),
                   SDO_ORDINATE_ARRAY(6,10,10,1,14,10,10,14,6,10))
Ring 2 out of 2 SDO_GEOMETRY(2003,NULL,NULL,
                   SDO_ELEM_INFO_ARRAY(1,1005,2,1,2,2,5,2,1),
                   SDO_ORDINATE_ARRAY(13,10,10,13,7,10,10,2,13,10))
Ring 1 out of 1 SDO_GEOMETRY(2003,NULL,NULL,
                   SDO_ELEM_INFO_ARRAY(1,1005,2,1,2,1,5,2,2),
                   SDO_ORDINATE_ARRAY(106,110,110,101,114,110,
                               110,114,106,110))
```

SDO_UTIL.Extract stops extracting at the level ring and does not process each compound ring's subelements. Accessing the circular arcs that describe a compound subelement (For example, SDO_ETYPE 1005/2005—compound rings) of a polygon is a useful thing to do when analyzing the correctness of an SDO_GEOMETRY object construction, or for reuse of an arc into another geometry. This ultimate fragmentation of an SDO_GEOMETRY object may seem abstract, but having such a capability is powerful and useful. How this can be done will now be examined.

Fragmenting a geometry – ST_ExplodeGeometry

As already seen, the SDO_UTIL.Extract function is useful only for extracting elements and subelements of simple SDO_GEOMETRY objects. However, to help the function extract the subelement rings of even a polygon, the ST_NumRings function was built. If a subelement is compound, SDO_UTIL.Extract does not allow us to extract its subelements (that is, the lines and circular arcs that describe it). Because no standard function is available that enables the counting of the compound elements of a ring or linestring; the ST_NumSubElems function was built that allows for the counting of the linestring and circular arc constituents of a compound element or subelement.

The functions introduced so far have given us knowledge of the structure of the SDO_ELEM_INFO and SDO_ORDINATES attributes, the SDO_UTIL.Extract function, and pipelining. These can now be used to implement a new function, ST_ExplodeGeometry, which breaks any object into its fundamental descriptive elements.

The actual code for `ST_ExplodeGeometry` will not be shown in this book due to its length and complexity. (This complexity is mostly due to it being a pipelined function that requires a lot of code duplication.)

```
Create Or Replace
Function ST_ExplodeGeometry(p_geom      In SDO_GEOMETRY,
                            p_tolerance In number default 0.005)
   Return sdo_geometries Pipelined;
```

The following SQL shows a number of combinations of types of geometries:

```
With geoms As (
Select row_number() over (order by 1) As geom_id, geom
   From (Select SDO_GEOMETRY(2004,null,null,
                  SDO_ELEM_INFO_ARRAY(1,1,1,    3,2,1,
                                      7,1003,3,11,2003,3),
                  SDO_ORDINATE_ARRAY (1,2,      1,1,2,1,
                                      3,3,4,4,3.1,3.1,3.9,3.9))
                As geom
          From dual Union All
         Select SDO_GEOMETRY(2002,NULL,NULL,
                  SDO_ELEM_INFO_ARRAY(1,4,2, 1,2,1, 3,2,2),
                  SDO_ORDINATE_ARRAY (0,0,
                                      70,70,50,70,50,50,
                                      55,16,13.5,13.5))
                As geom
          From dual Union All
         Select SDO_GEOMETRY(2005,NULL,NULL,
                  SDO_ELEM_INFO_ARRAY(1,1,2),
                  SDO_ORDINATE_ARRAY (1,2,3,4))
                As geom
          From dual Union All
         Select SDO_GEOMETRY(2007,NULL,NULL,
                  SDO_ELEM_INFO_ARRAY(1,1005,2,  1,2,1, 5,2,2,
                                      11,2005,2, 11,2,1,15,2,2,
                                      21,1005,2, 21,2,1,25,2,2),
                  SDO_ORDINATE_ARRAY(6,10,10,1, 14,10,10,14,6,10,
                                     13,10,10,2, 7,10,10,13,13,10,
                                     21,25,25,16,29,25,25,29,21,25))
                As geom
          From dual
   )
)
Select a.geom_id ||
       '/' ||
```

```
        (row_number() over (partition by a.geom_id order by 1)) ||
        '/' ||
        a.geom.get_gtype() as ELEMID,
        t.geom
  From geoms a,
        Table(ST_ExplodeGeometry(a.geom)) t
  Order By a.geom_id;


--Results
--a
ELEMID GEOM
------ -------------------------------------------------------------
1/1/4  SDO_GEOMETRY(2001,NULL,NULL,SDO_ELEM_INFO_ARRAY(1,1,1),
           SDO_ORDINATE_ARRAY(1,2))
1/2/4  SDO_GEOMETRY(2002,NULL,NULL,SDO_ELEM_INFO_ARRAY(1,2,1),
           SDO_ORDINATE_ARRAY(1,1,2,1))
1/3/4  SDO_GEOMETRY(2003,NULL,NULL,SDO_ELEM_INFO_ARRAY(1,1003,3),
           SDO_ORDINATE_ARRAY(3,3,4,4))
1/4/4  SDO_GEOMETRY(2003,NULL,NULL,SDO_ELEM_INFO_ARRAY(1,1003,3),
           SDO_ORDINATE_ARRAY(3.1,3.1,3.9,3.9))
2/1/2  SDO_GEOMETRY(2002,NULL,NULL,SDO_ELEM_INFO_ARRAY(1,2,1),
           SDO_ORDINATE_ARRAY(0,0,70,70))
2/2/2  SDO_GEOMETRY(2002,NULL,NULL,SDO_ELEM_INFO_ARRAY(1,2,2),
           SDO_ORDINATE_ARRAY(70,70,50,70,50,50))
2/3/2  SDO_GEOMETRY(2002,NULL,NULL,SDO_ELEM_INFO_ARRAY(1,2,2),
           SDO_ORDINATE_ARRAY(70,50,50,55,16,13.5))
3/1/5  SDO_GEOMETRY(2001,NULL,NULL,SDO_ELEM_INFO_ARRAY(1,1,1),
           SDO_ORDINATE_ARRAY(1,2))
3/2/5  SDO_GEOMETRY(2001,NULL,NULL,SDO_ELEM_INFO_ARRAY(1,1,1),
           SDO_ORDINATE_ARRAY(3,4))
4/1/7  SDO_GEOMETRY(2002,NULL,NULL,SDO_ELEM_INFO_ARRAY(1,2,1),
           SDO_ORDINATE_ARRAY(6,10,10,1,14,10))
4/2/7  SDO_GEOMETRY(2002,NULL,NULL,SDO_ELEM_INFO_ARRAY(1,2,2),
           SDO_ORDINATE_ARRAY(14,10,10,14,6,10))
4/3/7  SDO_GEOMETRY(2002,NULL,NULL,SDO_ELEM_INFO_ARRAY(1,2,2),
           SDO_ORDINATE_ARRAY(13,10,10,13,7,10))
4/4/7  SDO_GEOMETRY(2002,NULL,NULL,SDO_ELEM_INFO_ARRAY(1,2,1),
           SDO_ORDINATE_ARRAY(7,10,10,2,13,10))
4/5/7  SDO_GEOMETRY(2002,NULL,NULL,SDO_ELEM_INFO_ARRAY(1,2,1),
           SDO_ORDINATE_ARRAY(21,25,25,16,29,25))
4/6/7  SDO_GEOMETRY(2002,NULL,NULL,SDO_ELEM_INFO_ARRAY(1,2,2),
           SDO_ORDINATE_ARRAY(29,25,25,29,21,25))


 15 rows Selected
```

The function copes with all the compound elements and subelements described in the table "SDO_ETYPE and SDO_INTERPRETATION values relating to circular arcs."For further clarity, take a look at the following example, which is based on a single compound linestring:

```
SDO_GEOMETRY(2002,NULL,NULL,
        SDO_ELEM_INFO_ARRAY(1,4,2,
                            1,2,1,
                            3,2,2),
        SDO_ORDINATE_ARRAY (0,0,
            70,70,50,70,50,50,55,16,13.5,13.5))
```

This compound linestring (4) is made up of two subelements. The first element is a linestring (2) composed of two vertices connected by a straight line (1) segment: (0,0),(70,0). The second element is also a linestring (2), but described by a connected sequence of circular arcs (2). A circular arc is described by three coordinates. Note that the last point of a subelement (for example, the linestring in this case) is the first point of the next subelement (it is not to be repeated).

```
Select ST_NumSubElems(SDO_GEOMETRY(2002,NULL,NULL,
        SDO_ELEM_INFO_ARRAY(1,4,2,
                            1,2,1,
                            3,2,2)
        SDO_ORDINATE_ARRAY (0,0,
            70,70,50,70,50,50,55,16,13.5,13.5)))
        As SubElems
    From Dual;


SUBELEMS
--------
       3
```

Such a function could find use in building new adjoining polygons to an existing polygon that contains compound elements.

Vectorizing geometries with linestrings

This section shows you how to take any linestring or polygon geometry object and decompose it even further than ST_ExplodeGeometry; not just the extraction of individual elements and subelements, but right down to individual vectors. A vector is a directed two vertex linestring.

```
(X1,Y2) ---> (X2,Y2)
E.g. SDO_GEOMETRY(2002,NULL,NULL,
                SDO_ELEM_INFO_ARRAY(1,2,1),
                SDO_ORDINATE_ARRAY (1,1,2,1))
```

The only level lower than this is the individual vertex level!

Vectorization of an SDO_GEOMETRY object requires access to all the vertices of that object. The SDO_UTIL.GetVertices function can be used to return the vertices of an SDO_GEOMETRY object in a form that is easy and natural to work with. This function returns the MDSYS.VERTEX_SET_TYPE collection object, which is defined on the MDSYS.VERTEX_TYPE object:

```
Create Or Replace Type vertex_set_type As Table Of mdsys.vertex_type;
```

The MDSYS.VERTEX_TYPE object is defined (11*g*R1 and above) as follows:

```
Create Or Replace Type T_Vertex Is Object (
   x  Number,
   y  Number,
   z  Number,
   w  Number,
   id Number
   [ constructors, function details removed ]
);
```

This object has changed with every release of the Oracle Locator product since 9*i*, when it was first introduced. These changes are immaterial where we use functions that create objects for us c.f., SDO_UTIL.GetVertices. However, to use this object in user object types and PL/SQL functions across multiple product versions, a method that will deliver stability for those derived types and functions must be implemented.

This will be done by creating a user-defined vertex object, which will then be used to create a new vector object that will hold the start and end vertices that compose a vector. This is sensible in such a way that almost all uses for vectorization is at the 2D X, Y coordinate level (that is, the v5 to v11 numbers are never used).

This object will be called T_Vertex. A simplified type declaration (extended implementation details removed though available in the code that is shipped with this book) is as follows:

```
Create Or Replace Type T_Vertex Is Object (
   x  Number,
   y  Number,
   z  Number,
   w  Number,
   id Number
   [ constructors, function details removed ]
);
```

Implementing vectorization – ST_Vectorize

The T_Vertex object can now be used to create the T_Vector object. A simplified type declaration follows (again its extended implementation details have been removed):

```
Create Or Replace Type T_Vector Is Object (
  Id             Integer,
  element_id     Integer ,
  subelement_id  Integer ,
  startCoord     T_Vertex,
  EndCoord       T_Vertex
  [ constructor details removed ]
  Member Function ST_Dims    Return pls_integer Deterministic,
  Member Function ST_SdoGeometry(p_SRID in Number)
                             Return SDO_GEOMETRYDeterministic,
  Member Function ST_AsText Return varchar2 Deterministic
);
```

Now that these two objects are available, a pipelined function (because vectorization can create thousands of objects in one execution) can be created that will process vertex-connected linestrings and polygons only. This function is called ST_Vectorize and will return an array of vector objects. Therefore two additional types are required.

```
Create Or Replace Type T_VectorObj As Object (vector T_Vector);
```

T_VectorObj is needed so that the ST_Vectorize function will return a single T_Vector object per row and not the individual elements of the T_Vector object.

```
Create Or Replace Type T_Vectors As Table of T_VectorObj;
```

This type is the array type that ST_Vectorize will use to return one or more vectors.

Where a polygon ring is defined by an optimized rectangle, the rectangle will be converted to its vertex-connected equivalent before being processed. A special "helper" function, ST_Rectange2Polygon, is used to do this and is included with the code shipped with this book.

This version of the ST_Vectorize function will not process the SDO_GEOMETRY objects that include compound elements. This is because a circular arc is defined using three vertices, and it is invalid to construct two vectors from the vertices that define these elements (you would be returning the chord and not the circular arc, thus losing definition (A function, ST_ArcVectorize, which handles compound objects, is included in the T_GEOMETRY object that is shipped with this book.))

```
Select ST_Rectangle2Polygon(
         SDO_GEOMETRY(2003,null,null,
            SDO_ELEM_INFO_ARRAY(1,1003,3),
            SDO_ORDINATE_ARRAY (1,1,20,20))) as geom
  From Dual;
-- Results
--
GEOM
-----------------------------------------------------------
SDO_GEOMETRY(2003,NULL,NULL,
            SDO_ELEM_INFO_ARRAY(1,1003,1),
            SDO_ORDINATE_ARRAY(1,1,20,1,20,20,1,20,1,1))
```

ST_Vectorize in skeletal form looks like this:

```
Create Or Replace
Function ST_Vectorize(p_geometry in SDO_GEOMETRY)
  Return T_Vectors Pipelined
Is
  [ declarations removed ]
Begin
  [ geometry validity and suitability testing removed ]
  v_dims := p_geom.get_dims();
  <<Extract_all_elements>>
  For v_element_no In 1..SDO_UTIL.GetNumElem(p_geom) Loop
    v_element := SDO_UTIL.Extract(p_geom,v_element_no,0);
    If ( v_element Is not null ) Then
      If ( v_element.get_gtype() = 3) Then -- Polygons
        <<Exract_All_Rings>>
        For v_ring_no IN 1..ST_NumRings(v_element,0) Loop
          v_ring := SDO_UTIL.Extract(p_geom,
                                     v_element_no,v_ring_no);
          If ( v_ring Is not null ) Then
            -- Check If ring Is an optimized rectangles
            If ( ST_hasOptimizedRectangles(v_ring)=1 ) Then
              v_ring := ST_Rectangle2Polygon(v_ring);
            End If;
            -- Vectorize the ring
            For v_coord_no
                IN 1..((v_ring.sdo_ordinates.COUNT/v_dims)-1)
            Loop
              v_ord := ((v_coord_no - 1) * v_dims) + 1;
              PIPE ROW(T_VectorObj(T_Vector(
                  v_element_no,v_ring_no,v_coord_no,
                  T_Vertex(v_ring.sdo_ordinates(v_ord),
```

```
                        v_ring.sdo_ordinates(v_ord+1),
                        case when v_dims>2 Then
                                v_ring.sdo_ordinates(v_ord+2)
                            Else null End,
                        case when v_dims>3 Then
                                v_ring.sdo_ordinates(v_ord+3)
                            Else null End,
                        v_coord_no),
              T_Vertex(v_ring.sdo_ordinates(v_dims+v_ord),
                        v_ring.sdo_ordinates(v_dims+v_ord+1),
                        case when v_dims>2
                            Then v_ring.sdo_ordinates
                                        (v_dims+v_ord+2)
                            Else null End,
                        case when v_dims>3
                            Then v_ring.sdo_ordinates
                                        (v_dims+v_ord+3)
                            Else null End,
                        v_coord_no + 1) ) ) );
        End Loop;
      End If;
    End Loop Extract_All_Rings;
  ElsIf ( v_element.get_gtype() = 2) Then  -- Linestrings
    -- Vectorize the linestring
    For v_coord_no
        IN 1..((v_element.sdo_ordinates.COUNT/v_dims)-1)
    Loop
      v_ord := ((v_coord_no - 1) * v_dims) + 1;
      PIPE ROW(T_VectorObj(T_Vector(
          v_element_no,null,v_coord_no,
          T_Vertex(v_element.sdo_ordinates(v_ord),
                    v_element.sdo_ordinates(v_ord+1),
                    case when v_dims>2
                        Then v_element.sdo_ordinates(v_ord+2)
                        Else null End,
                    case when v_dims>3
                        Then v_element.sdo_ordinates(v_ord+3)
                        Else null End,
                    v_coord_no),
          T_Vertex(v_element.sdo_ordinates(v_dims+v_ord),
                    v_element.sdo_ordinates(v_dims+v_ord+1),
                    case when v_dims>2
                        Then v_element.sdo_ordinates
                                    (v_dims+v_ord+2)
```

```
                         Else null End,
             case when v_dims>3
                Then v_element.sdo_ordinates
                                    (v_dims+v_ord+3)
                Else null End,
             v_coord_no + 1) ) ) );
      End Loop;
    End If;
  End If;
  End Loop Extract_all_elements;
  Return;
End ST_Vectorize;
```

Simply put, ST_Vectorize identifies pairs of coordinates and from them constructs the vectors. The T_Vector structure allows for the source of the vector to be returned as well. Thus, if a vector came from the first two vertices of the first ring of a single polygon, element_id would be set to 1, subelement_id (ring) to 1 and the vector to 1.

```
With geoms As (
  Select 'Line'  As id,
         SDO_GEOMETRY(2002,null,null,
             SDO_ELEM_INFO_ARRAY(1,2,1),
             SDO_ORDINATE_ARRAY(1,1,20,20)) As geom
    From dual Union All
  Select 'Rect'  As id,
         SDO_GEOMETRY (2003,null,null,
             SDO_ELEM_INFO_ARRAY(1,1003,3),
             SDO_ORDINATE_ARRAY(1,1,20,20)) As geom
    From dual Union All
  Select 'Poly'  As id,
         SDO_GEOMETRY (2003,null,null,
             SDO_ELEM_INFO_ARRAY(1,1003,1,11,2003,1,21,2003,1),
             SDO_ORDINATE_ARRAY (0,0,20,0,20,20,0,20,0,0,
                                 10,10,10,11,11,11,11,10,10,10,
                                 5,5,5,7,7,7,7,5,5,5)) As geom
    From dual Union All
  Select 'MPly' As id,
```

```
                SDO_GEOMETRY (2007,null,null,
                    SDO_ELEM_INFO_ARRAY(1,1003,1,11,2003,1,21,2003,1,
                                31,1003,1),
                    SDO_ORDINATE_ARRAY(0,0,20,0,20,20,0,20,0,0,
                                10,10,10,11,11,11,11,10,10,10,
                                5,5,5,7,7,7,7,5,5,5,
                                40,40,50,40,50,50,40,50,40,40))
            As geom
      From dual
  )
  Select a.id,
         Cast(v.vector.element_id              || '/' ||
             NVL(v.vector.subelement_id,0)   || '/' ||
              v.vector.vector_id as Varchar2(8)) As ESV,
         v.vector.startCoord.ST_asText() As StartCoord,
         v.vector.EndCoord.ST_asText()    As EndCoord
    From geoms a,
         Table(ST_Vectorize(a.geom)) v
  Order By Decode(a.id,'Line',1,'Rect',2,'Poly',3,'MPly',4),
           v.vector.element_id,
           v.vector.subelement_id,
           v.vector.vector_id;


  -- Results
  --
  ID    ESV    STARTCOORD                    EndCOORD
  ----  -----  --------------------------    --------------------------
  Line 1/0/1 T_Vertex(1,1,NULL,NULL,1)    T_Vertex(20,20,NULL,NULL,2)
  Rect 1/1/1 T_Vertex(1,1,NULL,NULL,1)    T_Vertex(20,1,NULL,NULL,2)
  Rect 1/1/2 T_Vertex(20,1,NULL,NULL,2)   T_Vertex(20,20,NULL,NULL,3)
  Rect 1/1/3 T_Vertex(20,20,NULL,NULL,3)  T_Vertex(1,20,NULL,NULL,4)
  Rect 1/1/4 T_Vertex(1,20,NULL,NULL,4)   T_Vertex(1,1,NULL,NULL,5)
  Poly 1/1/1 T_Vertex(0,0,NULL,NULL,1)    T_Vertex(20,0,NULL,NULL,2)
  Poly 1/1/2 T_Vertex(20,0,NULL,NULL,2)   T_Vertex(20,20,NULL,NULL,3)
  Poly 1/1/3 T_Vertex(20,20,NULL,NULL,3)  T_Vertex(0,20,NULL,NULL,4)
  Poly 1/1/4 T_Vertex(0,20,NULL,NULL,4)   T_Vertex(0,0,NULL,NULL,5)
```

```
Poly 1/2/1  T_Vertex(10,10,NULL,NULL,1)  T_Vertex(11,10,NULL,NULL,2)
Poly 1/2/2  T_Vertex(11,10,NULL,NULL,2)  T_Vertex(11,11,NULL,NULL,3)
Poly 1/2/3  T_Vertex(11,11,NULL,NULL,3)  T_Vertex(10,11,NULL,NULL,4)
Poly 1/2/4  T_Vertex(10,11,NULL,NULL,4)  T_Vertex(10,10,NULL,NULL,5)
Poly 1/3/1  T_Vertex(5,5,NULL,NULL,1)    T_Vertex(7,5,NULL,NULL,2)
Poly 1/3/2  T_Vertex(7,5,NULL,NULL,2)    T_Vertex(7,7,NULL,NULL,3)
Poly 1/3/3  T_Vertex(7,7,NULL,NULL,3)    T_Vertex(5,7,NULL,NULL,4)
Poly 1/3/4  T_Vertex(5,7,NULL,NULL,4)    T_Vertex(5,5,NULL,NULL,5)
MPly 1/1/1  T_Vertex(0,0,NULL,NULL,1)    T_Vertex(20,0,NULL,NULL,2)
MPly 1/1/2  T_Vertex(20,0,NULL,NULL,2)   T_Vertex(20,20,NULL,NULL,3)
MPly 1/1/3  T_Vertex(20,20,NULL,NULL,3)  T_Vertex(0,20,NULL,NULL,4)
MPly 1/1/4  T_Vertex(0,20,NULL,NULL,4)   T_Vertex(0,0,NULL,NULL,5)
MPly 1/2/1  T_Vertex(10,10,NULL,NULL,1)  T_Vertex(11,10,NULL,NULL,2)
MPly 1/2/2  T_Vertex(11,10,NULL,NULL,2)  T_Vertex(11,11,NULL,NULL,3)
MPly 1/2/3  T_Vertex(11,11,NULL,NULL,3)  T_Vertex(10,11,NULL,NULL,4)
MPly 1/2/4  T_Vertex(10,11,NULL,NULL,4)  T_Vertex(10,10,NULL,NULL,5)
MPly 1/3/1  T_Vertex(5,5,NULL,NULL,1)    T_Vertex(7,5,NULL,NULL,2)
MPly 1/3/2  T_Vertex(7,5,NULL,NULL,2)    T_Vertex(7,7,NULL,NULL,3)
MPly 1/3/3  T_Vertex(7,7,NULL,NULL,3)    T_Vertex(5,7,NULL,NULL,4)
MPly 1/3/4  T_Vertex(5,7,NULL,NULL,4)    T_Vertex(5,5,NULL,NULL,5)
MPly 2/1/1  T_Vertex(40,40,NULL,NULL,1)  T_Vertex(50,40,NULL,NULL,2)
MPly 2/1/2  T_Vertex(50,40,NULL,NULL,2)  T_Vertex(50,50,NULL,NULL,3)
MPly 2/1/3  T_Vertex(50,50,NULL,NULL,3)  T_Vertex(40,50,NULL,NULL,4)
MPly 2/1/4  T_Vertex(40,50,NULL,NULL,4)  T_Vertex(40,40,NULL,NULL,5)
```

```
33 rows Selected
```

It is possible to visualize the vectors generated from a polygon using the `T_Vector` object's `ST_SdoGeometry` member function in the following SQL, which the SQL Developer extension, `GeoRaptor`, can draw:

```
Select v.vector.element_id    ||'/'||
       v.vector.subelement_id||'/'||
       v.vector.vector_id as vector,
       v.vector.ST_SdoGeometry(null) As geom
  From Table(ST_Vectorize(
             SDO_GEOMETRY(2003,null,null,
               SDO_ELEM_INFO_ARRAY(1,1003,1,11,2003,1,21,2003,1),
               SDO_ORDINATE_ARRAY(0,0,20,0,20,20,0,20,0,0,
```

```
                               10,10,10,11,11,11,11,10,10,10,
                                5,5,5,7,7,7,7,5,5,5)))) v;
```

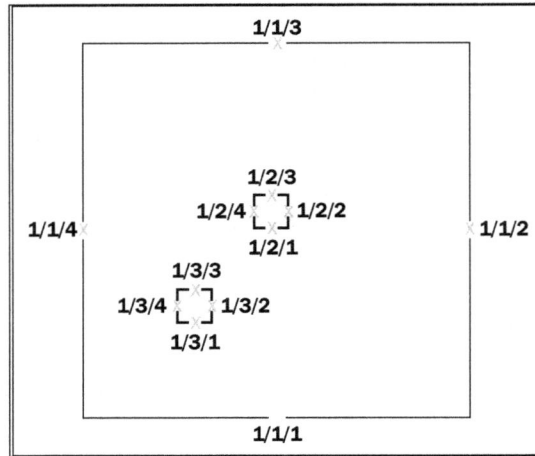

Using vectorization

So what is the value to the database spatial practitioner of a vectorizing function? One example of its use is in the implementation of a centroid algorithm that the author has written and made available on his website For many years. It has proven to be a much sought after function. A copy is included with this book.

In addition, a vectorizing function can be used to discover where gaps or overlaps exist between two geometries. This is done by breaking up the two polygons into their component vectors. Then the vectors are sorted and analyzed, looking for pairs. Where a pair exists, they represent the shared boundary between the two polygons; the pair can be safely discarded. Where a vector does not have a pair it is either a part of the polygon that does not share a boundary, or its pair is in a nearby location as both are a part of an overlap or gap. The intricacies of how this is done are not shown here, but are included with the code associated with this book. What is shown are the two SDO_GEOMETRY objects and the GeoRaptor image that shows the unpaired vectors:

```
Select SDO_GEOMETRY(2003,NULL,NULL,
            SDO_ELEM_INFO_ARRAY(1,1003,1),
            SDO_ORDINATE_ARRAY(0,0,10,0,9.98,5.23,9.87,9.51,
                                10.01,13.09,9.59,15.68,9.99,18.94,
                                10,20,0,20,0,0)
        ) As geom
    From dual
  Union All
  Select SDO_GEOMETRY(2003,NULL,NULL,
```

```
                SDO_ELEM_INFO_ARRAY(1,1003,1),
                SDO_ORDINATE_ARRAY(10,20,9.99,18.94,9.59,15.68,
                                   10.01,13.09,9.77,9.51,9.98,5.23,
                                   10,0,20,0,20,20,10,20)
          ) As geom
    From dual;
```

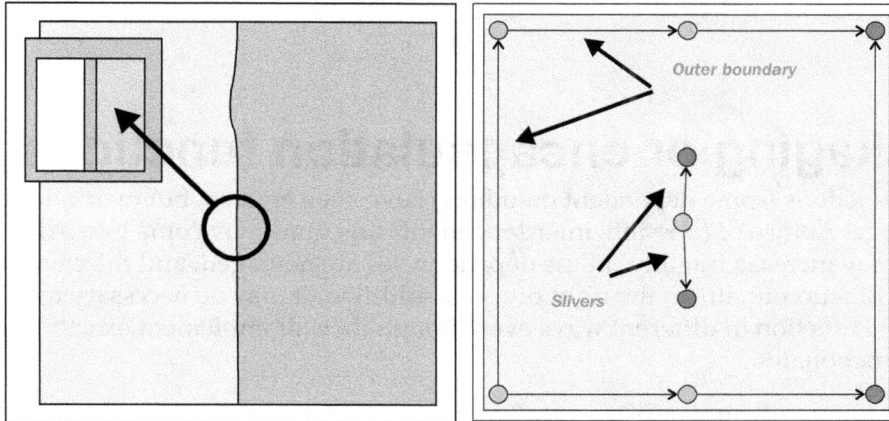

Finally, vectorization is useful for dimensioning the individual segments of a linestring or polygon, as shown in the following code:

```
Select Round(SDO_GEOM.Sdo_Length(t.vector.ST_SdoGeometry(),
                                 0.005),2) As dist,
       t.vector.ST_SdoGeometry() As segment
  From Table(ST_Vectorize(
             SDO_GEOMETRY(2002,NULL,NULL,
                 SDO_ELEM_INFO_ARRAY(1,2,1),
                 SDO_ORDINATE_ARRAY(1,1,10,1,10,5,5,10,1,1))))) t;
-- Results
--
DIST SEGMENT
---- -----------------------------------------------------------
   9 SDO_GEOMETRY(2002,NULL,NULL,SDO_ELEM_INFO_ARRAY(1,2,1),
        SDO_ORDINATE_ARRAY(1,1,10,1))
   4 SDO_GEOMETRY(2002,NULL,NULL,SDO_ELEM_INFO_ARRAY(1,2,1),
        SDO_ORDINATE_ARRAY(10,1,10,5))
7.07 SDO_GEOMETRY(2002,NULL,NULL,SDO_ELEM_INFO_ARRAY(1,2,1),
        SDO_ORDINATE_ARRAY(10,5,5,10))
9.85 SDO_GEOMETRY(2002,NULL,NULL,SDO_ELEM_INFO_ARRAY(1,2,1),
        SDO_ORDINATE_ARRAY(5,10,1,1))
```

Visually, it looks like this:

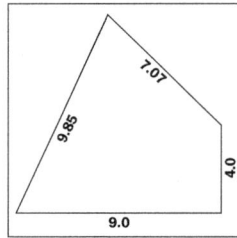

Packaging or encapsulation functions

Many functions (some dependent on others) have been created, but over time, an even larger library of (possibly interdependent) functions may come into existence, which may increase fragility unless dependencies are managed, and the compilation of each function occurs in the right order. In addition, it may be necessary to call a specific function in different ways even though they all implement exactly the same functionality.

Oracle offers us a number of ways to handle these situations.

Calling the same processing in different ways

The ST_Dimension function must be called with an SDO_GEOMETRY parameter. However, it may be more efficient to call the function with an SDO_GEOMETRY object's SDO_GTYPE numeric attribute; after all, the ST_Dimension function's processing is based entirely on this attribute. The current ST_Dimension function cannot be called with a non-SDO_GEOMETRY parameter value:

```
With testGeom As (
    Select SDO_GEOMETRY(3302,NULL,NULL,
                        Sdo_Elem_Info_Array(1,2,1),
                        SDO_ORDINATE_ARRAY(1,1,0, 2,2,1.414)
            ) As mGeom
      From dual
)
Select ST_Dimension(a.mGeom.sdo_gtype) As dim
  From testGeom a;

-- Results
--
Error report:
SQL Error: ORA-06553: PLS-306: wrong number or types of arguments in
call to 'ST_DIMENSION'
```

To fix this error, another version of the function needs to be created that takes an SDO_GEOMETRY object's SDO_GTYPE number value rather than SDO_GEOMETRY, through a mechanism called overloading:

Overloading is the ability for functions in a computer language like PL/SQL to have the same name but different parameters.

The function is written as:

```
Create Function ST_Dimension(p_gtype in Number)
  Return Integer Deterministic
AuthId Current_User
As
Begin
    If ( p_gtype Is null ) Then
       Return NULL;
    End If;
       If ( p_gtype in (1,5) ) Then Return 0;
    ElsIf ( p_gtype in (2,6) ) Then Return 1;
    ElsIf ( p_gtype in (3,7) ) Then Return 2;
    ElsIf ( p_gtype in (8,9) ) Then Return 3;
    ElsIf ( p_gtype = 4 )      Then Return -4;
    Else                            Return -1;
    End If;
End ST_Dimension;

--Results
--
Error report:
ORA-00955: name is already used by an existing object
```

While Oracle allows subprogram overloading (Standalone Subprograms", Oracle® Database PL/SQL Language Reference 11*g* Release 2 (11.2) Part Number E25519-05) it does not allow standalone subprograms to be overloaded (the "Or Replace" part of the function DDL was removed to make the statement fail).

A standalone subprogram is a subprogram created at the schema level (refer the `Procedures` or `Functions` nodes in a SQL Developer connection).

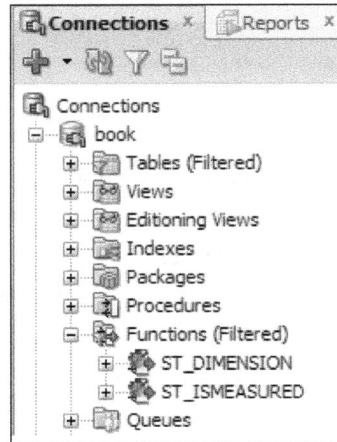

Implementing a PL/SQL package

How can the two functions (Oracle calls a these functions subprograms) be overloaded?

```
ST_Dimension(p_geom   In SDO_GEOMETRY)  Return Integer
ST_Dimension(p_gtype In number)        Return Integer
```

Oracle allows nested and package subprograms to be overloaded. For the purpose of this chapter, only the package subprogram approach will be examined.

What is a package and how can it be created (For a full description of packages, see "Chapter 10 PL/SQL Packages", Oracle® Database PL/SQL Language Reference 11*g* Release 2 (11.2) Part Number E25519-05)?

A package is a schema object that groups logically related PL/SQL types, variables, constants, subprograms, cursors, and exceptions.

A package always has a specification, which declares the public items that can be referenced from outside the package. [...] the package specification [constitutes an] application programming interface (API).

If the public items include [...] subprograms, then the package must also have a body.

Here are the functions so far presented collected together under a single package specification.

```
Create or Replace Package GEOM_PKG
AuthId Current_User
As
  Function ST_Dimension(p_gtype in Number)
    Return Integer Deterministic;
  Function ST_Dimension(p_geom in SDO_GEOMETRY)
    Return Integer Deterministic;
  Function ST_hasDimension(p_geom in SDO_GEOMETRY,
                           p_dim  in Integer  default 2)
    Return Integer Deterministic;
End GEOM_PKG;
```

This package specification is composed only of function declarations (no implementation details), and of these, the two ST_Dimension functions exhibit "overloading". Once the specification is defined, each function can then be implemented in the package body.

```
Create Or Replace Package Body GEOM_PKG
As

  Function ST_Dimension(p_geom in SDO_GEOMETRY)
    Return Integer
  As
     v_gtype pls_integer := 0;
  Begin
    If ( p_geom Is null or p_geom.get_gtype() Is null ) Then
       Return NULL;
    End If;
    v_gtype := MOD(p_geom.sdo_gtype,1000);
       If ( v_gtype in (1,5) ) Then Return 0;
    ElsIf ( v_gtype in (2,6) ) Then Return 1;
    ElsIf ( v_gtype in (3,7) ) Then Return 2;
    ElsIf ( v_gtype in (8,9) ) Then Return 3;
    ElsIf ( v_gtype = 4 )      Then Return -4;
    Else                            Return -1;
    End If;
  End ST_Dimension;

  Function ST_Dimension(p_gtype in Number)
    Return Integer
  As
  Begin
```

```
      If ( p_gtype Is null ) Then
         Return NULL;
      End If;
          If ( p_gtype in (1,5) ) Then Return 0;
      ElsIf ( p_gtype in (2,6) ) Then Return 1;
      ElsIf ( p_gtype in (3,7) ) Then Return 2;
      ElsIf ( p_gtype in (8,9) ) Then Return 3;
      ElsIf ( p_gtype = 4 )        Then Return -4;
      Else                              Return -1;
      End If;
End ST_Dimension;

Function ST_hasDimension(p_geom in SDO_GEOMETRY,
                          p_dim  in Integer  default 2)
   Return Integer
As
   v_etype pls_integer := 0;
   v_gtype pls_integer := -1;
Begin
   If ( p_geom Is null or p_geom.get_gtype() Is null ) Then
      Return 0;
   End If;
   v_gtype := MOD(p_geom.sdo_gtype,100);   -- DLTT
   If (  ( v_gtype in (1,5) and p_dim = 0 )
      or ( v_gtype in (2,6) and p_dim = 1 )
      or ( v_gtype in (3,7) and p_dim = 2 )
      or ( v_gtype in (8,9) and p_dim = 3 ) ) Then
         Return 1;
   End If;
   If ( v_gtype = 4 ) Then
     <<element_Extraction>>
     For v_i IN 0..((p_geom.sdo_elem_info.COUNT / 3) - 1) Loop
        v_etype := p_geom.sdo_elem_info(v_i * 3 + 2);
        -- Check etype value, until find required p_dim
        If ( (v_etype = 1                       and p_dim = 0)
          or (v_etype in (2,4)                  and p_dim = 1)
          or (v_etype in (1003,2003,1005,2005) and p_dim = 2)
          or (v_etype in (1007,1006,2006)      and p_dim = 3)
            ) Then
             Return 1;
          End If;
     End Loop element_Extraction;
     Return 0;
   Else
```

```
        Return 0;
    End If;
  End ST_hasDimension;

End GEOM_PKG;
```

Testing the functions in this package shows how the overloading works:

```
Select GEOM_PKG.ST_Dimension(a.mGeom)           As GeomDimension,
       GEOM_PKG.ST_Dimension(a.mGeom.sdo_gtype) As TypeDimension,
       GEOM_PKG.ST_Dimension(a.mGeom)           As hasDimimension
  From (Select SDO_GEOMETRY(3302,NULL,NULL,
               SDO_ELEM_INFO_ARRAY(1,2,1),
               SDO_ORDINATE_ARRAY(1,1,0, 2,2,1.414)
             ) As mGeom
         From dual ) a;

-- Results
--
GEOMDIMENSION TYPEDIMENSION HASDIMIMENSION
------------- ------------- --------------
           -1            -1             -1
```

Oracle describes the benefits of package-based development as:

- Modularity
- Easier application design
- Information hiding
- Added functionality
- Better performance

Implementing user object types

Since these functions inspect properties of an SDO_GEOMETRY object, why not add them as methods to the base object, rather than using a package?

The simple answer is that the SDO_GEOMETRY object type is declared as FINAL, that is, it can't be changed. If SDO_GEOMETRY was declared as NOT FINAL, it would not be a good idea to attempt to alter the type for the following reasons:

- It is not your object to modify (it is Oracle's)
- Objects depend on it

- Applying patch sets (or even bug fixing) might overwrite the user modifications creating major support problems

- Getting support via My Oracle Support would be impossible

- Migration (import/export/transportable tablespaces) and upgrades would be problematic due to type version number differences

- Inconsistencies when two different users alter the type separately yet want to share their work

- The wrapping of the associated SDO_GEOMETRY object type body makes implementation impossible

If it was possible to do so, the ST_Dimension function could be added as a new method as follows:

```
Alter Type SDO_GEOMETRY
Add Member Function ST_Dimension
    Return Integer Deterministic
Cascade;
Alter Type Body SDO_GEOMETRY
Add Member Function ST_Dimension
    Return Integer
Is
Begin
   If ( SELF.geom Is null or SELF.get_gtype() Is null ) Then
      Return NULL;
   End If;
   Return MdSys.ST_Geometry(SELF.geom).ST_Dimension();
End ST_Dimension;
```

While this neatly packages the additional methods with the host object, for the reasons outlined earlier, modifying the core SDO_GEOMETRY object should be avoided in every situation.

Note that, an object type based on SDO_GEOMETRY does not require the SDO_GEOMETRY it is acting upon be passed in as a parameter to each method, as each method operates on its encapsulated SDO_GEOMETRY attribute. Thus, there is no need to implement the overloaded ST_Dimension (p_gtype in number) as the SDO_GTYPE is an attribute of the object.

Creating a custom object type

It is possible, however, to create a custom object type and use it to collect Oracle and any custom functions together in one object. An Oracle object type is very much like a PL/SQL package as it contains a specification and a body, whereas a function in a package must be supplied with all the data it needs to process via parameters; a member function in an object works on the data encapsulated within the object.

The best way to learn is to jump right in and create a new object type called T_GEOMETRY based on a single SDO_GEOMETRY with two specific methods: ST_Dimension and ST_hasDimension.

```
Create Or Replace Type T_Geometry As Object (
   geom   SDO_GEOMETRY,
   Member Function ST_Dimension
          Return Integer Deterministic,
   Member Function ST_hasDimension(p_dim in Integer  default 2)
          Return Integer Deterministic
);


Create Or Replace Type Body T_Geometry
As
   Member Function ST_Dimension
          Return Integer
   Is
      v_gtype pls_integer;
   Begin
      If ( SELF.geom Is null ) Then
         Return NULL;
      End If;
      v_gtype := SELF.geom.get_gtype();
    /* Returns 0 For POINT,   1 For LINESTRING,
               2 For POLYGON, 3 For SOLID
     * and the largest dimension of the components
       of a GEOMETRYCOLLECTION.
     **/
      Return Case WHEN v_gtype in (1,5) Then 0
                  WHEN v_gtype in (2,6) Then 1
                  WHEN v_gtype in (3,7) Then 2
                  WHEN v_gtype =   4    Then -1 /* Layer */
                  WHEN v_gtype in (8,9) Then 3
                  Else -1
             End;
   End ST_Dimension;
```

```
      Member Function ST_hasDimension(p_dim in Integer default 2)
            Return Integer
  As
      v_etype pls_integer := 0;
      v_gtype pls_integer := -1;
  Begin
      If ( SELF.geom Is null or
          SELF.geom.get_gtype() Is null ) Then
        Return 0;
      End If;
      v_gtype := MOD(SELF.geom.sdo_gtype,100); -- DLTT
      If (  ( v_gtype in (1,5) and p_dim = 0 )
         or ( v_gtype in (2,6) and p_dim = 1 )
         or ( v_gtype in (3,7) and p_dim = 2 )
         or ( v_gtype in (8,9) and p_dim = 3 ) ) Then
            Return 1;
      End If;
      If ( v_gtype = 4 ) Then
        <<element_Extraction>>
        For v_i In 0..((SELF.geom.sdo_elem_info.COUNT/3)-1) Loop
          v_etype := SELF.geom.sdo_elem_info(v_i * 3 + 2);
          -- Check etype value, until find required p_dim
          If ( (v_etype = 1                       and p_dim = 0)
            or (v_etype in (2,4)                  and p_dim = 1)
            or (v_etype in (1003,2003,1005,2005) and p_dim = 2)
            or (v_etype in (1007,1006,2006)       and p_dim = 3)
             ) Then
             Return 1;
          End If;
        End Loop element_Extraction;
        Return 0;
      Else
        Return 0;
      End If;
  End ST_hasDimension;
End;


Grant Execute On T_Geometry To Public With Grant Option;
```

In the example that follows, the methods of the host SDO_GEOMETRY object can still be accessed.

```
With testGeom As (
   Select T_Geometry(SDO_GEOMETRY(3302,NULL,NULL,
                      SDO_ELEM_INFO_ARRAY(1,2,1),
                      SDO_ORDINATE_ARRAY(1,1,0, 2,2,1.414))
           ) As mGeom
     From dual
)
Select (a.mgeom.geom).get_gtype()  As gtype,
        a.mGeom.ST_Dimension()      As T_dim,
        a.mGeom.ST_hasDimension(1) As T_hasDim
   From testGeom a;

-- Results
--
GTYPE T_DIM T_HASDIM
----- ----- --------
    2     1        1
```

There is a certain additional cost involved in using a type this way, which is the need to access the underlying SDO_GEOMETRY object before executing the standard method against it. Against this is the benefits that a consistent approach provides, as shown in the following section.

Including default SDO_GEOMETRY functions

We can use our new object type to package, in a more consistent fashion, all existing functions or methods that Oracle provides thorough other types and packages. The two that will be exposed are the ST_GEOMETRY ST_GeometryType method and the SDO_UTIL.GetNumVertices function. Also presented is a ST_isMeasured function that is a wrapper over the SDO_GEOMETRY.Get_Lrs_Dim method: this wrapper shows how to present a different "view" of the data (simply "is this geometry measured?" not, "what is the current LRS measure value?").

In the example that follows, access to the methods of the host SDO_GEOMETRY object is a little cleaner, as it does not require access to the host geom object before executing the method; the type takes care of this. The extended type specification now looks like this:

```
Create Or Replace Type T_Geometry As Object (
   geom    SDO_GEOMETRY,
```

```
Member Function ST_Get_Gtype       Return Number    Deterministic,
Member Function ST_Get_Dims        Return Number    Deterministic,
Member Function ST_Get_Lrs_Dim     Return Number    Deterministic,
Member Function ST_Get_Wkb         Return Blob      Deterministic,
Member Function ST_Get_Wkt         Return Clob      Deterministic,
Member Function ST_CoordDim        Return Smallint  Deterministic,
Member Function ST_isMeasured      Return Integer   Deterministic,
Member Function ST_GeometryType    Return Varchar2  Deterministic,
Member Function ST_NumVertices     Return Integer   Deterministic,
Member Function ST_hasDimension (p_dim in Integer default 2)
                                   Return Integer   Deterministic,
Member Function ST_Dimension       Return Integer   Deterministic
);
/* Only the implementation of two new methods Is shown */
Create Or Replace Type Body T_Geometry
As
  Member Function ST_isMeasured
          Return Integer
  Is Begin
    Return Case When SELF.Geom.Get_Lrs_Dim() <> 0
               Then 1 Else 0 End;
  End ST_isMeasured;
  Member Function ST_GeometryType

          Return varchar2

  Is Begin

    Return mdsys.ST_Geometry(SELF.geom).ST_GeometryType();

  End;

  Member Function ST_NumVertices

          Return Integer

  Is Begin

    Return mdsys.SDO_UTIL.GetNumVertices(SELF.geom);

  End ST_NumVertices;

End;
```

The following example shows how neat, tidy, and consistent packaging can be:

```
SET NULL '(null)'
With testGeoms As (
  Select T_Geometry(
           SDO_GEOMETRY(2001,NULL,
```

```
                              SDO_POINT_TYPE(1,2,NULL),NULL,NULL)) As geom
                 SDO_POINT_TYPE(1,2,NULL),NULL,NULL)) As geom
       From dual Union All
     Select T_Geometry(
              SDO_GEOMETRY(2005,NULL,NULL,
                 SDO_ELEM_INFO_ARRAY(1,1,2),
                 SDO_ORDINATE_ARRAY(1,2,3,4))) As geom
       From dual Union All
     Select T_Geometry(
              SDO_GEOMETRY(2002,NULL,NULL,
                 SDO_ELEM_INFO_ARRAY(1,2,1),
                 SDO_ORDINATE_ARRAY(1,2,3,4))) As geom
       From dual Union All
     Select T_Geometry(
              SDO_GEOMETRY(2003,NULL,NULL,
                 SDO_ELEM_INFO_ARRAY(1,1003,3),
                 SDO_ORDINATE_ARRAY(1,1,2,2))) As geom
       From dual Union All
     Select T_Geometry(
              SDO_GEOMETRY(3302,NULL,NULL,
                 SDO_ELEM_INFO_ARRAY(1,2,1),
                 SDO_ORDINATE_ARRAY(1,1,0,2,2,1.4))) As geom
       From dual Union All
     Select T_Geometry(
              SDO_GEOMETRY(3302,NULL,NULL,SDO_ELEM_INFO_ARRAY(1,2,1),
                    SDO_ORDINATE_ARRAY(1,1,0,2,2,1.4))) As geom
       From dual
)
Select a.geom.ST_GeometryType()  As gType,
       a.geom.ST_NumVertices()   As nVerts,
       a.geom.ST_CoordDim()      As cDim,
       a.geom.ST_Dims()          As sDim,
       a.geom.ST_hasDimension(1) As T_hasD,
       a.geom.ST_Dimension()     As T_Dim
  From testGeoms a;

-- Results
--
GTYPE          NVERTS   CDIM   SDIM   T_HASM  T_DIM
------------- ------ ------ ------ ------- ------
ST_POINT       (null)    2      2       0       0
ST_MULTIPOINT    2       2      2       0       0
ST_LINESTRING    2       2      2       1       1
ST_POLYGON       2       2      2       0       2
ST_LINESTRING    2       3      3       1       1

 6 rows Selected
```

Sorting geometries

Before finishing this section on object types, one more powerful method will be added to the T_GEOMETRY type to demonstrate how to sort the result set of a SQL statement using geometry objects in the SQL's Order By clause.

This will demonstrate the basics of a powerful capability that is used elsewhere in this book for improving query performance by sorting data on disk in a way that implements *Waldo Tobler's* first law of geography, which is *"Everything is related to everything else, but near things are more related than distant things."*

An Oracle object (not a package) can include a special method called an Order Member method. This method allows the Oracle query engine to compare two objects and decide if one is less than the other. Now, for numbers, the concept of "less than" is easy. For geometries, such a concept is somewhat more difficult to conceive of.

One simplistic method is to extract the centroid (SDO_GEOM.Sdo_Centroid) of each geometry object and then compare the X and Y ordinates (and the others if you so desire) as per their normal ordering. If an ordinate is less than another, a negative number is returned to the sort engine, otherwise a positive value is returned; if equal, a 0 is returned. Comparing both X and Y ordinates instead of just one will provide a more reasonable spatial sorting that will reflect *Waldo Tobler's "* law of geography".

If our SDO_GEOMETRY data only contained points stored in the SDO_POINT attribute, then the sorting could be achieved as follows:

```
-- Where x or y < compared x y, Return negative Number
-- Where x or y > compared x y, Return positive Number
If ( V_Geom.Sdo_Point.X < V_Compare_Geom.Sdo_Point.X ) Then
    Return -1;
ElsIf ( V_Geom.Sdo_Point.X > V_Compare_Geom.Sdo_Point.X ) Then
    Return  1;
ElsIf ( V_Geom.Sdo_Point.Y < V_Compare_Geom.Sdo_Point.Y ) Then
    Return -1;
ElsIf ( V_Geom.Sdo_Point.Y > V_Compare_Geom.Sdo_Point.Y ) Then
    Return  1;
Else
    Return  0;
End If;
```

Implementing an ordering function

SDO_GEOMETRY objects can be of any type and can be described in many ways. For example, a polygon object may not have an SDO_POINT attribute, but in other situations (as is common in MapInfo installations) an SDO_POINT may exist and may be the centroid of the polygon. Our order function has to cope with all situations. The Order Member Function in the T_GEOMETRY of our object looks like this:

```
Order Member Function orderBy(p_compare_geom in T_Geometry)
Return Number
Is
    v_point            SDO_POINT_TYPE;
    v_compare_point SDO_POINT_TYPE;

    Procedure getSortPoint
    As
    Begin
        If ( SELF.geom.sdo_point Is not null ) Then
            v_point := SDO_POINT_TYPE(SELF.geom.sdo_point.x,
                                      SELF.geom.sdo_point.y,
                                      null);
        Else
            v_point := SDO_POINT_TYPE(SELF.geom.sdo_ordinates(1),
                                      SELF.geom.sdo_ordinates(2),
                                      null);
        End If;
        If ( p_compare_geom.geom.sdo_point Is not null ) Then
            v_compare_point :=
                SDO_POINT_TYPE(p_compare_geom.geom.sdo_point.x,
                               p_compare_geom.geom.sdo_point.y,
                               null);
        Else
            v_compare_point :=
                SDO_POINT_TYPE(p_compare_geom.geom.sdo_ordinates(1),
                               p_compare_geom.geom.sdo_ordinates(2),
                               null);
        End If;
    End getSortPoint;

Begin
        If (SELF.geom Is null)        Then Return -1;
    ElsIf (p_compare_geom Is null) Then Return 1;
    End If;
```

```
        getSortPoint;
        -- Where x or y < compared x y, Return negative Number
        -- Where x or y > compared x y, Return positive Number
           If ( V_Point.X < V_Compare_Point.X ) Then Return -1;
        ElsIf ( V_Point.X > V_Compare_Point.X ) Then Return  1;
        ElsIf ( V_Point.Y < V_Compare_Point.Y ) Then Return -1;
        ElsIf ( V_Point.Y > V_Compare_Point.Y ) Then Return  1;
        Else                                          Return 0;
        End If;
      End;
```

The orderBy function works as follows. Firstly it checks that the geometries to be sorted are not null. Then, an internal function, getSortPoint, is called that analyzes the two provided objects and constructs two simple SDO_POINT_TYPE structures that will hold the actual two points to be compared in the sorting. The ordinates of the two points are then compared to determine the order. The sorting is tested as follows. First a query against 12 points is executed, two of which are guaranteed to be equal without any sorting:

```
With points As (
Select level as point_id,
        Case When level in (11,12)
            Then SDO_GEOMETRY(2001,NULL,
                    SDO_POINT_TYPE(1000,2000,NULL),NULL,NULL)
            Else SDO_GEOMETRY(2001,NULL ,
                    SDO_POINT_TYPE(
                        round(dbms_random.value(10000,19999),3),
                        round(dbms_random.value(20000,29999),3),
                        NULL),NULL,NULL)
        End as pGeom
  From dual
Connect By Level <= 12
)
Select a.point_id as id,
        a.pGeom.sdo_point as Point
  From points a;

-- Results
--
```

```
        ID POINT
---------- ------------------------------------------------
         1 MDSYS.SDO_POINT_TYPE(10523.725,29750.711,NULL)
         2 MDSYS.SDO_POINT_TYPE(16241.119,28811.61,NULL)
         3 MDSYS.SDO_POINT_TYPE(12341.034,28588.506,NULL)
         4 MDSYS.SDO_POINT_TYPE(19936.363,23390.776,NULL)
         5 MDSYS.SDO_POINT_TYPE(19483.802,29401.266,NULL)
         6 MDSYS.SDO_POINT_TYPE(19848.487,29442.227,NULL)
         7 MDSYS.SDO_POINT_TYPE(10361.191,27151.295,NULL)
         8 MDSYS.SDO_POINT_TYPE(14030.977,23144.388,NULL)
         9 MDSYS.SDO_POINT_TYPE(10396.729,21152.735,NULL)
        10 MDSYS.SDO_POINT_TYPE(16257.586,25821.374,NULL)
        11 MDSYS.SDO_POINT_TYPE(1000,2000,NULL)
        12 MDSYS.SDO_POINT_TYPE(1000,2000,NULL)

 12 rows Selected
```

Then the same query (with the clause removed for brevity) will now be executed, but this time with an `Order By` clause that includes our geometries:

```
With points As (
[ data generation query removed ]
)
Select a.point_id as id,
       a.pGeom.sdo_point as Point
  From Points a
 Order By t_geometry(a.pGeom,0.005);

-- Results
--
        ID POINT
---------- ------------------------------------------------
        11 MDSYS.SDO_POINT_TYPE(1000,2000,NULL)
        12 MDSYS.SDO_POINT_TYPE(1000,2000,NULL)
         7 MDSYS.SDO_POINT_TYPE(10624.136,29752.641,NULL)
         2 MDSYS.SDO_POINT_TYPE(12382.42,20970.931,NULL)
         1 MDSYS.SDO_POINT_TYPE(12622.696,26435.449,NULL)
         9 MDSYS.SDO_POINT_TYPE(14508.88,27084.664,NULL)
        10 MDSYS.SDO_POINT_TYPE(14724.049,21432.018,NULL)
         4 MDSYS.SDO_POINT_TYPE(15819.915,20084.528,NULL)
         6 MDSYS.SDO_POINT_TYPE(16764.28,24377.218,NULL)
         5 MDSYS.SDO_POINT_TYPE(17196.955,24229.005,NULL)
         3 MDSYS.SDO_POINT_TYPE(18695.319,25840.929,NULL)
         8 MDSYS.SDO_POINT_TYPE(19388.327,24076.913,NULL)

 12 rows Selected
```

Another chapter will look at a more sophisticated method for spatial sorting and investigate the benefits of storing spatially sorted data.

Packaging summary

Which method of packaging our functions is the best?

- **Standalone functions**: These are great for developing and testing algorithms or creating something one off or standalone that can be given to others. But they could become unwieldy, if they grow too many interdependencies.

- **Packages**: These allow for shared package variables, which are related to some logical function, but are more loosely coupled in relation to each other or to the data stored in the package.

- **Types**: These allow for the consistent and homogeneous packaging of all the methods of an object with its persistent data. Methods in an object type tend to be tightly coupled with the object data as they provide the property inspectors and manipulators for them.

Programming with object types is clean and logical. While within Oracle it is possible to store any object type in a database table (persisting it across database sessions), storage of the T_GEOMETRY type is not recommended for the simple reason that it is presented only as a packaging mechanism for enduser programming purposes. In addition, T_GEOMETRY should not be stored in tables as Oracle already provides SDO_GEOMETRY for this purpose which is indexed, supported across multiple versions, and throughout its support channels (This book is about applying and extending Oracle SDO_GEOMETRY not replacing it!).

In the end there is no right answer to which method to use for packaging functionality. Sometimes packages are better, other times standalone functions, and sometimes object types. Whatever decision is taken, the full power of the Oracle PL/SQL programming framework and object type system is always there to use as required.

Using, extending, and applying Oracle Spatial is possible due to the openness of its storage, and the ability to use the whole of its database architecture to achieve what you want to achieve. However, to do this, one has to be prepared to learn more about what the Oracle database has to offer.

Chapter 1, Defining a Data Model for Spatial Data Storage, demonstrated the power of using Oracle's Advance Queue API's DBMS_AQADM and DBMS_AQ packages to implement spatial referential integrity checking.

This section exposed some of the power of the PL/SQL programming language through use of its functions, packages, and types. Familiarity with these non-spatial tools is vital to being able to deliver, for an organization, improved return on its investment in the Oracle database. It is therefore recommended that the Oracle documentation is always available to programmers during development; not just the spatial books but specifically the PL/SQL programming guide. Other parts of this book will show how other spatial and non-spatial database functionality can be used when developing solutions. In all cases the relevant documentation should be close at hand.

Summary

This chapter has been an introduction to the basics of programming new functionality based on the SDO_GEOMETRY data type's attributes and methods. New functions were created that showed you how to interpret and manipulate the SDO_ELEM_INFO and SDO_ORDINATES attributes both alone and together using SQL and PL/SQL.

New functionality is often built on top of existing or new functions. Such functionality needs to be organized so that it is understandable, maintainable, and accessible. This chapter therefore introduced different methods for packaging such functionality with an emphasis on the use of an object type called T_GEOMETRY. The chapter concluded with some practical functions for sorting and vectorizing geometries.

In *Chapter 7, Editing, TransForming, and Constructing Geometries*, everything learned in this chapter will be used when building more advanced functions For editing, converting, and transforming the SDO_GEOMETRY data.

7
Editing, Transforming, and Constructing Geometries

Desktop **GIS**, **CAD**, and **Extract Transform and Load** (ETL) software provide a rich set of tools that the experienced operator can use to construct, edit, or process geometric objects. But few realize that creating and applying such functionality within the database is also possible and can be more effective, efficient, and less complicated.

While the Oracle database SDO_GEOMETRY data type provides an excellent storage, search, and processing engine for spatial data, what users often overlook is its ability to provide geometry modification and processing capabilities. These can be used in database objects such as views, materialized views, or triggers for the implementation of specific business functionality; with that functionality being available to any software product that connects to the database.

The functions in *Chapter 6, Implementing New Functions*, were about highlighting the structure of an SDO_GEOMETRY object and its processing. The goal of this chapter is to build on, and extend, the concepts learned in the previous chapter, by creating functions that have a more direct application in the real world. All the examples chosen to illustrate usage and aid your understanding come from solving real world problems.

In this chapter, the following aspects of editing/modifying or processing geometry data will be covered:

- Insert/update/delete coordinates
- Extending/Shrinking a linestring
- Affine: translate, rotate, and scale geometries
- Moving a line parallel to itself

- One-sided and square buffering of linestrings
- Splitting geometries at known points
- Converting between dimensions (2D<->3D<->4D, and so on)
- Tiling vector objects and smoothing raster to vector data

Finally, this chapter will not include the implementation details of the functions it presents. All source code is available with the book.

Inserting, modifying, and deleting coordinates

The description of a spatial object is controlled by the **Open Geospatial Consortium** http://www.opengeospatial.org/ **(OGC)** and **International Organization for Standardization (ISO)** standard bodies. In particular, the **ISO/IEC 13249-3 Part 3: Spatial** (ISO 13249-3, Information technology - Database languages - SQL Multimedia and Application Packages - Part 3: Spatial.) and the related **OGC Simple Features Access – Part 2: SQL Option** (OpenGIS® Implementation Specification for Geographic information - Simple feature access - Part 2: SQL option, Version 1.1 and 1.2. Previously known as Simple Features–SQL) (known as SFA). These standards define how a polygon and its exterior and interior rings are defined and how it is to be encoded.

Similarly, these standards describe the functionality that must be made available for conforming implementations. But there are a number of functions not described or required by the current standards that are most useful when building server-side or database-centric processing solutions.

Three such non-standardized functions that will be implemented in this section are those that enable the inserting, updating, and deleting of specific points of a geometry object. These functions do not exist with Oracle's SDO_GEOMETRY object type or related functions, but do exist in competitor products as shown in the following table:

Function to be implemented	PostGIS 2.0	IBM Informix Spatial DataBlade 11.7
ST_InsertVertex	ST_AddPoint	SE_VertexAppend
ST_DeleteVertex	ST_RemovePoint	SE_VertexDelete
ST_UpdateVertex	ST_SetPoint	SE_VertexUpdate

Each of these functions will be introduced and their use outlined, via specific examples.

Identifying an arc point – ST_inCircularArc

When editing an SDO_GEOMETRY, one of the constraints which must be honored is that a point may not be added to or deleted from a circular arc segment.

The member functions in *Chapter 6, Implementing New Functions*, presented the functions `T_Geometry ST_NumSubElems` and `ST_hasCircularArcs` that identify the circular arcs within a geometry. Here is an additional member function that detects if a supplied coordinate number is a part of a circular arc/circle definition:

```
Create Or Replace Type Body T_Geometry
[...]
  -- Returns position in circular arc
  -- 0 means not in a circular arc
  -- 1 means Is first point in circular arc
  -- 2 means Is second point in circular arc
  -- 3 means Is third point in circular arc
  --
  Member Function ST_inCircularArc(p_point_number In Integer)
        Return Integer
  Is
    v_arc_start_point pls_integer;
    v_etype           pls_integer;
    v_offset          pls_integer;
    v_interpretation  pls_integer;
    v_elements        pls_integer;
    v_elem_info       SDO_ELEM_INFO_ARRAY;
  Begin
    If ( SELF.ST_Dimension() = 0 ) Then
      Return 0;
    End If;
    v_elements := ( ( SELF.geom.SDO_ELEM_INFO.COUNT / 3 ) - 1 );
    v_elem_info := SELF.geom.SDO_ELEM_INFO;
    <<element_extraction>>
    For v_i In 0 .. v_elements Loop
      v_offset         := (v_i * 3) + 1;
      v_etype          := v_offset + 1;
      v_interpretation := v_etype  + 1;
      If (    ( v_elem_info(v_etype) = 2 AND
                v_elem_info(v_interpretation) = 2 )
          Or ( v_elem_info(v_etype) In (1003,2003) AND
```

```
                    v_elem_info(v_interpretation) = 4 ) ) then
          v_arc_start_point := ((v_elem_info(v_offset) - 1) /
                                SELF.ST_dims()) + 1;
          If ( p_point_number between v_arc_start_point
                              and v_arc_start_point + 2 ) Then
              Return (p_point_number - v_arc_start_point) + 1;
          End If;
        End If;
      End Loop element_extraction;
      Return 0;
    End ST_inCircularArc;
    [...]
End;
```

The following SQL tests each vertex to see which one is in a circular arc or not:

```
Select level as point_num,
       a.geom.ST_inCircularArc(level) as inCircularArc
  From (Select T_Geometry(SDO_GEOMETRY(2002,null,null,
               SDO_ELEM_INFO_ARRAY(1,4,2, 1,2,1, 7,2,2, 11,2,1),
               SDO_ORDINATE_ARRAY (10,78, 10,75, 20,75,
                                   20,78, 15,80,
                                   10,78, 5,78  ))) as geom
        From dual ) a
      Connect By Level <= a.geom.ST_ST_NumVertices();

-- Results
--
POINT_NUM INCIRCULARARC
--------- -------------
        1             0
        2             0
        3             0
        4             1
        5             2
        6             3
        7             0

  6 rows selected
```

Implementing ST_InsertVertex

Adding a vertex to an existing geometry object requires consideration of the type of change being requested and the current description of the geometry object itself. For example, the following will be respected in our implementation:

1. If a request to add a vertex to a single point is made, then the point will cease to be a single point, rather it will become a multipoint.

2. Adding a vertex will not result in an invalid geometry object.

3. An added vertex may not cause a duplicate vertex error (ORA-13356).

4. A single new vertex may not be added into an existing three vertex circular arc (a minimum of two must be added)

5. A vertex may not be added where a change to the definition of the geometry is required; for example, a vertex added to the end of a circular arc element, if that vertex is expected to be part of a new vertex-connected linestring element.

The first situation is independent of an ordinate precision (or tolerance), whereas the third situation is very much dependent on tolerance to ensure two vertexes are not treated as being the same.

> All functions in this chapter will be added as new member methods to the `T_Geometry` object presented in *Chapter 6, Implementing New Functions*. The T_GEOMETRY object with all its member functions will be shipped with the source code for this book.

Now that the `ST_inCircularArc` member function exists, the `ST_InsertVertex` member function (code removed for brevity) can be created:

```
Create Or Replace Type Body T_Geometry
[...]
   -- Supplied p_vertex must have Z and W coded correctly
   -- p_vertex.id Is position for inserting new vertex.
   -- All existing vertexes shuffled down ie
   -- Insert Is "add before" except end.
   --
   Member Function
   ST_InsertVertex(p_vertex in mdsys.vertex_type)
   Return T_Geometry Deterministic,
```

The function processes the existing geometry data as follows:

1. Handles easiest cases of adding a vertex to existing single point objects.

2. Checks if the inserted point would fall within a circular arc.

3. For non-single point geometries, the vertex is then inserted into the sdo_ordinate array. Note that adding a vertex to the end of the sdo_ordinate array is handled separately as it is faster. Also, the existing vertexes must be "shuffled down" to create a gap into which a vertex can be written.

4. Modifies the offset at the relevant SDO_ELEM_INFO (everywhere) triplet.

5. Tests the changes to ensure that they are not invalid.

Modifying SDO_ELEM_INFO could have been done by directly processing SDO_ELEM_INFO_ARRAY via a for loop. However, in this situation (while slower), the processing is done via a single SQL statement. This is done to demonstrate how to implement an algorithm declaratively and not programmatically. The SQL was developed outside of the PL/SQL source code in a normal SQL session. Once the SQL is working, it is inserted back into the code. Often, SQL-based implementation is a great way to build a solution that can be optimized later on via pure PL/SQL programming. Here performance is less of an issue; the readability and ease of testing of pure SQL is often preferable.

Testing ST_InsertVertex

The following tests demonstrate how some specific cases are handled by the ST_InsertVertex function:

- Change a point (variously encoded) to a multipoint:

```
-- Insert vertex into single vertex stored in sdo_point
Select T_Geometry(SDO_GEOMETRY(2001,null,
                 SDO_POINT_TYPE(1000,2000,null),
                 null,null))
          .ST_InsertVertex(T_Vertex(0,0,1).ST_VertexType()
       ).geom as geom
  From Dual Union All
-- Insert vertex into single vertex stored in sdo_ordinates
Select T_Geometry(SDO_GEOMETRY(2001,null,null,
                 SDO_ELEM_INFO_ARRAY(1,1,1),
                 SDO_ORDINATE_ARRAY(1000,2000)))
          .ST_InsertVertex(T_Vertex(0,0,-1).ST_VertexType()
          ).geom as geom
  From Dual Union All
```

```
-- Insert vertex into multi-point with 1 vertex in sdo_ordinates
Select T_Geometry(SDO_GEOMETRY(2005,null,null,
                             SDO_ELEM_INFO_ARRAY(1,1,1),
                             SDO_ORDINATE_ARRAY(1000,2000)))
          .ST_InsertVertex(T_Vertex(0,0,-1).ST_VertexType()
          ).geom as geom
  From dual;

-- Results
--
GEOM
----------------------------------------------------------
SDO_GEOMETRY(2005,NULL,NULL,SDO_ELEM_INFO_ARRAY(1,1,2),
          SDO_ORDINATE_ARRAY(0.0,0.0, 1000.0,2000.0))
SDO_GEOMETRY(2005,NULL,NULL,SDO_ELEM_INFO_ARRAY(1,1,2),
          SDO_ORDINATE_ARRAY(1000.0,2000.0, 0.0,0.0))
SDO_GEOMETRY(2005,NULL,NULL,SDO_ELEM_INFO_ARRAY(1,1,2),
          SDO_ORDINATE_ARRAY(1000.0,2000.0, 0.0,0.0))
```

- Insert vertexes into a multipoint (notice how the additions can be stacked):

```
Select T_Geometry(
          SDO_GEOMETRY(2005,NULL,NULL,
             SDO_ELEM_INFO_ARRAY(1,1,3),
             SDO_ORDINATE_ARRAY(6,10, 10,1, 10,14)))
       .ST_InsertVertex(
          T_Vertex(14,10,3).ST_VertexType()
       ).ST_InsertVertex(
          T_Vertex(6,10,-1).ST_VertexType()
       ).geom as Geom
  From dual;

-- Results
--
GEOM
----------------------------------------------------------
SDO_GEOMETRY(2005,NULL,NULL,SDO_ELEM_INFO_ARRAY(1,1,5),
          SDO_ORDINATE_ARRAY(6,10,10,1,14,10,10,14,6,10))
```

- Insert a new first vertex, a new third vertex, and then add another two in the end:

```
Select T_Geometry(SDO_GEOMETRY(2002,NULL,NULL,
            SDO_ELEM_INFO_ARRAY(1,2,1),
            SDO_ORDINATE_ARRAY(6,10,10,1,14,10,10,14,6,10))
        ).ST_InsertVertex(
            T_Vertex(2,6,1).ST_VertexType()
        ).ST_InsertVertex(
            T_Vertex(7.7,6.2,3).ST_VertexType()
        ).ST_InsertVertex(
            T_Vertex(2,6,-1).ST_VertexType()
        ).ST_InsertVertex(
            T_Vertex(-2,10,-1).ST_VertexType()
        ).geom as Geom
  From dual;

-- Results
--
GEOM
------------------------------------------------------------
SDO_GEOMETRY(2002,NULL,NULL,SDO_ELEM_INFO_ARRAY(1,2,1),
            SDO_ORDINATE_ARRAY(2,6, 6,10,7.7,6.2,10,1,
                               14,10,10,14,6,10,2,6,-2,10))
```

- Try and add a new arc to the end of an existing one:

```
Select T_Geometry(
            SDO_GEOMETRY(2002,NULL,NULL,
            SDO_ELEM_INFO_ARRAY(1,4,2, 1,2,1, 5,2,2),
            SDO_ORDINATE_ARRAY(6,10,10,1,14,10,10,14,6,10))
            ,0.005
        ).ST_InsertVertex(
            T_Vertex( 2, 6,-1).ST_VertexType()
        ).ST_InsertVertex(
            T_Vertex(-2,10,-1).ST_VertexType()
        ).geom as Geom
  From dual;

-- Results
--
GEOM
------------------------------------------------------------
SDO_GEOMETRY(2002,NULL,NULL,
    SDO_ELEM_INFO_ARRAY(1,4,2,1,2,1,5,2,2),
    SDO_ORDINATE_ARRAY(6,10,10,1,14,10,10,14,6,10,2,6,-2,10)
    )
```

- Create a loop (ORA-13349) in the outer boundary of a polygon:

```
Select T_Geometry(
          SDO_GEOMETRY(2003,NULL,NULL,
            SDO_ELEM_INFO_ARRAY(1,1005,2,1,2,1,5,2,2),
            SDO_ORDINATE_ARRAY(6,10,10,1,14,10,10,14,6,10))
        ).ST_InsertVertex(
            T_Vertex(0,0,3).ST_VertexType()
        ).geom as newGeom
   From geoms a;

-- Result
--
ORA-20124: Vertex insert invalidated geometry. Reason: ORA-13349
```

- Break the polygon by inserting a new vertex at the start of a ring (ORA-13348):

```
Select T_Geometry(
          SDO_GEOMETRY(2003,NULL,NULL,
            SDO_ELEM_INFO_ARRAY(1,1005,2, 1,2,1, 5,2,2),
            SDO_ORDINATE_ARRAY(6,10,10,1,14,10,10,14,6,10))
        ).ST_InsertVertex(
            T_Vertex(0,0,1).ST_VertexType()
        ).geom as geom
   From dual;

-- Result
--
ORA-20124: Vertex insert invalidated geometry. Reason: ORA-13348
```

> ORA-13348: polygon boundary is not closed.

- Fix the polygon (ORA-13348) by inserting a new vertex at the end of a ring:

```
Select T_Geometry(
          SDO_GEOMETRY(2003,NULL,NULL,
            SDO_ELEM_INFO_ARRAY(1,1003,1),
            SDO_ORDINATE_ARRAY(6,10,10,1,14,10,10,14))
        ).ST_InsertVertex(
            T_Vertex(6,10,-1).ST_VertexType()
        ).geom as geom
   From dual;

-- Result
--
```

```
GEOM
----------------------------------------------------------
SDO_GEOMETRY(2003,NULL,NULL,
          SDO_ELEM_INFO_ARRAY(1,1003,1),
          SDO_ORDINATE_ARRAY(6,10,10,1,14,10,10,14,6,10))
```

- Insert the second point in a polygon with a compound outer ring:

```
Select T_Geometry(
          SDO_GEOMETRY(2003,NULL,NULL,
          SDO_ELEM_INFO_ARRAY(1,1005,2,1,2,1,5,2,2),
          SDO_ORDINATE_ARRAY(6,10,10,1,14,10,10,14,6,10))
        ).ST_InsertVertex(
            T_Vertex(7.7,6.2,2).ST_VertexType()
        ).geom as geom
  From dual a;
-- Result
--
GEOM
----------------------------------------------------------
SDO_GEOMETRY(2003,NULL,NULL,
        SDO_ELEM_INFO_ARRAY(1,1005,2,1,2,1,7,2,2),
        SDO_ORDINATE_ARRAY(6.0,10.0, 7.7, 6.2,10.0, 1.0,
                          14.0,10.0,10.0,14.0, 6.0,10.0))
```

Implementing ST_UpdateVertex

Updating a vertex within an existing object is simpler than adding a vertex. The main issues to be aware of are:

- An update of a defining vertex of a circular arc must not break that arc (for example, making the three points collinear; collapsing two vertexes so that they are the same)

- Updating the first or last vertex in the ring of a polygon such that the ring no longer closes

- Updating a vertex in a ring of a polygon such that a loop is created in the boundary (ORA-13349)

Oracle already implements the necessary validation mathematics in its SDO_GEOM. Validate_Geometry function. As such, the algorithm will:

- Apply the update without concerning ourselves with the previous issues

- Check the modified geometry via VALIDATE_GEOMETRY, returning the modified geometry only if it is correct

The `ST_UpdateVertex` member function of the `T_Geometry` type is as follows. Note that updating the identification of the particular vertex is expected in the `ID` attribute of the `VERTEX_SET` object:

```
Create Or Replace Type Body T_Geometry
[...]
   -- Supplied p_vertex must have Z and W coded correctly
   -- p_vertex.id Is position of vertex to be updated.
   Member Function
   ST_UpdateVertex (p_vertex In mdsys.vertex_type)
   Return T_Geometry Determinstic,
```

Testing ST_UpdateVertex

`ST_UpdateVertex` can be tested in the following ways:

- Update a single point stored in a variety of ways:
```
-- Update single point stored in sdo_point
Select T_Geometry(
         SDO_GEOMETRY(2001,null,
                      SDO_POINT_TYPE(1000,2000,null),
                      null,null)
         ).ST_UpdateVertex(
             T_Vertex(0,0,1).ST_VertexType()
         ).geom as geom
 From Dual Union All
-- Update single point stored in sdo_ordinates
Select T_Geometry(
         SDO_GEOMETRY(2001,null,null,
                      SDO_ELEM_INFO_ARRAY(1,1,1),
                      SDO_ORDINATE_ARRAY(1000,2000))
         ).ST_UpdateVertex(
             T_Vertex(0,0,-1).ST_VertexType()
         ).geom as geom
 From Dual Union All
 -- Single point in multipoint
Select T_Geometry(
         SDO_GEOMETRY(2005,null,null,
                      SDO_ELEM_INFO_ARRAY(1,1,1),
                      SDO_ORDINATE_ARRAY(1000,2000))
         ).ST_UpdateVertex(
             T_Vertex(0,0,-1).ST_VertexType()
         ).geom as geom
 From dual;

-- Results
```

```
--
GEOM
-----------------------------------------------------------
SDO_GEOMETRY(2001,NULL,
             SDO_POINT_TYPE(0.0,0.0,NULL),NULL,NULL)
SDO_GEOMETRY(2001,NULL,NULL,
             SDO_ELEM_INFO_ARRAY(1,1,1),
             SDO_ORDINATE_ARRAY(0.0,0.0))
SDO_GEOMETRY(2005,NULL,NULL,
             SDO_ELEM_INFO_ARRAY(1,1,1),
             SDO_ORDINATE_ARRAY(0.0,0.0))
```

- Update the second and last vertex of a multipoint:

```
Select T_Geometry(
          SDO_GEOMETRY(2005,NULL,NULL,
               SDO_ELEM_INFO_ARRAY(1,1,3),
               SDO_ORDINATE_ARRAY(6,10, 10,1, 10,14))
       ).ST_UpdateVertex(
           T_Vertex(14,10,2).ST_VertexType()
       ).ST_UpdateVertex(
           T_Vertex(6,10,-1).ST_VertexType()
       ).geom as geom
  From dual;


-- Results
--
GEOM
-----------------------------------------------------------
SDO_GEOMETRY(2005,NULL,NULL,SDO_ELEM_INFO_ARRAY(1,1,3),
             SDO_ORDINATE_ARRAY(6,10, 14,10, 6,10))
```

- Invalid circular arc change (ORA-13346:collinear):

```
Select T_Geometry(
          SDO_GEOMETRY(2002,null,null,
             SDO_ELEM_INFO_ARRAY(1,4,2, 1,2,1, 7,2,2),
             SDO_ORDINATE_ARRAY(10,78,/*1*/ 10,75,/*2*/
                                20,75,/*3*/ 20,78,/*4*/
                                15,80,/*5*/ 10,78 /*6*/))
       ).ST_UpdateVertex(
           T_Vertex(10,82,-1).ST_VertexType()
       ).geom as Geom
  From dual;

-- Results
--
ORA-20123: Vertex update invalidated geometry. Reason: ORA-13346
```

- Invalid polygon ring change (ORA-13348: last vertex not equal to first):

```
Select T_Geometry(
        SDO_GEOMETRY(2003,NULL,NULL,
            SDO_ELEM_INFO_ARRAY(1,1003,1),
            SDO_ORDINATE_ARRAY (0,0,10,0,10,10,0,10,0,0))
        ).ST_UpdateVertex(
            T_Vertex_type(5,5,-1).ST_VertexType()
        ).geom as geom
 From dual;


-- Results
--
ORA-20123: Vertex update invalidated geometry. Reason: ORA-13348
```

Variations: A number of important variations of this function exist:

1. Replace all the instances of a particular vertex with another one:

```
Member Function
  ST_UpdateVertex (p_old_point IN MDSYS.Vertex_Type,
                   p_new_point IN MDSYS.Vertex_Type)
```

2. Provide a list of vertexes to insert, update, or delete (set X and Y ordinates to NULL) and let the function make the changes in one call:

```
Member Function
  ST_UpdateVertex (p_vertexes IN MDSYS.Vertex_Type_Set)
```

These variations are not presented in this book, but are available in the associated source code.

Implementing ST_DeleteVertex

The deletion of vertexes is somewhat simpler than adding them. Single points could be downgraded to SDO_GEOMETRY objects with NULL SDO_POINT/SDO_ORDINATE elements. These would still be reported as invalid (for example, ORA-13032: Invalid NULL SDO_GEOMETRY object), so the algorithm will turn them into NULL geometries. The deletion of a vertex in a multipoint with one vertex will also return a NULL SDO_GEOMETRY object.

Deletion of the vertex that describes a circular arc will not be allowed. In all other situations, the vertex will be deleted and the SDO_ELEM_INFO array's appropriate offset values decremented. The modified geometry will be validated before being returned.

The `ST_DeleteVertex` member function is declared as follows:

```
Create Or Replace Type Body T_Geometry
[...]
  -- p_vertex_id Is position of vertex to be deleted.
  Member Function
  ST_DeleteVertex(p_vertex_id In Integer)
  Return T_Geometry Deterministic.
```

Testing ST_DeleteVertex

The following tests are applied:

- Single point deletion:

```
SET NULL NULL
-- Delete point in single point coded in sdo_point
Select T_Geometry(
        SDO_GEOMETRY(2001,null,
                     SDO_POINT_TYPE(1000,2000,null),
                     null,null)
       ).ST_DeleteVertex(1).geom as geom
 From Dual Union All
-- Delete point in single point coded in sdo_ordinates
Select T_Geometry(
        SDO_GEOMETRY(2001,null,null,
                     SDO_ELEM_INFO_ARRAY(1,1,1),
                     SDO_ORDINATE_ARRAY(1000,2000))
       ).ST_DeleteVertex(-1).geom as geom
  From Dual Union All
-- Or single point in multipoint sdo_ordinates
Select T_Geometry(
        SDO_GEOMETRY(2005,null,null,
                     SDO_ELEM_INFO_ARRAY(1,1,1),
                     SDO_ORDINATE_ARRAY(1000,2000))
       ).ST_DeleteVertex(-1).geom as geom
  From dual;

-- Results
--
GEOM
----
NULL
NULL
NULL
```

- **Remove third and last vertex of a multipoint:**

```
Select T_Geometry(
         SDO_GEOMETRY(2005,NULL,NULL,
            SDO_ELEM_INFO_ARRAY(1,1,5),
            SDO_ORDINATE_ARRAY(6,10, 10,1, 14,10,
                                   10,14, 6,10))
       ).ST_DeleteVertex(3)
        .ST_DeleteVertex(-1).geom as geom
  From dual;

-- Results
--
GEOM
---------------------------------------------------------
SDO_GEOMETRY(2005,NULL,NULL,SDO_ELEM_INFO_ARRAY(1,1,3),
         SDO_ORDINATE_ARRAY(6,10,10,1,10,14))
```

- **Remove third and last vertex of a linestring:**

```
Select T_Geometry(
         SDO_GEOMETRY(2002,NULL,NULL,
            SDO_ELEM_INFO_ARRAY(1,2,1),
            SDO_ORDINATE_ARRAY(6,10,10,1,14,10,10,14,6,10))
       ).ST_DeleteVertex(3)
        .ST_DeleteVertex(-1).geom
        as geom
  From dual;

-- Results
--
GEOM
----------------------------------------------------------------
SDO_GEOMETRY(2002,NULL,NULL,SDO_ELEM_INFO_ARRAY(1,2,1),
         SDO_ORDINATE_ARRAY(6,10, 10,1, 10,14))
```

- **Break the circular arc in a compound ring:**

```
Select T_Geometry(
         SDO_GEOMETRY(2003,NULL,NULL,
            SDO_ELEM_INFO_ARRAY(1,1005,2, 1,2,1, 5,2,2),
            SDO_ORDINATE_ARRAY(6,10,10,1,14,10,10,14,6,10))
       ).ST_DeleteVertex(3).geom as geom
  From dual;

-- Results
--
SQL Error: ORA-20122: Deletion of vertex within an existing
circular arc not allowed.
```

- Break a polygon by deleting the vertex at the start of a ring:

```
Select T_Geometry(
        SDO_GEOMETRY(2003,NULL,NULL,
           SDO_ELEM_INFO_ARRAY(1,1005,2, 1,2,1, 5,2,2),
           SDO_ORDINATE_ARRAY(6,10,10,1,14,10,10,14,6,10))
        ).ST_DeleteVertex(1).geom as geom
  From dual;
-- Results
--
SQL Error: ORA-20123: Vertex delete invalidated geometry. Reason:
ORA-13348
```

- Delete the second point in a polygon with a compound outer ring:

```
Select T_Geometry(
        SDO_GEOMETRY(2003,NULL,NULL,
           SDO_ELEM_INFO_ARRAY(1,1005,2, 1,2,1, 5,2,2),
           SDO_ORDINATE_ARRAY(6,10,10,1,14,10,10,14,6,10))
        ).ST_DeleteVertex(2).geom as geom
  From dual a;

-- Results
--
GEOM
-----------------------------------------------------------
SDO_GEOMETRY(2003,NULL,NULL,
           SDO_ELEM_INFO_ARRAY(1,1005,2, 1,2,1, 3,2,2),
           SDO_ORDINATE_ARRAY(6,10,
                              14,10, 10,14, 6,10))
```

Real world example

While some of the test examples just seen are drawn from real world examples, a number of others will demonstrate what can be done with a richer API. For example, it is not uncommon to see an external software load polygons that are not "closed"; that is, the last vertex is not the same as the first vertex.

To fix this, the first and last vertices of the polygon need to be extracted or inspected. This can be done by coding a vertex "inspection" member function, ST_GetVertex, and then creating two "wrapper" functions, ST_StartVertex and ST_EndVertex, that call it.

```
Create Or Replace Type T_Geometry As Object (
[...]
  /* Vertex Inspectors */
  Member Function ST_GetVertex(p_vertex In Integer)
         Return mdsys.vertex_type Deterministic,
  Member Function ST_StartVertex
         Return mdsys.vertex_type Deterministic,
  Member Function ST_EndVertex
         Return mdsys.vertex_type Deterministic,
```

Examples of their use can be seen as follows:

*   Inspecting the first vertex:

```
Select a.geom.ST_StartVertex() as svertex
  From (Select T_Geometry(
                 SDO_GEOMETRY(2005,NULL,NULL,
                     SDO_ELEM_INFO_ARRAY(1,1,3),
                     SDO_ORDINATE_ARRAY(6,10, 10,1, 10,14)))
               as geom
         From dual) a;

-- Results
--
SVERTEX
-------------------------------------
VERTEX_TYPE(6,10,NULL,NULL,NULL,NULL,
        NULL,NULL,NULL,NULL,NULL,1)
```

- Inspecting the last vertex:

```
Select a.geom.ST_EndVertex() as svertex
  From (Select T_Geometry(
                SDO_GEOMETRY(2005,NULL,NULL,
                  SDO_ELEM_INFO_ARRAY(1,1,3),
                  SDO_ORDINATE_ARRAY(6,10, 10,1, 10,14)))
              as geom
          From dual) a;


-- Results
--
SVERTEX
---------------------------------------
VERTEX_TYPE(10,14,NULL,NULL,NULL,NULL,
        NULL,NULL,NULL,NULL,NULL,3)
```

- Inspecting all vertexes:

```
Select a.geom.ST_getVertex(b.vertex) as vertex
  From (Select T_Geometry(
                SDO_GEOMETRY(2005,NULL,NULL,
                  SDO_ELEM_INFO_ARRAY(1,1,3),
                  SDO_ORDINATE_ARRAY(6,10, 10,1, 10,14)))
              as geom
          From dual) a,
        (Select level as vertex
          From dual
        Connect By Level <= 3) b;


-- Results
--
VERTEX
---------------------------------------
VERTEX_TYPE(6,10,NULL,NULL,NULL,NULL,
        NULL,NULL,NULL,NULL,NULL,1)
VERTEX_TYPE(10,1,NULL,NULL,NULL,NULL,
        NULL,NULL,NULL,NULL,NULL,2)
VERTEX_TYPE(10,14,NULL,NULL,NULL,NULL,
        NULL,NULL,NULL,NULL,NULL,3)
```

Assume that a polygon is not closed and that the last vertex, which should be equal to the first, is missing. The polygons can be closed by selecting the starting vertex of the polygon using ST_StartVertex and then inserting it at the end (-1) of the polygon using ST_InsertVertex.

```
Select b.geom
       .ST_InsertVertex(
           T_Vertex(b.vertex.x,b.vertex.y,-1).ST_VertexType()
       ).geom as geom
  From (Select a.geom, a.geom.ST_StartVertex() as vertex
        From (Select T_Geometry(
                  SDO_GEOMETRY(2003,NULL,NULL,
                     SDO_ELEM_INFO_ARRAY(1,1003,1),
                     SDO_ORDINATE_ARRAY(6,10, 10,1,
                                        14,10, 10,14))) as geom
              From dual
            ) a
       ) b;

-- Results
--
GEOM
-------------------------------------------------------------
SDO_GEOMETRY(2003,NULL,NULL,
             SDO_ELEM_INFO_ARRAY(1,1003,1),
             SDO_ORDINATE_ARRAY(6,10,10,1,14,10,10,14,6,10))
```

Extending a linestring

The requirement to extend a linestring is quite common in the world of a practitioner. This can be done by coding a single ST_Extend member function. The rules that will govern this function are:

- Operates only on linestrings
- Allows for shortening via provision of a negative extension distance
- Can be applied to the start, end, or both ends at the same time
- Negative extension distances that could collapse the linestring to nothing need to be captured

The function's declaration in the `T_Geometry` type is as follows:

```
Create Or Replace TYPE T_Geometry As Object (
[...]
  Member Function
  ST_Extend(p_extend_dist In number,
            p_start_end   In varchar2 default 'START',
            p_unit        In varchar2 default null)
      Return T_Geometry Deterministic,
```

The `ST_Extend` function's `p_extend_dist` parameter can be applied in two ways. Firstly, if the supplied value is negative, the underlying linestring is shortened; if positive, the linestring is extended.

The `p_start_end` parameter must be one of either START, END, or BOTH. If it is START, the linestring is extended from the beginning of the linestring in the reverse direction of the first vector. If it is END, the extension is at the end in the direction of the last vector. If it is BOTH, the line is extended/shortened at both ends.

The `p_unit` parameter allows for the specification of the extension distance in units that are compatible with the SRID. Generally, it is considered a best practice to always provide the unit parameter to any **Oracle Spatial** function that requires it. The value provided for `p_extend_dist` must be expressed in the same units as described by `p_unit`.

The following example demonstrates how this function works:

A driveway (the dotted black line in the top-right) is proposed between the buildings with FIDs of 9161 and 9265. What is actually required is to take the existing road centerline with FID 1334 and extend it by 90ft at its start so as to connect to the proposed driveway.

The SQL to create the extension at the start of the line is as follows:

```
Select rl.fid,
       T_Geometry(rl.geom,0.005)
                 .ST_Extend(90,'START','unit=U.S. Foot')
                 .ST_RoundOrdinates(2)
                 .geom as sGeom
  From road_clines rl
 Where rl.fid = 1334;

-- Results
--
FID SGEOM
--- ------------------------------------------------------------
334 SDO_GEOMETRY(2002,2872,NULL,SDO_ELEM_INFO_ARRAY(1,2,1),
        SDO_ORDINATE_ARRAY(5995356.7,2105963.84,
                           5995094.87,2105871.28,
                           5995044.46,2105492.8,
                           5995033.03,2105411.14,
                           5995012.6,2105371.13,
                           5994950.04,2105332.47))
```

Visually, the result looks like this:

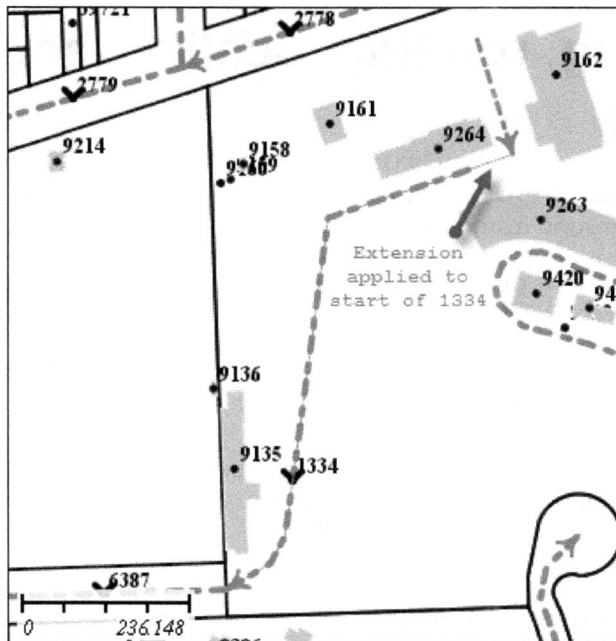

For completeness, the extend operation should also involve a snapping of the road centerline's modified start vertex with the end vertex of the proposed driveway. The provision of this capability is covered in a chapter on the use of the **JTS Topology Suite (JTS)** within the Oracle database's JVM.

Later on in this chapter, another example is provided where ST_Extend is used in consort with ST_LineShift to process the vectors that describe a land parcel.

Finally, the following SQL shows how to shorten a linestring at both ends by an amount that is greater than the length:

```
Select T_Geometry(
        SDO_GEOMETRY(2002,null,null,
                     SDO_ELEM_INFO_ARRAY(1,2,1),
                     SDO_ORDINATE_ARRAY(1,1,2,2)),0.05)
       .ST_Extend(-0.708,'BOTH')
       .ST_RoundOrdinates(2).geom as geom
  From dual;

-- Result
--
ORA-20124: Reducing geometry of length (1.4142135623731) by (.708)
at both ends would result in a zero length geometry.
```

Translating, rotating, scaling, and reflecting

This section examines a collection of commonly used transformation functions that provide the ability to: move or translate a geometry from one location to another; rotate a geometry around a supplied point; scale or enlarge/shrink a geometry.

The standard method for implementing these is via the application of an affine transformation matrix.

Oracle Spatial did not provide such capability until 11*g*, when it provided the SDO_UTIL.AffineTransforms function. In 10*g*, Oracle released the UTL_NLA built-in SYS package (including UTL_NLA_ARRAY_DBL and UTL_NLA_ARRAY_INT types), which can be used to construct the appropriate affine transformation matrices to implement spatial transformations. (for Oracle 9*i*, implementation can be done with custom PL/SQL code, but it is not covered in this book.)

Both the SDO_UTIL.AffineTransforms and UTL_NLA-based approaches are provided in the source code associated with this book.

Introducing a set of transformation member functions

The following member functions have been added to the `T_Geometry` type. These member functions include overloaded functions that implement a simplified set of parameters:

```
Create Or Replace Type T_Geometry As Object (
[...]
  /* Transformers */
  Member Function ST_Rotate      (p_angle_rad In Number,
                                  p_dir       In pls_integer,
                                  p_rotate_pt In SDO_GEOMETRY,
                                  p_line1     In SDO_GEOMETRY)
          Return T_Geometry deterministic,
  Member Function ST_Rotate      (p_angle_rad In Number,
                                  p_rx        In Number,
                                  p_ry        In Number)
          Return T_Geometry deterministic,
  Member Function ST_Rotate      (p_angle_rad In Number,
                                  p_rotate_pt In SDO_GEOMETRY)
          Return T_Geometry deterministic,
  Member Function ST_Rotate      (p_angle_rad In Number)
          Return T_Geometry deterministic,
  Member Function ST_Scale       (p_sx        In Number,
                                  p_sy        In Number,
                                  p_sz        In Number,
                                  p_scale_pt  In SDO_GEOMETRY)
          Return T_Geometry deterministic,
  Member Function ST_Scale       (p_sx        In Number,
                                  p_sy        In Number,
                                  p_sz        In Number)
          Return T_Geometry deterministic,
  Member Function ST_Scale       (p_sx        In Number,
                                  p_sy        In Number)
          Return T_Geometry deterministic,
  Member Function ST_Translate (p_tx          In Number,
                                  p_ty        In Number,
                                  p_tz        In Number)
          Return T_Geometry deterministic,
  Member Function ST_Translate (p_tx          In Number,
                                  p_ty        In Number)
          Return T_Geometry deterministic,
  Member Function ST_Reflect(p_reflect_geom In SDO_GEOMETRY,
                             p_reflect_plane In Number default -1)
          Return T_Geometry deterministic,
```

```
Member Function ST_Affine  (p_a    In Number,
                            p_b    In Number,
                            p_c    In Number,
                            p_d    In Number,
                            p_e    In Number,
                            p_f    In Number,
                            p_g    In Number,
                            p_h    In Number,
                            p_i    In Number,
                            p_xoff In Number,
                            p_yoff In Number,
                            p_zoff In Number)
        Return T_Geometry deterministic,
```

Example one – shifting a geometry's position

An organization in Tasmania, Australia, for many years stored the spatial data it needed for its business operations in single precision format. In addition, the nature of their mapping software saw them incapable of producing a single state-wide map at 1:500,000 scale, while King Island (see the following figure) was in its correct geographic position. Therefore, all King Island data was physically shifted 8,950m south from its correct position and 66,680 meters to the east:

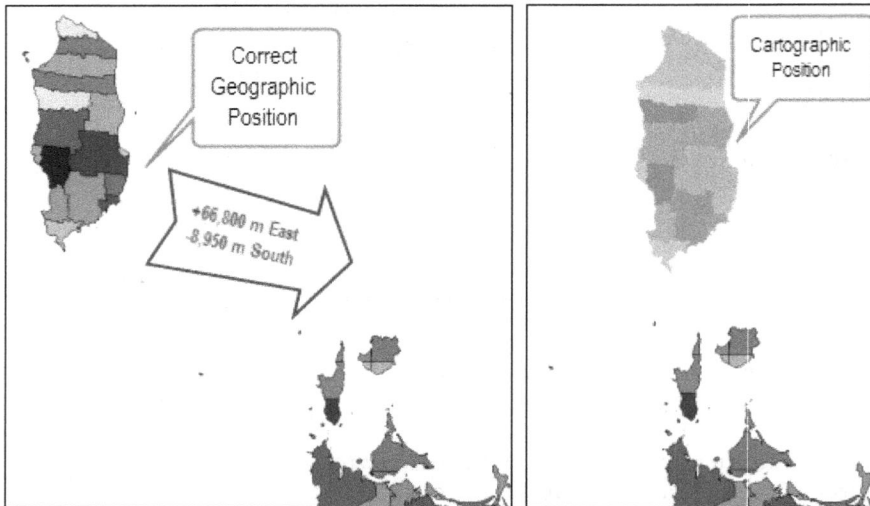

Later, with the purchase of Oracle Spatial (8*i*) and more modern mapping and GPS software, there arose the need to transform all the data in the database from cartographic position to their correct geographic position. So that the fully-integrated spatial information and asset management databases that were being constructed would no longer include this historical artifact.

To move all the King Island data within the Oracle database back to their correct geographic position, a translation algorithm was written in PL/SQL that provided the name of a table and an SDO_GEOMETRY column within it:

- Spatially searched for all data in the King Island location
- Modified all the coordinates such that 66,680 meters were subtracted from each X ordinate, and 8,950 meters added to each Y ordinate that described the geometry object
- Committed the changes

To affect such a move of geometry objects, the ST_Geometry.ST_Translate wrapper over the SDO_UTIL.AffineTransformations package function, can be used. The following example shows, how moving King Island from its cartographic to its geographic position can be achieved:

```
WITH cartoFence as (
   Select SDO_GEOMETRY(2003,28355,NULL,
                       SDO_ELEM_INFO_ARRAY(1,1003,3),
                       SDO_ORDINATE_ARRAY(292901.564,5539682.144,
                                          327337.044,5608340.539))
             as geom
      From dual
)
Select book.T_Geometry(b.geom,0.0005)
          .ST_Translate(-66800,8950)
          .ST_RoundOrdinates(3).geom
        as geom
   From cartoFence a,
        locality b
  Where sdo_anyinteract(b.geom,a.geom) = 'TRUE';
```

This technique was used by the same organization in the year 2000, when in Australia; the central mapping council required that all Australian spatial data and maps be converted from the Australian Geodetic Datum (1966 AGD66, 1984 AGD84) to a new Geocentric Datum of Australia (1994 GDA94). The former was a localized datum, whereas the latter was aligned with the WGS84. All the data was held as Australian Map Grid (AMG) Zone 55 data (based on AGD66). The move had to be to Map Grid of Australia (MGA) Zone 55 (based on GDA94).

That organization had large amounts of low accuracy point data (AGD) in its Oracle database, and large amounts of point data stored as Northing and Easting columns in Microsoft Access and Spreadsheet formats. For these datasets, it was decided that moving the data could be achieved via a simple translation in which 112 meters was added to each Easting and 183 meters to each Northing.

Example two – duplicating a road centerline using reflection

A common scenario in spatial data processing is the creation of new geometry data based on existing data. Consider an example where a surveyor is working in a housing development where all the houses have identical footprints. After a complete survey of the footprint of one building, a baseline is surveyed along the other buildings via a set of copy, translate, and rotate operations, then the footprint is copied to create the others.

In what follows, a design for a proposed road centerline will be created from an existing centerline via a rotate and translation operation using the ST_Rotate and ST_Translate member functions.

The following figure shows an area in the San Francisco dataset in which an existing road will be selected and mirrored around an imaginary centerline (dotted) to create a new centerline. To create the new design centerline, the existing road will be rotated by 180 degrees around its starting vertex (ST_StartVertex) and then translated so that the starting vertex becomes the ending vertex of the original road (the starting/ending vertex pair forms the rotation axis):

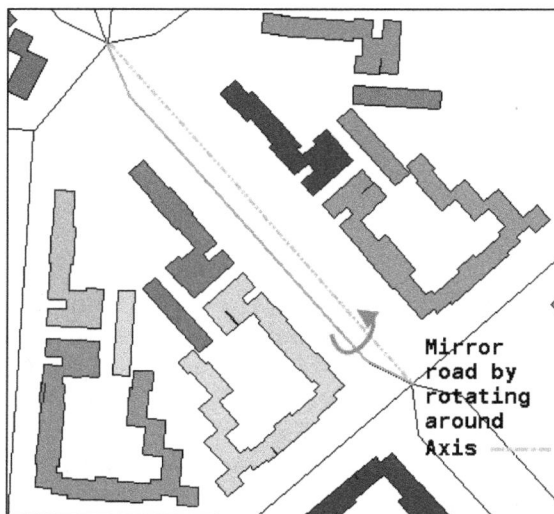

The table holding the road centerline data is ROAD_CLINES. The source road has an FID of 9328. An existing record already exists in ROAD_CLINES that will hold our new geometry; its FID is 9329. Our update will occur as follows:

```
Update road_clines a
Set a.geom = (
  With testLine As (
  Select SDO_GEOMETRY(2002,b.geom.ST_Srid(),null,
              SDO_ELEM_INFO_ARRAY(1,2,1),
              SDO_ORDINATE_ARRAY(b.startV.x,b.startV.y,
                                  b.endV.x,  b.endV.y))
            as rAxis, b.geom
      From (Select a.geom.ST_StartVertex() as startV,
                  a.geom.ST_EndVertex()    as endV,
                  geom
              From (Select T_Geometry(r.geom,0.05) as geom
                    From road_clines r
                    Where r.fid = 9328
                  ) a
          ) b
  )
  Select a.geom.ST_Reflect(a.rAxis,-1).geom as geom    From testLine a
  )
Where a.fid = 9329;
Commit;
```

The creation of the new road centerline could also be done using the ST_Rotate and ST_Translate member functions as follows:

```
Update road_clines a
Set a.geom = (
Select (22/7) as rAngle,
        SDO_GEOMETRY(2002,b.geom.ST_Srid(),null,
            SDO_ELEM_INFO_ARRAY(1,2,1),
            SDO_ORDINATE_ARRAY(b.startV.x,b.startV.y,
                                b.endV.x,b.endV.y))
          as rAxis, b.startV, b.endV, b.geom
  From (Select a.geom.ST_StartVertex() as startV,
              a.geom.ST_EndVertex()    as endV,
              geom
          From (Select T_Geometry(r.geom,0.05) as geom
                From road_clines r
                Where r.fid = 9329
```

```
                        ) a
            ) b
    )
    Select a.geom.ST_Rotate(a.rAngle,a.startV.x,a.startV.y)
                .ST_Translate((a.endV.x-a.startV.x),
                            (a.endV.y-a.startV.y)).geom as geom
      From testLine a;
    )
    Where a.fid = 9329;
    commit;
```

The new road centerline can be seen in the following figure:

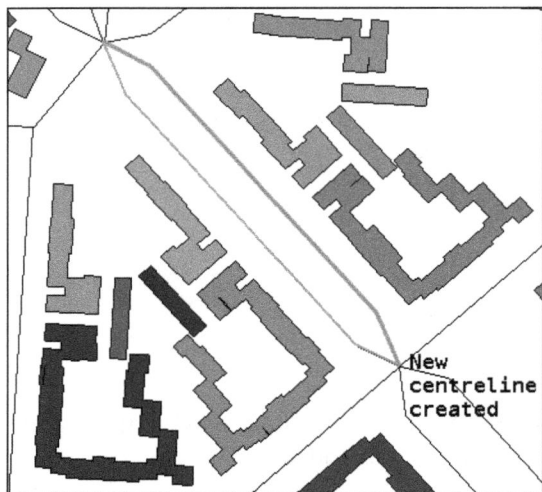

Splitting linestring geometries – ST_Split

Building on from this road centerline example, imagine a situation where the newly created centerline has to be split, so that a new driveway can be constructed to service the buildings that are in front of the street it describes. In the following figure the dotted line is the proposed driveway:

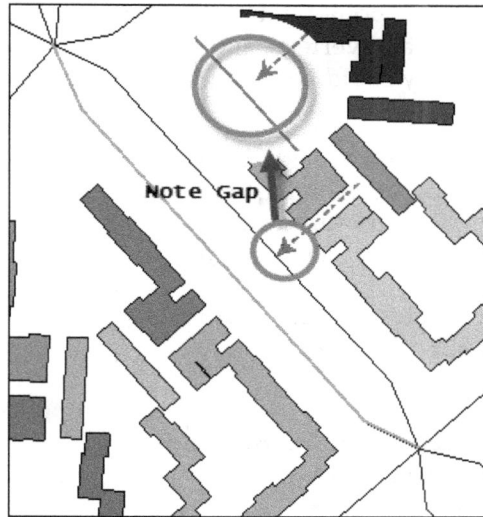

To split the new road centerline at the intersection point, a new member function that allows us to split an existing (multi) linestring using a single point needs to be constructed. The algorithm does not require the point to actually fall exactly on the line, rather it can be nearby (though the algorithm will not implement a test that limits the distance the point can be from the line – this can be done as an exercise later on).

The member function is described as follows:

```
Create or Replace T_Geometry As Object { T_Geometry As Object (
[...]
    Member Function ST_Split(p_point In SDO_GEOMETRY,
                             p_snap  In pls_integer default 0)
      Return t_Geometries pipelined
```

The p_point parameter is the estimated nearest point to the split point. The p_snap parameter is a Boolean parameter for which 0 (false) means that the line is split and the parts returned, whereas a positive value (true) means that the split point should be snapped to the actual position on the line and then returned. Therefore, the function returns one (the split point may fall at the start/end vertex, thus only the linestring itself is returned) or more linestrings or the adjusted snap point.

To split the road, the centerline point on the proposed driveway nearest to the road is extracted – in this case, the end vertex – and passed to the function. The results are then mapped (though they could be saved back to the database).

```
With testData As (
  Select T_Geometry(a.geom,0.005) as geom,
         T_Geometry(SDO_GEOMETRY(2002,2872,NULL,
                         SDO_ELEM_INFO_ARRAY(1,2,1),
                         SDO_ORDINATE_ARRAY(5989028.63,2090807.20,
                                            5988948.76,2090732.13)),
                    0.005) as driveLine
    From road_clines a
   Where a.fid = 9329
)
Select T_Geometry(s.geometry,s.tolerance)
         .ST_RoundOrdinates(2)
         .geom as geom
  From testData a,
       Table(a.geom.ST_Split(T_Vertex(a.driveLine.ST_EndVertex()),
                             'unit=U.S. Foot')) s;

-- Results
--
GEOM
-------------------------------------------------------
SDO_GEOMETRY(2002,2872,NULL,SDO_ELEM_INFO_ARRAY(1,2,1),
            SDO_ORDINATE_ARRAY(5989091.25,2090531.43,
                               5989071.35,2090588.77,
                               5988946.62,2090730.24))
SDO_GEOMETRY(2002,2872,NULL,SDO_ELEM_INFO_ARRAY(1,2,1),
            SDO_ORDINATE_ARRAY(5988946.62,2090730.24,
                               5988788.12,2090909.99,
                               5988722.74,2090950.65))
```

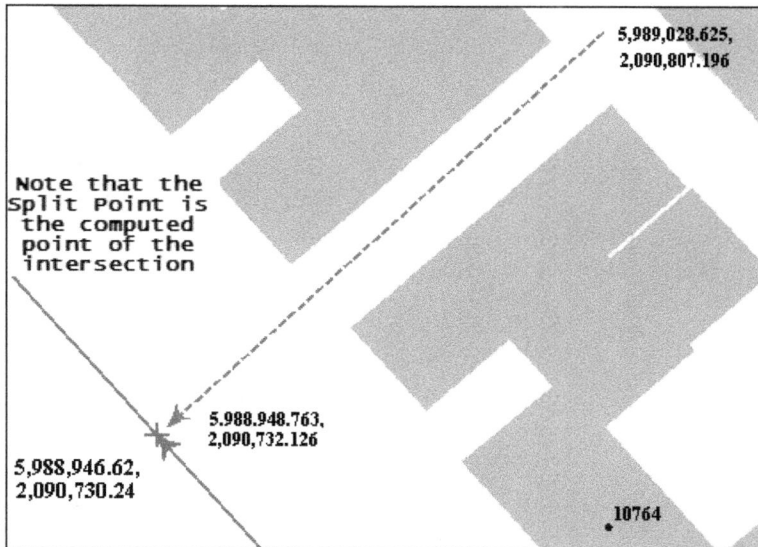

Note that the Split Point is the computed point of the intersection

5,989,028.625,
2,090,807.196

5.988.948.763,
2,090,732.126

5,988,946.62,
2,090,730.24

10764

To return the actual adjusted snap point (and to round the ordinates to two decimal places), a function called ST_Snap is introduced that snaps the end point of the proposed centerline to the existing centerline (the bold coordinates show the pre-snap and post-snap ordinate values of the point).

```
Create or Replace Type T_Geometry As Object (
[...]
  Member Function
  ST_Snap(p_point In mdsys.SDO_GEOMETRY,
          p_unit  In varchar2 default null)
  Return BOOK.T_geometries
```

Snapping the two lines is carried out by executing the following SQL statement:

```
With testData As (
Select T_Geometry(a.geom,0.005) as geom,
       T_Geometry(
         SDO_GEOMETRY(2002,2872,NULL,
```

```
                         SDO_ELEM_INFO_ARRAY(1,2,1),
                         SDO_ORDINATE_ARRAY(5989028.63,2090807.20,
                                            5988948.76,2090732.136)),
                0.005) as driveLine
     From road_clines a
    Where a.fid = 9329
    )
    Select T_Geometry(t.geometry,t.tolerance)
            .ST_RoundOrdinates(2).geom as geom
      From testData a,
           Table(a.geom
                    .ST_Snap( T_Vertex( a.driveLine.ST_EndVertex() )
                        .ST_SdoGeometry(a.geom.ST_CoordDim(),
                                        a.geom.ST_Srid() ),
                        'unit=U.S. Foot')
                ) t;

    -- Results
    --
    SPOINT
    -----------------------------------------------------------------
    SDO_GEOMETRY(2001,2872,
                SDO_POINT_TYPE(5988946.62,2090730.24,NULL),NULL,NULL)
```

Instead of splitting the linestring, the snapped point could be inserted into the linestring creating a vertex where the driveway linestring's end vertex ends. This is not covered in this book, but it could be constructed using the ST_InsertVertex member function or via modification of the ST_Split function.

Moving/shifting lines parallel to the original object

A common requirement for spatial data processing is to move a linear object parallel to itself. Examples include the desire to create a linestring that is parallel to the boundary of a land parcel, as is the case with rights of carriageway, often defined as being 10ft from an existing boundary. Similarly, one often hears of the need to create a line parallel to an existing road centerline; perhaps to define the boundaries of the road reserve or to present linear data graphically adjacent to the existing data.

Shifting a line sideways – ST_LineShift

A common and simple solution for moving a line parallel to itself is to extract the first and last vertex in a linestring, compute a single offset at right angles to an imaginary line composed of these vertexes, and then apply this to all the vertexes in the line. This is called line shifting and is implemented in ST_LineShift.

```
Create Or Replace Type T_Geometry As Object (
[...]
    Member Function ST_LineShift        (p_distance In Number)
                                        Return T_Geometry deterministic,
```

Our first example is drawn from a road centerline in the San Francisco dataset that is composed only of two vertexes:

```
Select rl.fid,
       T_Geometry(rl.geom,0.05)
          .ST_LineShift(-20)
          .ST_RoundOrdinates(2).geom as geom
   From road_clines rl
Where rl.fid = 638;
```

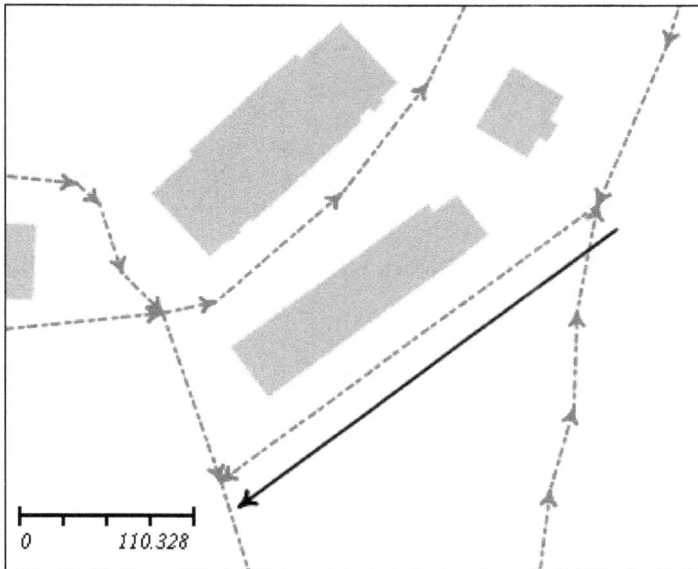

The offset road centerline is correct. This is because the constructed offset vector is the same as the actual linestring.

However, any road centerline with more than two vertexes will lose its shape if processed in this manner. This can be seen especially where any two vertex segments within the line have a similar alignment to the perpendicular, as seen in the following example:

```
Select fid,
       T_Geometry(rl.geom,0.005)
       .ST_LineShift(-20)
       .ST_RoundOrdinates(2)
       .geom as sGeom
  From road_clines rl
 Where rl.fid = 898;
```

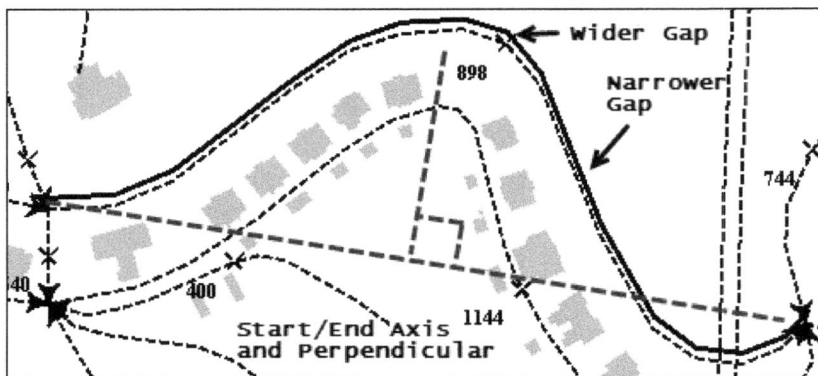

Finally, line shifting simply doesn't work where the linestring closes back upon itself as seen in the following example:

```
Select rl.fid,
       T_Geometry(rl.geom,0.005)
         .ST_LineShift(-10)
         .ST_RoundOrdinates(2)
         .geom as sGeom
  From road_clines rl
 Where rl.fid = 404;
```

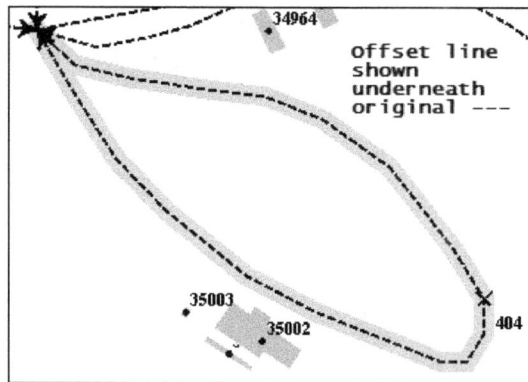

So, why have ST_LineShift at all? The following right of carriageway example shows just such a use.

Right of carriageway alongside land parcel boundary

Another real need for spatial processing for legal land parcels is the ability to create a boundary parallel to an existing one so that a right of carriageway (10ft wide) can be constructed to allow access to a building at the back of an existing land parcel. The black dashed line in the following figure shows the required boundary line over the existing top boundary of land parcel with a FID of 60143:

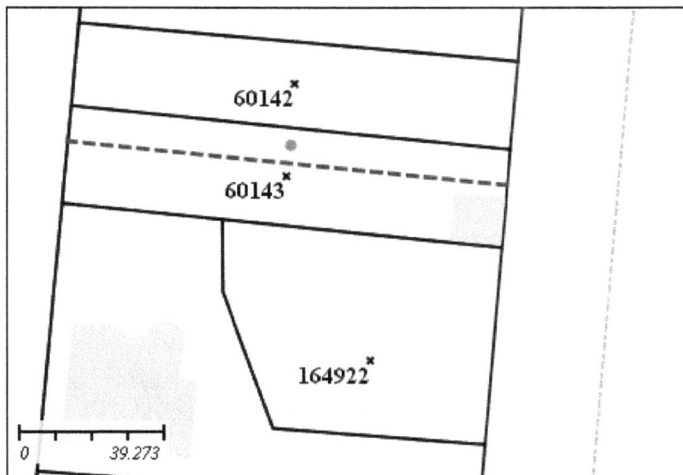

The SQL that can be used to extract and create this boundary line is as follows. The SQL first takes advantage of the simple four-sided nature of the land parcel by extracting the four boundaries via the ST_Vectorize member function. Then, to discover which boundary is to be used, a search point is created by clicking on the map display near the required boundary. The nearest boundary to the point is found via the ST_Distance member function. Finally, a parallel line is constructed via the ST_LineShift function.

```
    Select c.boundaryLine
           .ST_LineShift(-10/*10ft*/)
           .ST_RoundOrdinates(2).geom as rightOfWayLine
      From (Select b.boundaryLine.ST_Distance(b.searchPoint)
                       as distFeet,
                   Min(b.boundaryLine.ST_Distance(b.searchPoint))
                       Over (Order By 1) as minDistFeet,
                   b.boundaryLine
             From (Select T_Geometry(
                       t.vector.ST_SdoGeometry(2,lp.geom.sdo_srid),
                       0.005) as boundaryLine,
                       SDO_GEOMETRY(2001,2872,
                                     SDO_POINT_TYPE(5996463.162,
                                                    2105513.837,
                                                    NULL),
                                     NULL,NULL) as searchPoint
                   From land_parcels lp,
                       Table(T_Geometry(lp.geom,0.005)
                               .ST_Vectorize()) t
                   Where lp.fid = 60143
                 ) b
           ) c
    Where c.distFeet = c.minDistFeet;

-- Results
--
RIGHTOFWAYLINE
-------------------------------------------------------------
SDO_GEOMETRY(2002,2872,NULL,
             SDO_ELEM_INFO_ARRAY(1,2,1),
             SDO_ORDINATE_ARRAY(5996521.84,2105502.73,
                                5996402.45,2105514.96))
```

The previous figure was generated based on this result so you can see that it is correct.

The following example not only applies a line shift of 10ft, but also uses ST_Extend to shorten each end of the generated offset vectors by 10ft as well:

```
Select b.boundaryLine
        .ST_LineShift(-10)
        .ST_Extend(-10,'BOTH','unit=U.S. Foot')
        .ST_RoundOrdinates(2).geom as geom
  From (Select T_Geometry(
               t.vector.ST_SdoGeometry(2,a.geom.ST_Srid()),
                                  a.geom.tolerance)
            as boundaryLine
        From (Select T_Geometry(lp.geom,0.005) as geom
              From land_parcels lp
              Where lp.fid = 60142 ) a,
            Table(a.geom.ST_Vectorize()) t
      ) b;

-- Results
--
GEOM
-----------------------------------------------------------------
SDO_GEOMETRY(2002,2872,NULL,SDO_ELEM_INFO_ARRAY(1,2,1),
  SDO_ORDINATE_ARRAY(5996513.76,2105523.5,5996514.15,2105528.07))
SDO_GEOMETRY(2002,2872,NULL,SDO_ELEM_INFO_ARRAY(1,2,1),
  SDO_ORDINATE_ARRAY(5996514.12,2105528.11,5996414.65,2105537.04))
SDO_GEOMETRY(2002,2872,NULL,SDO_ELEM_INFO_ARRAY(1,2,1),
  SDO_ORDINATE_ARRAY(5996414.63,2105537.02,5996414.34,2105533.94))
SDO_GEOMETRY(2002,2872,NULL,SDO_ELEM_INFO_ARRAY(1,2,1),
  SDO_ORDINATE_ARRAY(5996414.43,2105533.83,5996513.93,2105523.64))
```

The result can be seen in the following figure:

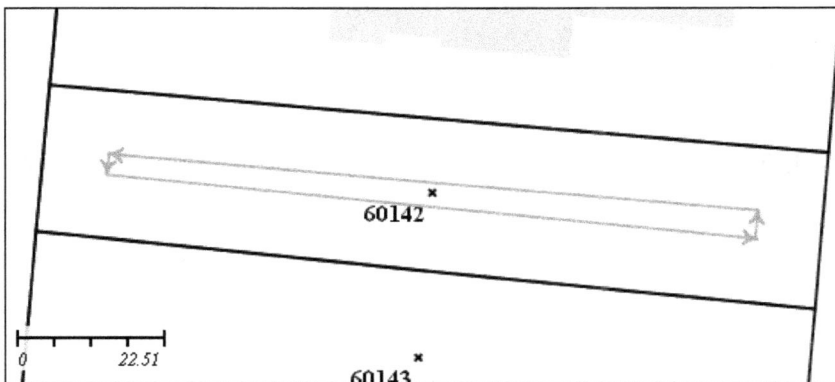

Creating truly parallel lines – ST_Parallel

The only way to get around the limited application and warping effect of `ST_LineShift` is to examine each pair of segments within the linestring. The construct the necessary offset vectors for each, and apply them by keeping in mind the extension or reduction in each segment that is necessary to ensure the modified vectors remain connected. Optionally (`p_curved=1`), where the bend between two segments is greater than 180 degrees, smooth the transition via the creation of a circular arc. The supplied distance (`p_distance`) must be expressed in the same units as described by the `p_unit` parameter.

```
Create Or Replace Type T_Geometry As Object(
[...]
  Member Function
  ST_Parallel(p_distance In Number,
              p_curved   In Number default 0,
              p_unit     In varchar2 default null)
  Return T_Geometry deterministic,
```

Processing a fabricated linestring will show what `ST_Parallel` can do. For that linestring, parallel lines 10ft to the left and another to the right will be created. In doing so, the function will be asked to "smooth" the linestrings at obtuse angles. The SQL code and the resultant figure follows:

```
With lineGeom As (
Select SDO_GEOMETRY(2002,null,null,
            SDO_ELEM_INFO_ARRAY(1,2,1),
            SDO_ORDINATE_ARRAY(0,0,45,45,90,0,135,45,
                    180,0,180,-45,90.84,-44.855,91.181,-90.221,
                    123.586,-107.277, 98.685,-74.872,
                    160.766,-111.37, 159.06, -62.933))
        as geom
    From dual
)
Select T_Geometry(a.geom,0.005).ST_Parallel(5,1).geom as RightGeom
    From lineGeom a
Union All
Select T_Geometry(a.geom,0.005).ST_Parallel(-5,1).geom as LeftGeom
    From lineGeom a;
```

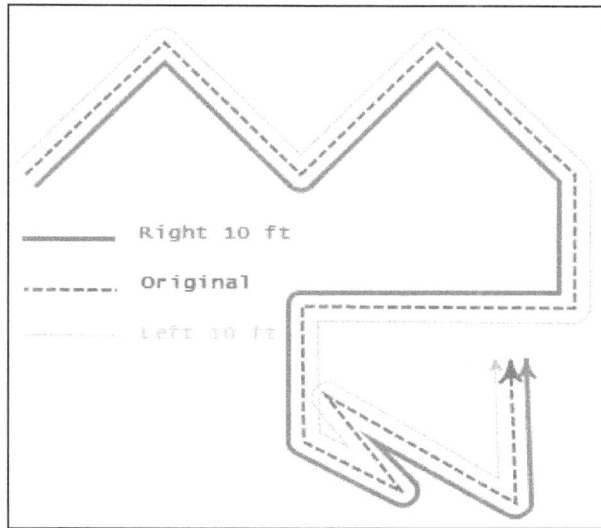

Now let's revisit the road in our San Francisco ROAD_CLINES table with FID = 898 used in the previous ST_LineShift example. Here is the SQL, except that parallel lines on both sides will be created which allows adding circular arcs for smoothing on obtuse angle bends:

```
Select a.fid,
       a.geom.ST_Parallel(-10,1,'unit=U.S. Foot').geom as geomLft,
       a.geom.ST_Parallel( 10,1,'unit=U.S. Foot').geom as geomRight
  From (Select fid,T_Geometry(rl.geom,0.005) as geom
          From road_clines rl
         Where rl.fid = 898) a;
```

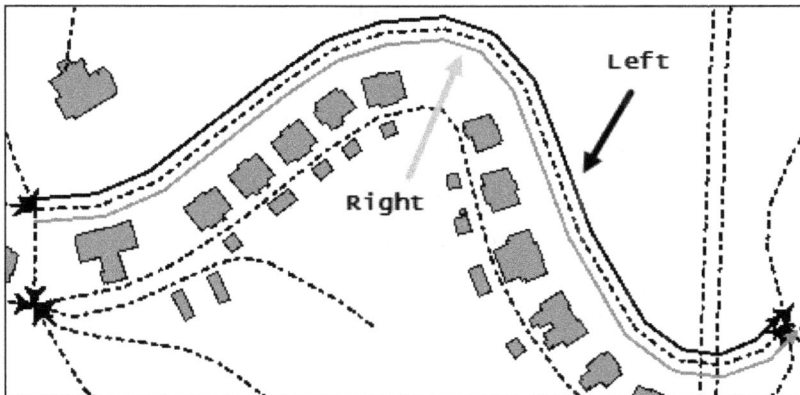

The operation of the `p_curved` parameter smoothing obtuse angles with circular arcs can be seen by looking at some of the vertices more closely:

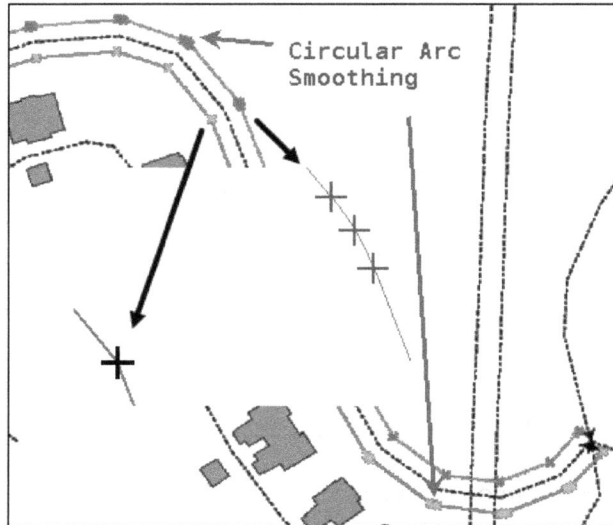

Buffering one side of a linestring – ST_ OneSidedBuffer

Two practical uses of `ST_Parallel` occur which are not difficult to create. The first is the ability to use the parallel line generated on one side of a line with the original line to create a square buffer on that side. This is done by:

- Taking the original linestring (or each linestring in a multiline string in turn) and creating a parallel line to the left or right

- If a parallel line is generated on the left, then reverse that line (`SDO_UTIL.Reverse_Linestring`) to ensure the correct vertex rotation of the polygon outer ring that is generated

- Append the reversed parallel line to the original line

- To close the polygon ring, add the first vertex of the original line on to the appended line

- Changing the `sdo_gtype` of the new geometry to that of a polygon

- If no smoothing has occurred, create SDO_ELEM_INFO_ARRAY containing 1,1003,1 only; otherwise, prefix the appended linestring's sdo_elem_info array that contains circular arcs with 1,1005, n, where n is the number of elements in the outer ring's description. The example SQL that generated the preceding figure follows this:

```
Select rl.fid,
       T_Geometry(rl.geom,0.005)
          .ST_OneSidedBuffer(p_distance=>15,
                             p_curved => 0,
                             p_unit   => 'unit=U.S. Foot')
          .ST_RoundOrdinates(2)
          .geom as rightSideBuff
  From road_clines rl
 Where rl.fid = 5039;
```

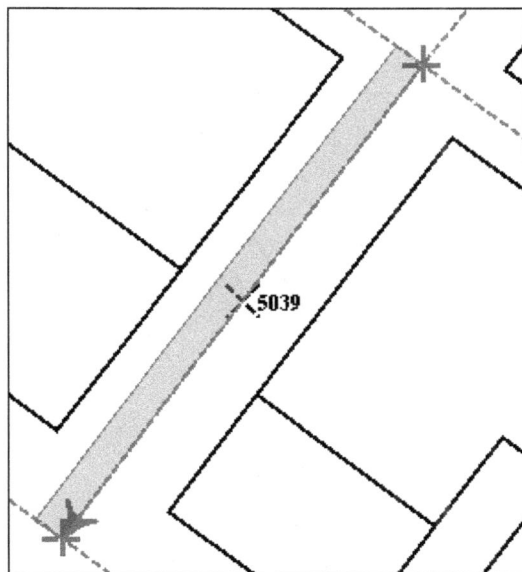

Generating a square buffer – ST_SquareBuffer

Creating a square buffered polygon on one side of a linestring is one of the uses for the ST_Parallel function; another is the creation of a single square buffer that covers both side of the linestring. Oracle's SDO_GEOM.Sdo_Buffer does this, but does not cut off the buffer at the end of the linestring, rather it creates a round end. The process is very similar to that described for ST_OneSidedBuffer so it won't be repeated. The SQL for generating the square buffer and the figure showing the result follows this:

```
Select  T_Geometry(rl.geom,0.005)
          .ST_SquareBuffer(p_distance=>15,
                           p_curved => 0,
                           p_unit   => 'unit=U.S. Foot')
          .ST_RoundOrdinates(2)
          .geom as sqBuff
   From road_clines rl
  Where rl.fid = 5039;
```

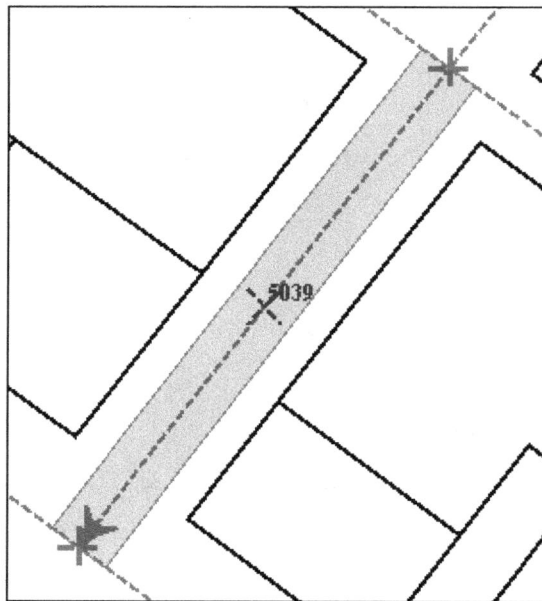

The following SQL compares the square buffer geometry with that generated by Oracle's SDO_BUFFER function:

```
Select SDO_GEOM.sdo_buffer(rl.geom,15,0.005,'unit=U.S. Foot') as geom
   From road_clines rl
  Where rl.fid = 5039;
```

The resulting figure visually compares the two outcomes:

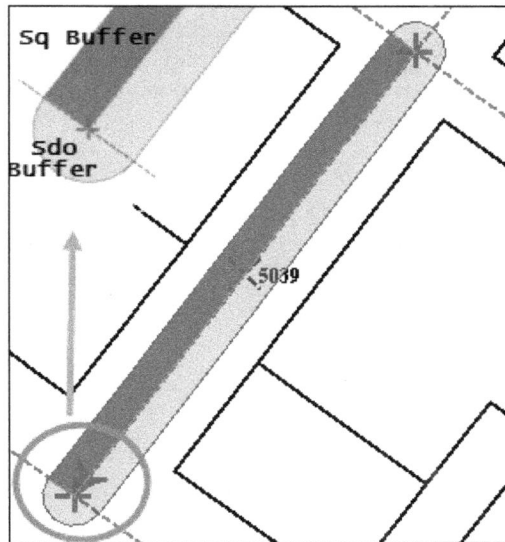

Tiling a vector geometry – ST_Tile

A common requirement for processing vector data is to tile (or tesselate) a vector object into a series of regular, non-overlapping squares or rectangles so that the entire object is covered. A function to do this is described as follows:

```
Create Or Replace T_Geometry As Object(
[...]
   Member Function ST_Tile(p_Tile_X    In Number,
                           p_Tile_Y    In Number,
                           p_grid_type In varchar2 Default 'TILE',
                           p_touch     In Integer default 0,
                           p_clip      In Integer default 0)
            Return T_Grids Pipelined,
```

This function requires the width of the tile in X and Y dimensions to be provided. They do not have to be the same. They, however, have to be expressed in the default units of the geometry's SRID. In addition, the p_grid_type parameter allows the user to request TILE that polygon tiles (expressed as optimized rectangles) to be returned, or POINT that only the centroid of the tile be returned, or BOTH, a polygon tile with its sdo_point structure filled in with the centroid of the tile be returned.

The p_touch parameter allows the user to request all the tiles that cover the **Minimum Bounding Rectangle (MBR)** of the input geometry be returned, including those, which fall outside the actual geometry, or only those tiles that touch (ANYINTERACT) the source geometry to be returned. Finally, the p_clip parameter allows the user to request any tile polygon that crosses the boundary of the object be clipped, so that only the part, which falls within the object is returned (this is for tiling polygon geometries only).

The member function returns a collection (T_Grids) of T_GRID objects, one per tile, via a pipeline. The required types for this function are as follows:

```
Create Or Replace Type T_Grid As Object (
    gcol   Number,
    grow   Number,
    geom   SDO_GEOMETRY
);
/
show errors
Create Or Replace Type T_Grids Is Table of T_Grid;
/
show errors
```

Applications of tiling

The applications of tiling area are actually greater than one might think. The following spring to mind:

- Creation of grid cells for an archeological site.

- Definition of patches across a road pavement for the purpose of measuring faults.

- Tiling of large polygons constructed from a large-scale (high precision) data (for example, the coast line of a legal area and boundary of vegetation data) can aid in query (for example, identification and assignment of vegetation type to a fauna observation via a before insert trigger), display, and processing.

- Determining optimum tile sizes when tiling very large raster images covering a large area.

- Use of tiles to support the extrapolation of treatments calculated from soil samples data across a field in a farm.

This section will discuss the first two applications.

Case 1 – creating grid cells over an archaeological site

It is common practice on archeological digs, to lay out a grid of squares to act as baselines that facilitate the recording of the finds. Let's assume that a building has been demolished on the land parcel with a FID of 88774.

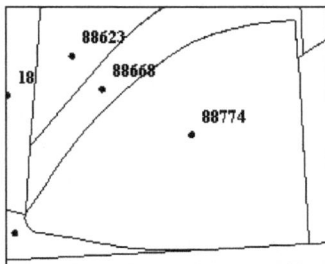

Before any new construction is allowed, an archeological survey needs to occur. To facilitate the survey a 15ft x 15ft square grid needs to be created over the land parcel.

Where a generate cell overlaps the boundary of the land parcel, it must be clipped by the boundary. The tiling and clipping is done by executing the following SQL:

```
Select t.geom as geom
   From land_parcels LP,
        Table(T_Geometry(lp.geom,0.005)
              .ST_Tile(p_Tile_X => 15,
                       p_Tile_Y => 15,
                       p_grid_type => 'TILE',
                       p_option    => 'CLIP',
                       p_unit => 'unit=U.S. Foot'
                      )
             ) t
   Where lp.fid = 88774;
```

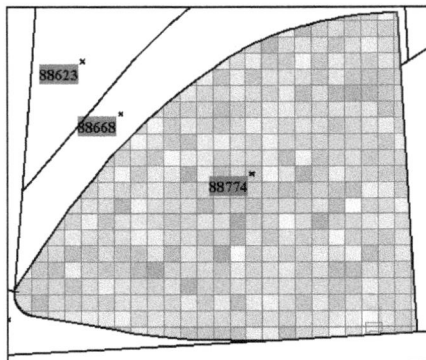

Case 2 – creating sampling patches over a road pavement

Another common use for tiling an object is for the place, where sample grid cells or patches are required to be created over a road surface. The road centerline that will be used is FID 5039.

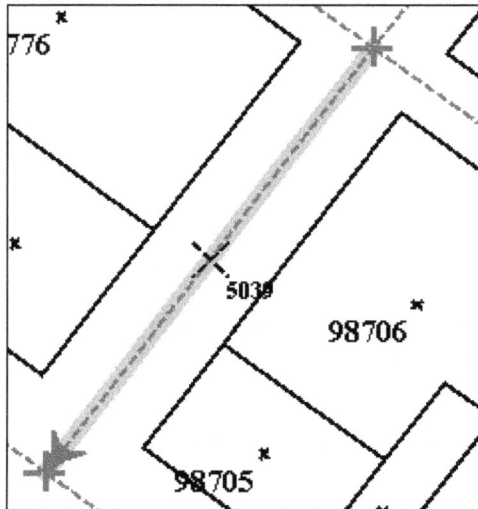

Firstly, let's grid the road centerline and return only those cells that touch the centerline. The road surface is 30ft, which is 15ft either side of the road centerline.

```
Select T_Geometry(t.geom,0.005).ST_RoundOrdinates(2).geom as geom
  From road_clines rc,
       Table(T_Geometry(rc.geom,0.005)
             .ST_Tile(p_Tile_X    => 30,
                      p_Tile_Y    => 30,
                      p_grid_type => 'TILE',
                      p_option    => 'TOUCH',
                      p_unit      => 'unit=U.S. Foot'
                     )
            ) t
  Where rc.fid = 5039;
```

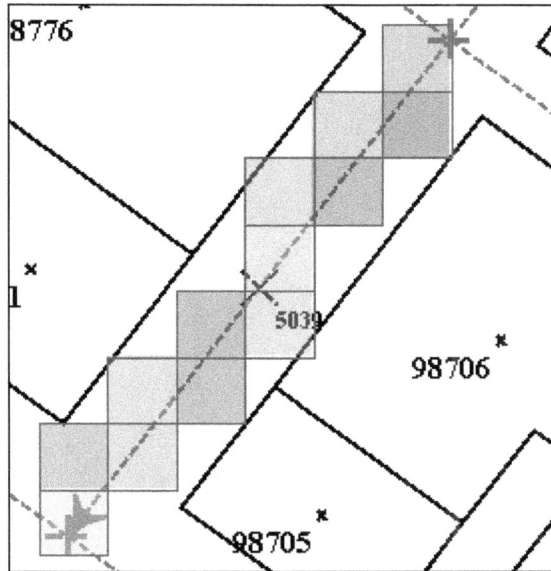

Note that the tiling is not oriented along the direction of the road centerline. This can be done via the use of our affine transformations by:

- Translating the line to 0,0
- Rotate it so that it is horizontal (this uses a static method, **Radians**, and the ST_Bearing method on the T_Vertex type)
- Tile the horizontal line
- Convert optimized rectangle tiles to 5-vertex polygons
- Shift the 5-vertex tile cell so it is halfway over the centerline
- Rotate the tiles back to the original line angle
- Translate the tiles back to the original location

The SQL for doing this is as follows:

```
With testLine As (
Select T_Vertex.Radians(90) -
        T_Vertex(b.endV)
          .ST_Bearing(T_Vertex(b.startV),b.geom.ST_Srid())
          as rotation,
        b.endV,
        b.geom
   From (Select a.geom.ST_StartVertex() as startV,
                a.geom.ST_EndVertex()   as endV,
                a.geom
          From (Select T_Geometry(rc.geom,0.005) as geom
                 From road_clines rc
                 Where rc.fid = 5039
                ) a
       ) b
)
Select T_Geometry(b.geom,a.geom.tolerance)
        .ST_Rectangle2Polygon()  /* Convert rectangle to polygon */
        .ST_Translate(0,-15)     /* Shift so half-way over line */
        .ST_Rotate(a.rotation) /* Rotate back to orig line angle */
        .ST_Translate(a.endV.x,
                      a.endV.y)  /* Move tile back to line start */
        .ST_RoundOrdinates(2)
        .geom as rgeom
   From testLine a,
        Table(a.geom
               .ST_Translate(0-a.endV.x,0-a.endV.y) /*Move to 0,0*/
               .ST_Rotate(0-rotation)    /* Rotate to horizontal */
               .ST_Tile(p_Tile_X    => 30,
                        p_Tile_Y    => 30,
                        p_grid_type => 'TILE',
                        p_option    => 'TOUCH',
                        p_unit      => 'unit=U.S. Foot'
               )                           /* Tile the horizontal line */
             ) b;
```

The result can be seen in the following figure:

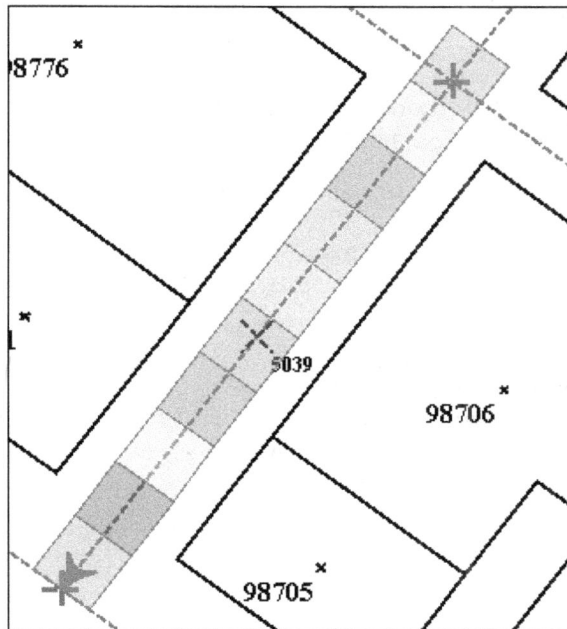

Removing steps in raster to vector data – ST_SmoothTile

A related concept to the tiling of a vector object is the generation of the vector object from raster data. It is not uncommon to see the vector data loaded into Oracle that obviously came from processed satellite observations. Satellite data is grid or raster data, that is, a value is observed, and assigned based on a single observational rectangular pixel. GIS processing often sees new pixel values or whole raster datasets being derived by applying filtering, resampling, aggregation, and other algorithms. Once the processing has occurred, raster data is sometimes converted to vector for use with other vector data.

For example, a new vegetation map could be created from satellite observations with the final result being exported, as a collection of vector polygons and loaded into Oracle. A dataset of beehive locations may be required by a business, in order to monitor the types of honey being produced to look for correlations based on vegetation. As the hive's location is changed, the vegetation type must be calculated automatically. A suitably constructed trigger on the beehive table could derive and assign the associated vegetation type via a simple point in a polygon operation between the point representing the hive and the polygon representing the vegetation data.

Raster to vector processing should include some measure of smoothing, and therefore coordinate a reduction of the representational polygons. This can aid in the reduction of the amount of storage needed for each vegetation polygon. But this does not always happen.

The following example is based on processing an arbitrary circle via a new member function called ST_SmoothTile:

```
Create Or Replace Type T_Geometry As Object (
[...]
  Member Function ST_SmoothTile
    Return T_Geometry Deterministic,
```

The circle that is "observed" is as follows:

```
Select SDO_GEOMETRY(2003,NULL,NULL,
          SDO_ELEM_INFO_ARRAY(1,1003,4),
          SDO_ORDINATE_ARRAY (4,50, 50,5, 95,50)) as Circle
   From dual;
```

A set of observational tiles are created that represent the satellite pixels:

```
Select t.geom as tile
  From Table(T_Geometry(
            SDO_GEOMETRY(2003,NULL,NULL,
                        SDO_ELEM_INFO_ARRAY(1,1003,4),
                        SDO_ORDINATE_ARRAY(4,50,50,5,95,50)),
            0.005
          ).ST_Tile(p_Tile_X     =>  5,
                    p_Tile_Y     =>  5,
                    p_grid_type  => 'TILE',
                    p_option     => 'TOUCH',
```

```
            p_unit        => 'unit=U.S. Foot'
    ) ) t;
```

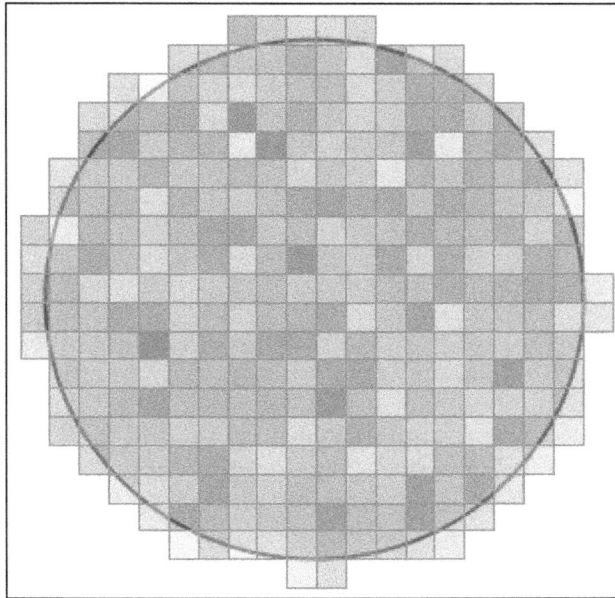

The GIS processing of the satellite data results in a single polygon per vegetation type. The following SQL creates a possible resultant polygon. Note the residual vertexes left in the vegetation polygon:

```
With tiles as (
Select t.geom as tile
  From Table(T_Geometry(
            SDO_GEOMETRY(2003,NULL,NULL,
                      SDO_ELEM_INFO_ARRAY(1,1003,4),
                      SDO_ORDINATE_ARRAY(4,50,50,5,95,50)),
            0.005
            ).ST_Tile(p_Tile_X    =>  5,
                    p_Tile_Y     =>  5,
                    p_grid_type => 'TILE',
                    p_option     => 'TOUCH',
                    p_unit       => 'unit=U.S. Foot'
            ) ) t
)
```

```
Select sdo_aggr_union(sdoaggrtype(a.tile,0.005)) as tile
  From tiles a;
```

It is not uncommon to see such data in databases in which the data perhaps should have been further smoothed or processed before the storage is given. But if it has not; is it possible to do something about it in this situation? The following SQL demonstrates, if this is the case.

Firstly, a new member function is created called ST_SmoothTile:

```
Create Or Replace Type T_Geometry As Object (
[...]
  Member Function ST_SmoothTile
  Return T_Geometry Deterministic,
[...]
```

Then, this function is applied as follows:

```
With tiles as (
Select t.geom as tile
  From Table(T_Geometry(
              SDO_GEOMETRY(2003,NULL,NULL,
                SDO_ELEM_INFO_ARRAY(1,1003,4),
                SDO_ORDINATE_ARRAY(4,50, 50,5, 95,50)),
              0.005
            ).ST_Tile(p_Tile_X    =>  5,
                      p_Tile_Y    =>  5,
                      p_grid_type => 'TILE',
                      p_option    => 'TOUCH',
                      p_unit      => 'unit=U.S. Foot'
            ) ) t
)
Select T_Geometry(sdo_aggr_union(sdoaggrtype(a.tile,0.005)),
                  0.005).ST_SmoothTile().geom as tile
  From tiles a;
```

> For Locator users, see ST_AggrUnionPolygons in *Chapter 10, Integrating Java Technologies with Oracle Spatial.*

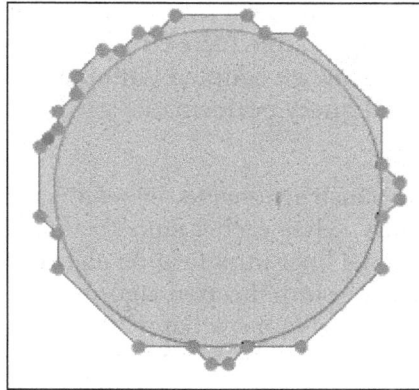

Finally, the savings in the number of vertexes can be calculated as follows:

```
With tiles as (
Select t.geom as tile
  From Table(T_Geometry(
              SDO_GEOMETRY(2003,NULL,NULL,
                SDO_ELEM_INFO_ARRAY(1,1003,4),
                SDO_ORDINATE_ARRAY(4,50, 50,5, 95,50)),
              0.005
              ).ST_Tile(p_Tile_X   => 5,
                        p_Tile_Y   => 5,
                        p_grid_type => 'TILE',
                        p_option   => 'TOUCH',
                        p_unit     => 'unit=U.S. Foot'
              ) ) t
)
Select smoothCount, originalCount,
       Round(((originalCount-smoothCount)/originalCount)*100,0)
         as saving
  From (Select b.geom.ST_SmoothTile()
                 .ST_NumVertices() as smoothCount,
               b.geom.ST_NumVertices() as originalCount
          From (Select T_Geometry(sdo_aggr_union(
                                    sdoaggrtype(a.tile,0.005)),
                                  0.005) as geom
                  From tiles a) b
        ) c;

-- Result
--
```

```
SMOOTHCOUNT ORIGINALCOUNT SAVING
----------- ------------- ------
         30            81     63
         30            81     63
```

As seen from the preceding code, the total saving in the storage by application of some in-database tile smoothing, resulted in a 63 percent saving in the number of vertexes. These sorts of savings, when coupled with the ordinate rounding, can improve storage efficiency and query performance as shown in *Chapter 2, Importing and Exporting Spatial Data*.

Before leaving this section, it is instructional to see what happens to the two-gridded polygons that share a boundary, when each is smoothed by this method and then compare the output to traditional line simplification algorithms (for example, SDO_UTIL.SIMPLIFY). Traditional line simplification algorithms tend to "pull apart" the boundaries shared between two polygons: what will happen with the ST_SmoothTile algorithm, given that it is designed to take advantage of the stepped nature of gridded polygons? ST_SmoothTile cannot be used as a generalized line simplifier.

The following SQL starts by tiling the circle as has already been done. It then compares each tile against an ad-hoc polygon generating a classification attribute of 0, if disjoint or 1, if not. The tiles are then aggregated by the classification attribute to create two polygons, each of which shares a common boundary. The two polygons are then simplified via application of SDO_UTIL.Simplify and ST_SmoothTile. Simplification using SIMPLIFY is only done with a value of 5. The reader is welcome to experiment with different values:

```
With tiles as (
Select case when SDO_GEOM.Relate(
             t.geom,
             'DETERMINE',
             SDO_GEOMETRY(2003,NULL,NULL,
                 SDO_ELEM_INFO_ARRAY(1,1003,1),
                 SDO_ORDINATE_ARRAY(
                     1.19,7.47, 34.5,27.58, 53.29,49.02,
                     49.01,65.5, 50.0,76.06, 61.87,78.7,
                     66.48,86.28, 71.76,85.29, 79.67,79.03,
                     99.13,84.3, 119.91,36.82, 85.28,-3.75,
                     6.47,0.87, 1.19,7.47)),
                 0.005) = 'DISJOINT'
        then 0 Else 1 end as inout,
      t.geom as tile
```

```
    From Table(T_Geometry(
                SDO_GEOMETRY(2003,NULL,NULL,
                        SDO_ELEM_INFO_ARRAY(1,1003,4),
                        SDO_ORDINATE_ARRAY(4,50,50,5,95,50)),
                0.005
                ).ST_Tile(p_Tile_X     => 5,
                        p_Tile_Y     => 5,
                        p_grid_type => 'TILE',
                        p_option    => 'TOUCH',
                        p_unit      => 'unit=U.S. Foot'
                ) ) t
    )
    Select SDO_UTIL.Simplify(b.poly,5.0,0.005) as uPoly,
           T_Geometry(b.poly,0.005)
                    .ST_SmoothTile()
                    .ST_RoundOrdinates(2)
                    .geom
                as sPoly
      From (Select sdo_aggr_union(sdoaggrtype(a.tile,0.005)) as poly
             From tiles a
            Group By inout
            ) b;
```

The results of the SIMPLIFY and ST_SmoothTile algorithms are shown in the following figures:

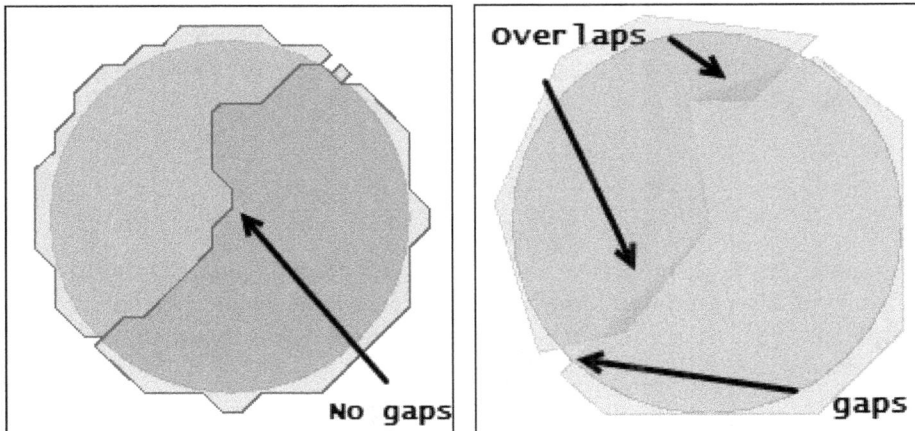

[309]

Adjusting coordinate dimensions

Some software products can create 3D SDO_GEOMETRY objects where the Z ordinate is NULL or 0. To use some of the Oracle Spatial processing functionality, the Z ordinate must be removed. While basic functions to do this are covered in the book *Pro Oracle Spatial*, variations are often needed. The following are introduced here:

```
Create Or Replace Type Body T_Geometry
[...]
  /* Dimensional Adjustment */
  Member Function ST_To2D Return T_Geometry Deterministic,
  Member Function ST_To3D(p_zordtokeep In Integer)
                          Return T_Geometry Deterministic,
  Member Function ST_To3D(p_start_z    In Number,
                          p_end_z      In Number,
                          p_unit In varchar2 default null)
                          Return T_Geometry Deterministic,
  Member Function ST_FixZ(p_default_z  In Number := -9999 )
                          Return T_Geometry Deterministic,
```

Reducing 3D to 2D – ST_To2D

One very tangible example of reducing a 3D SDO_GEOMETRY object to 2D is in the case where a CAD product has defined a table's SDO_GEOMETRY column to contain Z values, but none of them actually exist. While this may be appropriate for some data (for example, pipelines), it may not be present for traditional 2D data like vegetation polygons. This type of storage is quite a common situation where older CAD products are still being used to edit data stored in Oracle. An example of such type of geometry is as follows:

```
SDO_GEOMETRY(3002,NULL,NULL,
    SDO_ELEM_INFO_ARRAY(1,2,1),
    SDO_ORDINATE_ARRAY(6,10,0,10,1,0,14,10,0,10,14,0,6,10,0))
```

The existence of the additional (unused) Z ordinate increases storage space and can inhibit some 2D-based database-based geoprocessing, especially before 11*g*; for example, SDO_UTIL.Concat_Lines cannot process 3D linestrings, only 2D linestrings.

The correction of such a situation may be a natural part of a software upgrade of the CAD package. However, the ability to modify the data within the database independently of client software is an option that data administrators should have available to them. The previous 3D CAD linestring can be reduced to a 2D representation as follows:

```
Select T_Geometry(
         SDO_GEOMETRY(3002,NULL,NULL,
          SDO_ELEM_INFO_ARRAY(1,2,1),
          SDO_ORDINATE_ARRAY(6,10,0,10,1,0,14,10,0,10,14,0,6,10,0))
         ).ST_To2D()
         .geom as geom
   From dual a;

-- Results
--
GEOM
-----------------------------------------------------------
SDO_GEOMETRY(2002,NULL,NULL,SDO_ELEM_INFO_ARRAY(1,2,1),
            SDO_ORDINATE_ARRAY(6,10,10,1,14,10,10,14,6,10))
```

Fixing Z ordinate problems – ST_ FixZ

A particular organization stored hydrographic sounding data as a single 3D linestring in Oracle. In certain situations, the continuous set of observations did not include a depth, even though the longitude and latitude of the sounding was known. Quite rightly, they decided to store a NULL for the Z ordinate, where one was not known.

Little did they know that the GIS product they were using could not read such a linestring. Any query that included such a type of geometry was terminated. The software product's actions could not be modified; an alternate solution was required.

Finally, the solution decided, was to create a view over the hydrographic soundings, but to use a functional geometry, in which all NULL Z ordinates were replaced with a "marker" value of -9999. (A function-based index was then created over the geometry produced by the function.) This function used in the view that generates the correct geometry is called ST_FixZ, which can be used as follows:

```
Select T_Geometry(
        SDO_GEOMETRY(3002,8265,NULL,
            SDO_ELEM_INFO_ARRAY(1,2,1),
            SDO_ORDINATE_ARRAY(-79.230383,35.836761,-1292.0,
                                -79.230381,35.836795,-2347.7,
                                -79.23078, 35.837087,NULL,
                                -79.230765,35.837038,-3045.0)),
        0.05).ST_FixZ(-9999).geom as geom
  From dual a;

-- Results
--
GEOM
-------------------------------------------------------------------
SDO_GEOMETRY(3002,8265,NULL,
            SDO_ELEM_INFO_ARRAY(1,2,1),
            SDO_ORDINATE_ARRAY(-79.230383,35.836761,-1292.0,
                                -79.230381,35.836795,-2347.7,
                                -79.23078, 35.837087,-999.0,
                                -79.230765,35.837038,-3045.0))
```

Summary

Desktop GIS products provide an amazing range of editing functionality for geometric data, further creating a rich environment for the operator. While nothing in this chapter says that such functionality is not important or unnecessary, the processing of geometric data inside an Oracle database in response to clearly described and defined business processes is an important deployment option that should be available to all architects and developers. To do such deployment requires a rich set of functionality. This chapter extended the functions built in *Chapter 6, Implementing New Functions*, by creating functions for: editing vertices; transforming, scaling, rotating, and reflecting geometries; tiling, splitting, extending, shortening, moving sideways, and square buffering. Finally, the functions for converting from 2D to 3D, with examples, completed the chapter's presentation of what is possible for geometry editing, transformation, and construction.

The next chapter will continue to build on the functions created in *Chapter 6, Implementing New Functions* and in this chapter, by creating a complete set of functions for processing linear data.

8
Using and Imitating Linear Referencing Functions

The goal of this chapter is to use the functions created in the previous chapters to build new functions that can be used to solve business problems relating to managing linear assets. The main business problems that need this functionality are road, cycle way or track management, geocoding street addresses, survey, inventory, condition assessment, and water management applications.

Oracle Spatial has a robust linear referencing package (SDO_LRS) that can be applied to all these problems. The SDO_LRS package can only be licensed for use by purchasing the Spatial package and deploying it within an Oracle Enterprise database. In addition, SDO_LRS cannot be purchased separately from the whole Spatial package. Finally SDO_LRS cannot be used with Locator.

The functions created in this chapter will provide SDO_LRS functionality where licensing of the SDO_LRS package is beyond the user's resources. These functions will support simple linear processing against measured and non-measured geometries (a normal 2D linestring is a non-measured geometry). The examples used in this chapter will include real-world situations to demonstrate the power of developing and using these Types of functions. The SQL statements that demonstrate the use of the functions developed in this chapter are available in an SDO_LRS equivalent form in the SQL scripts shipped with this book for this chapter; they will not be included in the actual chapter.

This chapter will cover the following uses of linear processing:

- Snapping a point to line (refer the previous chapter for: ST_Snap)
- Splitting a line using a known point that is on or off the line (refer the previous chapter for: ST_Split)

- Adding, modifying, and removing measures to and from a linestring
- Finding linear centroids
- Creating a point at a known distance along, and possibly offset from, a line
- Extracting segments of linear geometries
- Linear analysis of point data

Understanding linear referencing and measures

All geographic representations stored in the `SDO_GEOMETRY` object are just that: representations. The linestring describing a pipe is not the pipe; the linestring describing a road in a GPS in-car navigation unit is not the road: they are scale-dependent representations of the middle of the pipe (an imaginary line drawn down the center of the pipe) or the middle of a marked lane for traffic on a road.

By scale-dependent is meant that the position of each vertex in the linestring has a locational "vagueness". The amount of location vagueness depends on the source of its measurement. A position measured by a surveyor using a modern electronic theodolite may be accurate to +/- 5mm; a GPS observation +/- 10cm; an old 1:1000 map from a roads department, +/- 2m.

Regardless as to capture method, each linestring's representation has an implicit length (measured by `SDO_GEOM.Sdo_Length` or the `T_GEOMETRY.ST_Length` method) property, but this length is also scale or measurement dependent: it is not the real-world length.

For linear asset management, the gap between representation and reality is bridged by not relying on the graphic length of the representation in the computer but by recording the position of an asset described, measured or positioned relative to a road, from a known point in the real world. Imagine a mark at the surveyed start of the road which is given a value of 0.0. Then, the surveyor places another mark at the end of the road along with a variable number of marks in the surface of the road at discernible changes of direction or grade. Each of these marks is assigned a surveyed distance (slope or planar) from the first mark at the origin of the road. This in-field measurement provides a reference system that allows a controlled world-to-computer representational system.

This is what is called a **Linear Referencing System** (LRS).

The following image shows a graphical representation of Sherman Road in San Francisco which has an in-database length of 298.36 meters, yet the imaginary surveyed marks in the actual road surface show a different length of 299.45 meters.

From maps.yahoo.com

While graphic length is an inherent property of a linestring, the surveyed reference measurement is not. Oracle allows this to be stored in its SDO_GEOMETRY object in either the Z or M ordinate. If a linestring has both X, Y, and Z (elevation as quasi-3D) ordinate Values, a measure must be stored in the fourth ordinate position; if it has only X and Y ordinates (2D) then the measure must be stored in the third ordinate position.

Sherman Road can be represented in an Oracle database as follows:

```
SDO_GEOMETRY(3302,2872,NULL,SDO_ELEM_INFO_ARRAY(1,2,1),
SDO_ORDINATE_ARRAY(5998875.284,2118417.957,0.0,
5998794.455,2118412.144,18.61, 5998733.653,2118410.486,38.14,
5998669.879,2118413.469,59.75, 5998599.64,2118421.085,86.99,
5998512.834,2118440.954,122.2, 5998401.496,2118470.252,166.51,
5998259.374,2118498.243,204.07, 5998137.78,2118515.132,225.59,
5998067.653,2118520.68,238.77, 5998025.462,2118529.372,250.96,
5997989.68,2118546.923,261.28, 5997965.493,2118570.435,274.66,
5997930.423,2118596.542,299.45))
```

While nowadays **GPS** is the field data capture technology of choice, this was not always the case. In addition, for the management of linear assets, GPS data may need to be supplemented with other representational methods, where analysis is concerned. For example, assume that San Francisco City Council passed a new by-law that requires that a public trash can must be placed, on average, every 200 feet along its streets; or, that street lights must be placed at a minimum, every 150 feet; how can such a measure be calculated? With a linear referencing system, plus suitable functions, GPS observed trash can or light points can be turned into linear measures that can then be analyzed.

Even if no surveyed measurement system exists, linear processing can be used to solve problems by assuming that the implicit length of a linestring is sufficient as a measure.

This chapter will show how to develop and use linear measurement technology to answer a range of problems such as developing a measure of trash cans per foot!

Linear referencing functions to be developed

A number of linear referencing functions will be developed in this chapter. These functions will draw upon or extend functions presented in the previous chapters, most notably ST_Split and ST_Snap. A recap of what these do follows:

1. ST_Split: Splits a linestring at one or more known points placed on or off the line.

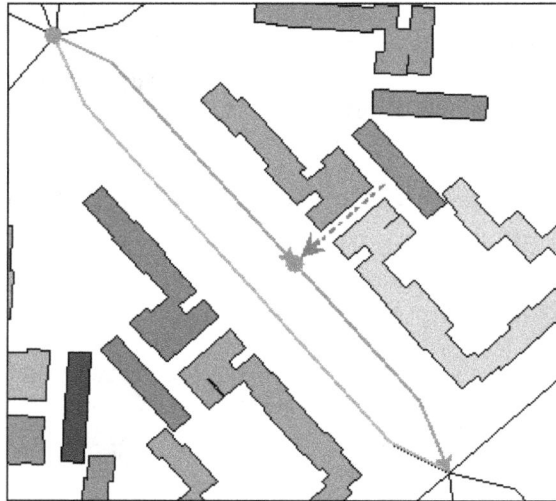

2. ST_Snap: Snaps (projects) a known point onto a linestring to determine its measure (or length) from beginning.

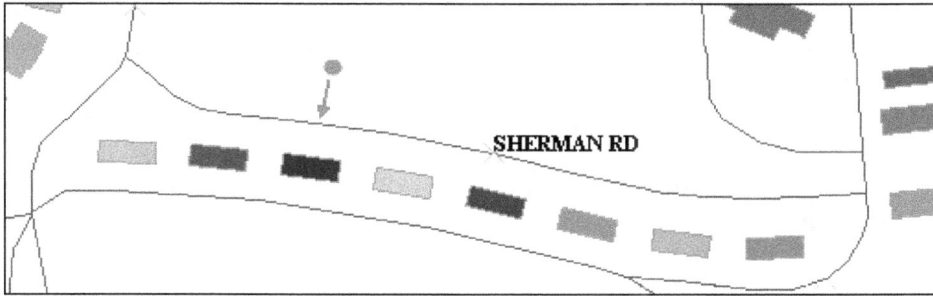

The following list outlines those linear referencing functions that will be implemented:

3. ST_Find_Measure: Given a point near a measured linestring, return the nearest measure.

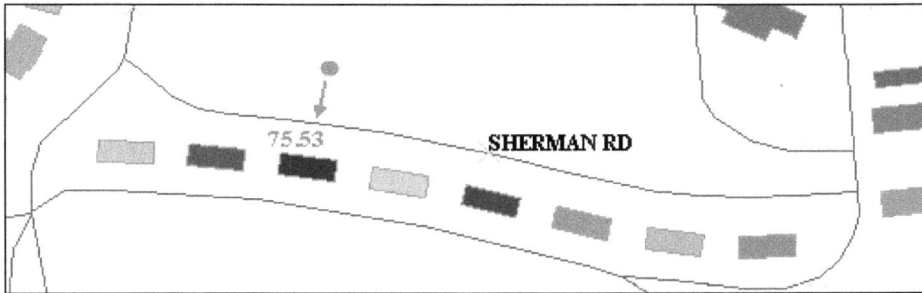

4. ST_Scale_Measures: Rescales the measures of a linestring and optionally applies an offset. Thus an existing measured linestring whose measure starts at 0 m and finishes at 100 m may be changed to start at 5 m (offset) and end at 106.5 m (rescale).

5. ST_Add_Measure: Takes an existing unmeasured linestring and a start and end measure, and returns a linestring where all vertices have a measure computed by linear interpolation between the start and end points.

6. ST_Start_Measure: Examines a linestring and returns measure of first vertex.

7. ST_End_Measure: Examines a linestring and returns measure of last vertex.

8. ST_Measure_Range: Examines a linestring and returns the difference between the last and first vertex measures. For non-measured linestrings this will be the length of the linestring.

9. `ST_Is_Measure_Increasing`: If a measured linestring's measure Values increase over the length of the linestring with the first being less than the second etc up to the last, then this function will return true.

10. `ST_Is_Measure_Decreasing`: If a measured linestring's measure Values decrease over the over the length of the linestring, with the first being greater than the second, and so on, up to the last, then this function will return true. This can occur especially if the linestring's direction is reversed from what is expected.

11. `ST_Measure_To_Percentage`: A function that converts a supplied measure value to an equivalent percentage. For example, if the linestring has measures that start at 0 m and end at 100 m, the a supplied measure of 50 m will return a percentage of 50.

12. `ST_Percentage_To_Measure`: A function that converts a supplied percentage value to its equivalent measure. For example, if the linestring has measures that start at 0 m and end at 100 m , then a supplied percentage of 50% will return a measure of 50 m.

13. `ST_Project_Point`: Find closest point on a linestring given any point. Is synonym for `ST_Snap`.

14. `ST_Find_Offset`: Given a point near a linestring, return the smallest (perpendicular) offset distance from the supplied point to the nearest point on the linestring.

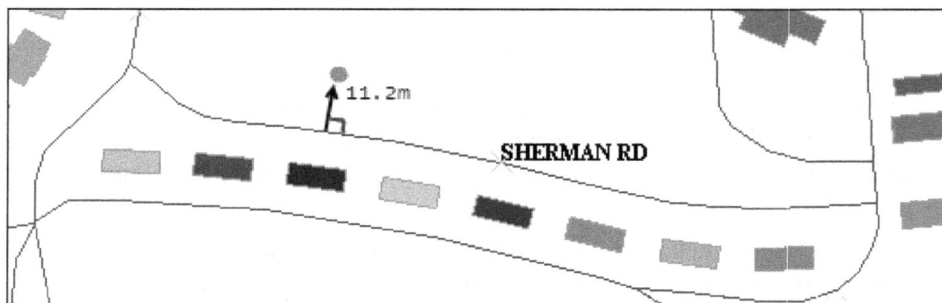

15. `ST_Locate_Measure` and `ST_Locate_Point`: Both take a measure and optional offset distance and create a new point at the referenced location. Where absolute measure Values are not known, a percentage value, expressed as a fraction between 0 and 1, can be used instead of the measure.

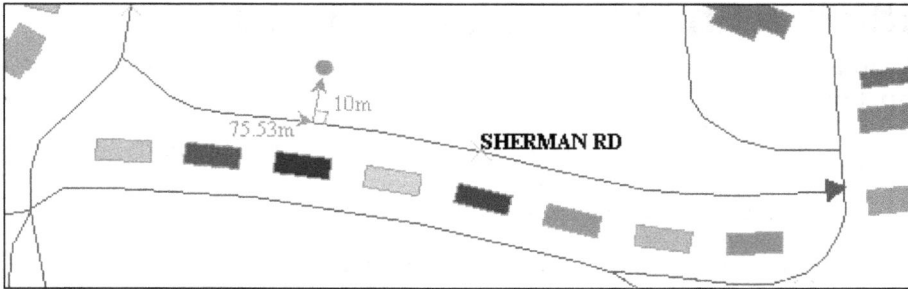

16. ST_Locate_Measures (no offset): Take a pair of measures and create a new linestring at the referenced location. Where absolute measure Values are not known, a percentage value, expressed as a fraction between 0 and 1, can be used.

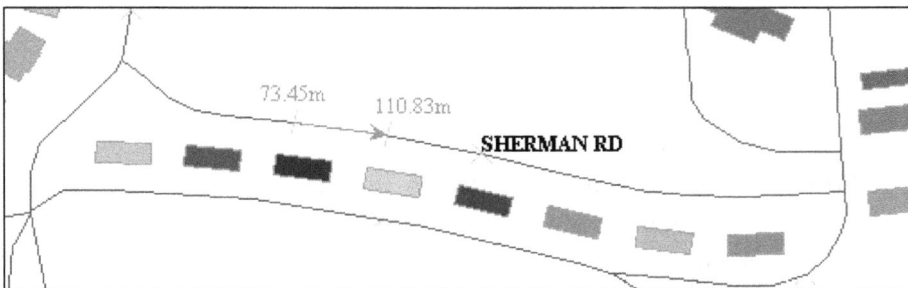

17. ST_Locate_Measures (with offset): Take a pair of measures and an offset distance and create a new linestring at the referenced location. Where absolute measure Values are not known, a percentage value, expressed as a fraction between 0 and 1, can be used.

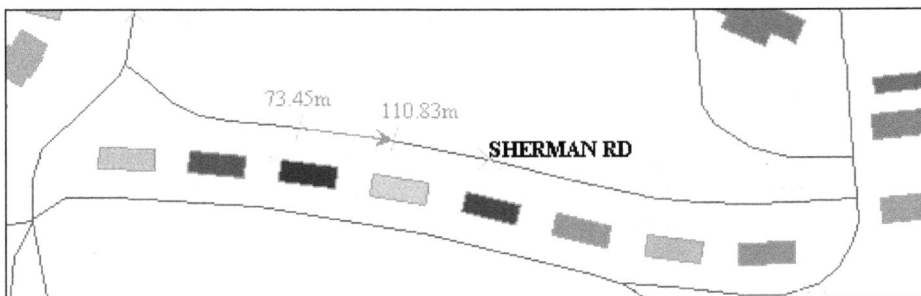

This chapter will not include the implementation details of the functions it presents. All source code is available with the book.

Splitting a line at a point – ST_Split

The previous chapter introduced the `ST_Split` method for our `T_GEOMETRY` object. To recap, this function looks like this:

```
Member Function ST_Split(p_vertex in MDSYSMDSYS.VERTEX_Type,
                         p_unit   in Varchar2 default null)
       Return T_Geometries pipelined
```

However, this member function had two limitations:

1. It only split a linestring at the first vector (a 2-vertex directed segment or a 3-vertex directed circular arc) it found. It is possible for a point to lie closest to more than one segment of a linestring as this image shows.

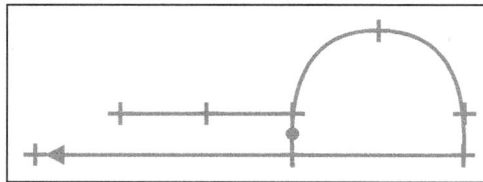

2. It did not handle compound linestrings, that is, a linestring with three point circular arcs as well as straight line connected segments.

For the applications required by this chapter, the member function has been modified to implement both these situations.

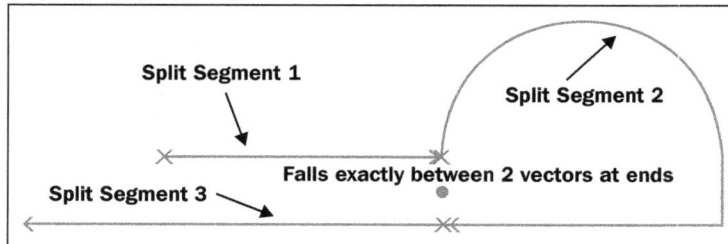

Previously, in situation 1, splitting at the circular point exactly midway between two vertices of different vectors sees the new `ST_Split` return three geometries while the old one would have returned 2.

```
Select 'Falls exactly between 2 vectors at ends' as test,
       g.geometry as splitLineSegment,
       a.point    as originalPoint
  From (Select SDO_GEOMETRY('POINT(3.0 -0.25)',null) as point,
               T_Geometry(
                 SDO_GEOMETRY(2002,null,null,
```

```
                    SDO_ELEM_INFO_ARRAY(1,4,3,1,2,1,5,2,2,9,2,1),
                    SDO_ORDINATE_ARRAY(1,0,2,0,3,0,4,1,5,0,
                                5,-0.5,3.0,-0.5,0.0,-0.5)),
               0.005) as geom
      From dual
     ) a,
     Table(a.geom.ST_Split(
              book.T_Vertex(a.point.sdo_point),null)) g;
```

In situation 2, splitting at a point near the circular arc now splits it, forming two new circular arcs with the same radius as before.

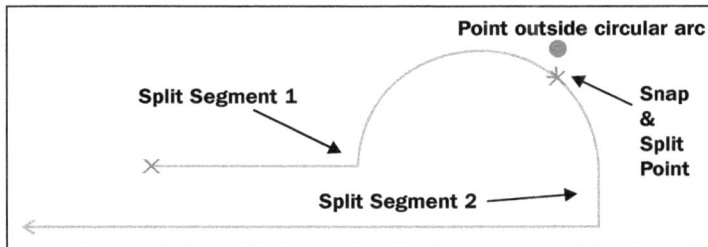

```
Select 'Point outside circular arc' as test,
       g.geometry as splitLineSegment,
       a.point     as originalPoint
  From (Select SDO_GEOMETRY('POINT(4.872 1.019)',null) as point,
              T_Geometry(
                SDO_GEOMETRY(2002,null,null,
                  SDO_ELEM_INFO_ARRAY(1,4,3,1,2,1,5,2,2,9,2,1),
                  SDO_ORDINATE_ARRAY(1,0,2,0,3,0,4,1,5,0,
                              5,-0.5,3.0,-0.5,0.0,-0.5)),
               0.005) as geom
      From dual ) a,
     Table(a.geom.ST_Split(
              book.T_Vertex(a.point.sdo_point),null)) g;
```

The SDO_LRS equivalent requires use of a PL/SQL procedure, Split_Geom_Segment. To use as a function would require the writing of a simple wrapper function. This procedure can only act on a measured linestring splitting at a known measure so use SDO_LRS's Convert_To_LRS_Geom and Split_Geom_Segment are required.

Snapping a point to a line – ST_Snap

ST_Snap takes a known point and snaps or moves it onto a linestring at its nearest point. Once a point has been snapped, it is then possible to determine its measure (or length) from the beginning of the line.

The signature for the `ST_Snap` member function is:

```
Member Function ST_Snap (p_point in mdsys.SDO_GEOMETRY,
                         p_unit  in Varchar2 default null)
      Return T_Geometries Pipelined,
```

This can be used as follows:

```
With geoms As (
Select 1 as id, 'At start of whole line' as test,
       SDO_GEOMETRY('POINT(1.0 0)',null) as point
  From Dual Union All
Select 2 as id, 'At end of whole line'  as test,
       SDO_GEOMETRY('POINT(1.75 -1.25)',null) as point
  From Dual Union All
Select 3 as id, 'Near end of whole line' as test,
       SDO_GEOMETRY('POINT(1.9 -1.5)',null) as point
  From Dual Union All
Select 4 as id, 'Falls exactly between 2 vectors at ends' as test,
       SDO_GEOMETRY('POINT(3.0 -0.25)',null) as point
  From Dual Union All
Select 5 as id,
       'Falls exactly between 2 vectors not at ends' as test,
       SDO_GEOMETRY('POINT(1.5 -0.25)',null) as point
  From Dual Union All
Select 6 as id, 'Falls exactly on existing vertex' as test,
       SDO_GEOMETRY('POINT(2 0)',null) as point
  From Dual Union All
Select 7 as id,
       'Falls exactly mid-angle inside two lines' as test,
       SDO_GEOMETRY('POINT(4.75 -0.25)',null)    as point
  From Dual Union All
Select 8 as id,
       'Falls exactly mid-angle outside two lines' as test,
       SDO_GEOMETRY('POINT(5.25 -0.75)',null) as point
  From Dual Union All
Select 9 as id, 'Falls off vector but in its middle' as test,
       SDO_GEOMETRY('POINT(2.47 0.21)',null) as point
  From Dual Union All
Select 10 as id, 'Point outside circular arc' as test,
       SDO_GEOMETRY('POINT(4.872 1.019)',null) as point
  From Dual Union All
Select 11 as id, 'Point outside mid-point circular arc' as test,
       SDO_GEOMETRY('POINT(4.0 1.2)',null) as point
  From Dual Union All
Select 12 as id, 'Point ON circular arc' as test,
       SDO_GEOMETRY('POINT(4.6798 .73339)',null) as point
```

```
      From Dual Union All
  Select 13 as id, 'Point inside circular arc' as test,
         SDO_GEOMETRY('POINT(3.528 0.562)',null) as point
    From dual
  )
  Select id, row_Number() over (order by 1) as rin, a.test,
         g.geometry as snappedPoint,
         a.point as originalPoint
    From geoms a,
         Table(T_Geometry(
               SDO_GEOMETRY(2002,null,null,
                   SDO_ELEM_INFO_ARRAY(1,4,3,1,2,1,5,2,2,9,2,1),
                   SDO_ORDINATE_ARRAY(1,0,2,0,3,0,4,1,5,0,5,
                                      -0.5,3.0,-0.5,0.0,-0.5)),0.005)
               .ST_Snap(a.point,null)) g;
```

Executing the query in SQL Developer and rendering the results using the
GeoRaptor extension produces the following image:

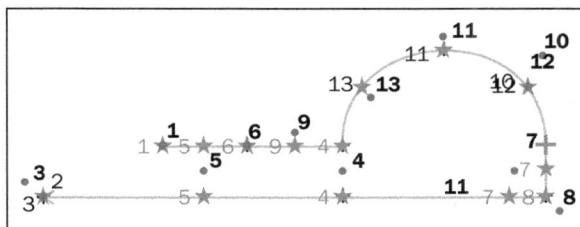

The circles are the original points and the stars are the resultant snapped points. Note
that original points 4, 5, and 7 snap to two different points on the original linestring.

> The same result can be achieved using two SDO_LRS functions.
> Firstly, a standard 2D geometry must be converted to a
> measured geometry using Convert_To_LRS_Geom followed by
> Project_Pt.

Finding the measure at a point – ST_Find_Measure

Find measure, in many ways, is a simplified version of ST_Snap. Simplified? ST_Snap
returns the input point snapped to the linestring. If the linestring is measured then
the snapped point will contain a suitable measure.

Since ST_Find_Measure only returns the measure of the snapped point, as long as ST_Snap is returning correctly snapped and measures points, ST_Find_Measure becomes a "wrapper" over ST_Snap.

ST_Find_Measure is declared as follows:

```
Member Function ST_Find_Measure (p_point in SDO_GEOMETRY,
                                 p_unit  in Varchar2 default null)
    Return MDSYS.SDO_ORDINATE_ARRAY Pipelined
```

This version of ST_Find_Measure is declared as returning a pipelined table because of the possibility that a point may fall perfectly between two vectors/segments of a linestring. This does make calling more convoluted.

> The equivalent SDO_LRS function is Find_Measure. This function is slightly simpler because it only returns the first of the two possible measure Values.

```
With geoms As (
  Select 4 as id,
         'Falls exactly between 2 vectors at ends' as test,
         SDO_GEOMETRY('POINT(3.0 -0.25)',null) as point
  From dual
)
Select id,
       row_Number() over (order by 1) as rin,
       a.test,
       Round(g.column_value,3) as measure
  From geoms a,
       Table(T_Geometry(
               SDO_GEOMETRY(3302,NULL,NULL,
                 SDO_ELEM_INFO_ARRAY(1,4,3, 1,2,1,
                                     7,2,2, 13,2,1),
                 SDO_ORDINATE_ARRAY(
                         1.0,0.0,0.0,      2.0,0.0,1.162,
                         3.0,0.0,2.324,    4.0,1.0,3.967,
                         5.0,0.0,5.61,     5.0,-0.5,6.191,
                         3.0,-0.5,8.514,   0.0,-0.5,12,
                         0.75,-2.0,13.677, 1.75,-1.25,14.927)),
                 0.005)
               .ST_Find_Measure(a.point,null)) g
  Order By 4;
```

```
-- Results
--
ID RIN TEST                                      MEASURE
-- --- ------------------------------------      -------
 4   1 Falls exactly between 2 vectors at ends   2.324
 4   2 Falls exactly between 2 vectors at ends   8.514
```

The result looks like this:

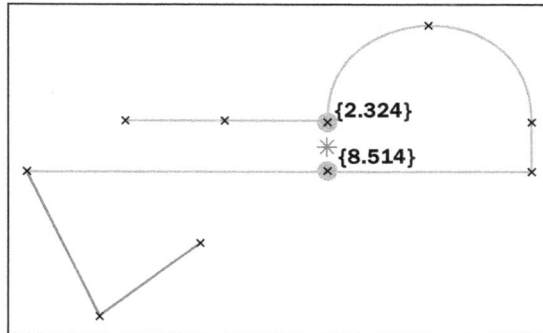

Calculating the offset from a line to a point – ST_Find_Offset

Before applying that which has been learned so far, another useful function (ST_Find_Offset) is needed. This function will generate the offset distance from a point to a linestring (does not have to be measured).

Offsets are expressed as either left or right of the road center-line's direction. Left is coded with a negative value and right with a positive value. A function to correctly compute the offset of a point asset is as follows:

```
Member Function ST_Find_Offset (p_geom in SDO_GEOMETRY,
                                p_unit in Varchar2   default null)
        Return number Deterministic;
```

This version of ST_Find_Offset is declared as returning a single number (The SDO_LRS equivalent is Find_Offset). While the possibility that a point may fall perfectly between two vectors/segments of a linestring exists, this situation has not been handled and is left up to the reader to implement.

```
With Geoms As (
   Select 'Falls off vector but in its middle' as test,
          SDO_GEOMETRY('POINT(2.47 0.21)',null) as point
```

```
                From Dual Union All
         Select 'Falls exactly between 2 vectors at ends' as test,
                SDO_GEOMETRY('POINT(3.0 -0.25)',null) as point
           From dual
     )
   Select b.test,
          Round(a.geom.ST_Find_Offset(b.point,'unit=U.S. Foot'),3) as
   offset
     From (Select T_Geometry(
                    SDO_GEOMETRY(3302,NULL,NULL,
                      SDO_ELEM_INFO_ARRAY(1,4,3, 1,2,1,
                                            7,2,2, 13,2,1),
                      SDO_ORDINATE_ARRAY(
                             1.0,0.0,0.0,   2.0,0.0,1.162,
                             3.0,0.0,2.324, 4.0,1.0,3.967,
                             5.0,0.0,5.61,  5.0,-0.5,6.191,
                             3.0,-0.5,8.514,0.0,-0.5,12,
                             0.75,-2.0,13.677,
                             1.75,-1.25,14.927)),
                    0.005) as geom
             From dual
           ) a,
           geoms b;

   -- Results
   --
   TEST                                               OFFSET
   ------------------------------------------- ----------
   Falls off vector but in its middle              -0.21
   Falls exactly between 2 vectors at ends          0.25
```

Applying measurement and offset in a real-world example

The San Francisco data includes the recording of city furniture: lamp posts, benches and trash cans. These data are recorded only as easting/northing SDO_GEOMETRY objects. To be able to answer the question of the density of these objects per meter, they need to be converted into linear measured observations. To do this the current unmeasured road centerline linestring, need to have measures added. The existing center-line data is not in a form that allows for linear measurement. This is because a street such as BATTERY ST is represented by a set of independent road centerline linestrings as can be seen in the associated screenshot:

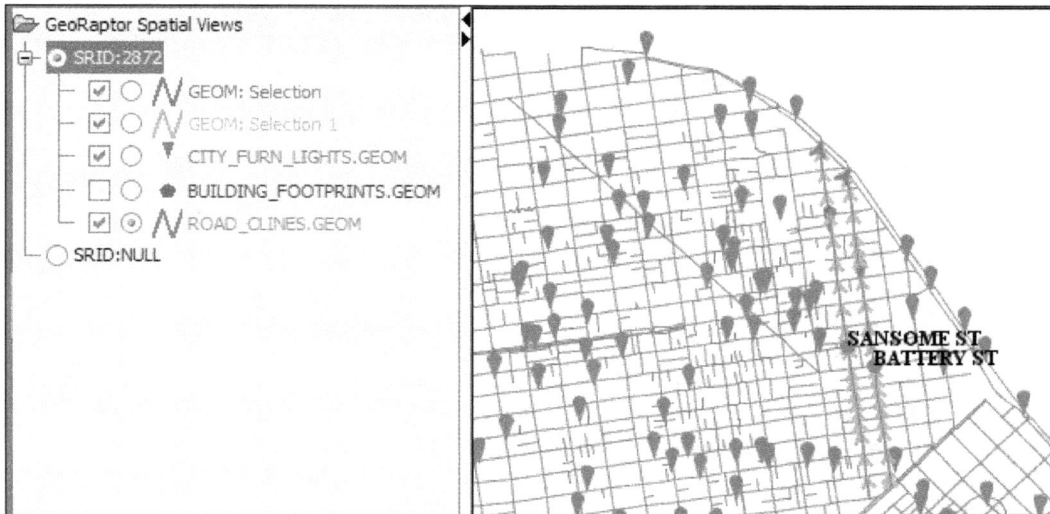

Individual road center-lines composed of two vertex segments is a difficult way of managing linear measures. The two tables in the San Francisco data model shipped with this book can be visualized using SQL Developer's Data Modeling tool as follows:

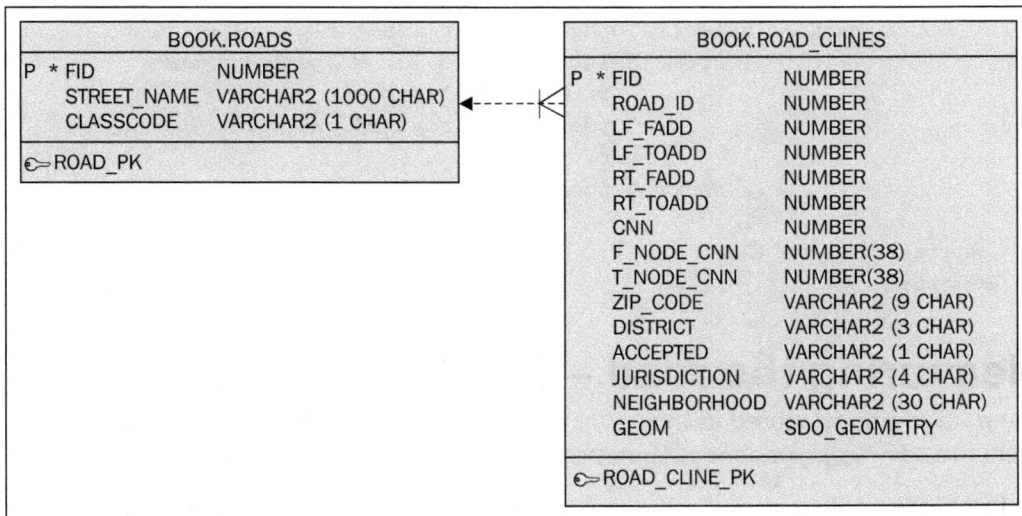

To process data at the road level (ROADS) and not the segment level (ROAD_CLINES) the data model needs to be modified to allow us to create an LRS measured geometry in the ROADS entity as follows:

```
Alter Table roads Add (geom SDO_GEOMETRY);
Delete From user_sdo_geom_metadata where table_name = 'ROADS';
Commit;
Insert Into user_sdo_geom_metadata (
    table_name,column_name,diminfo,srid
) Values (
  'ROADS', 'GEOM',
  MDSYS.SDO_DIM_ARRAY(
    MDSYS.SDO_DIM_ELEMENT('X',5900000,6100000,0.05),
    MDSYS.SDO_DIM_ELEMENT('Y',2000000,2200000,0.05),
    MDSYS.SDO_DIM_ELEMENT('M',-100,100000,0.05)),
  2872);
Commit;
```

To populate the ROADS GEOM column all the individual descriptive linestrings (stored in ROAD_CLINES.GEOM) for each independent road need to be concatenated in the correct order. This can be achieved in a variety of ways: the two example streets processed so far will be concatenated as follows:

```
Update roads r
  Set r.geom = (Select sdo_aggr_concat_lines(f.geom) as geom
                  From (Select rc.road_id, rc.geom as geom
                        From road_clines rc
                        Order By rc.road_id, rc.cnn
                       ) f
                  Where f.road_id = r.fid
               )
  Where r.fid in (2510,2274);
Commit;
```

Measuring the road – ST_Add_Measure

The road segments stored in ROAD_CLINES are defined as 2D (XY) linestrings. As such, the Sdo_Aggr_Concat_Lines processing previously generated 2D linestrings.

To be able to analyze data via linear processing measures need to be added to all the linestring geometries in ROADS.GEOM. Normally this would be done via an edit process that was based on real engineering and surveying data, but here the assignment will be via a new ST_Add_Measure function (the SDO_LRS equivalent function is Convert_To_LRS_Geom), which will populate the measure dimension using 2D line length as a surrogate.

```
Member Function
ST_Add_Measure(p_start_measure IN NUMBER Default NULL,
               p_end_measure   IN NUMBER Default NULL,
               p_unit          IN Varchar2 Default NULL)
      Return T_Geometry deterministic;

Update roads r
   Set r.geom = T_Geometry(r.geom,0.005)
                  .ST_Add_Measure(0,
                                   T_Geometry(r.geom,0.005)
                                     .ST_Length('Unit=U.S. Foot'),
                                   'Unit=U.S. Foot')
                  .ST_RoundOrdinates(2,2,2,2).geom
  Where r.fid in (2510,2274);
Commit;
```

The result looks as follows in SQL developer's table viewer:

Now that the two roads' linestrings are measured the measure of all the city furniture associated with those streets can be calculated. This can be done as follows:

```
Create Table city_furniture_event (
   road_id       Number,
   cf_id         Number,
   from_measure  Number not null,
   to_measure    Number,
   from_offset   Number,
   to_offset     Number
);
Alter Table city_furniture_event
   Add Constraint city_furniture_event_rd_id_fk
      Foreign Key (road_id)
      References book.roads (fid);
Alter Table city_furniture_event
   Add Constraint city_furniture_event_cf_id_fk
      Foreign Key (cf_id)
      References book.city_furniture(fid);
```

The table changes and relationships can be visualized using SQL developer's data modeling tool as follows:

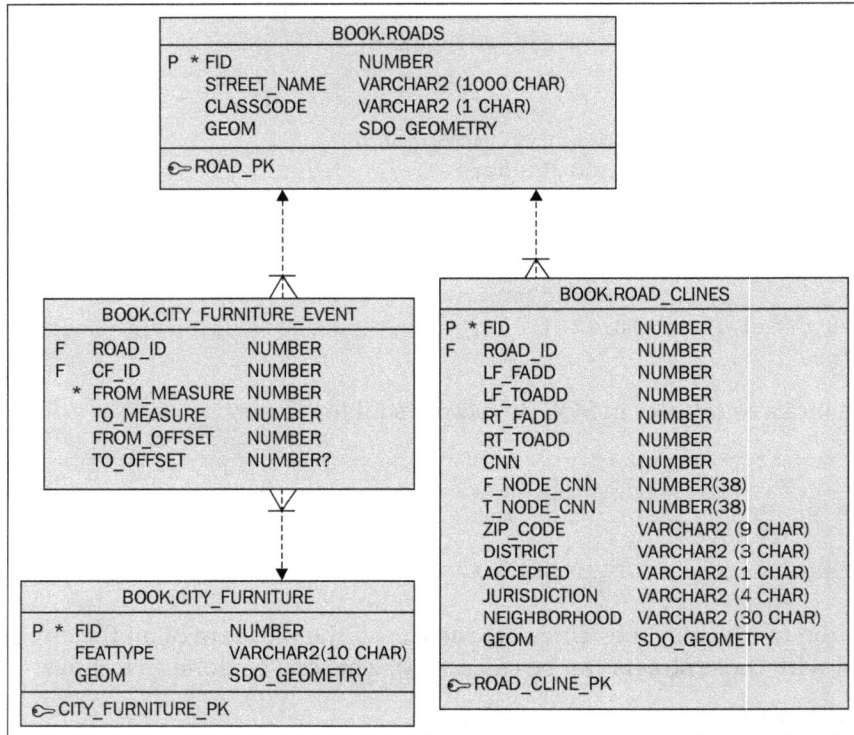

Populating the tables

The city furniture geometry objects can now be processed to create linear reference entries in our `CITY_FURNITURE_TABLE` as follows:

```
Insert Into city_furniture_event (
  road_id, cf_id, from_measure, from_offset
)
Select /*+ORDERED Index(r ROAD_CLINES_GEOM_SPIX)*/
       r.fid  as road_id,
       cf.fid as cf_id,
       Round(t.column_value,2) as from_measure,
       Round(T_Geometry(r.geom,0.005)
```

```
                    .ST_Find_Offset(cf.geom,'unit=U.S. Foot'),2)
              as from_offset
  From roads r,
         city_furniture cf,
         Table(T_Geometry(r.geom,0.05)
                    .ST_Find_Measure(cf.geom,'unit=U.S. Foot')) t
  Where r.fid in (2510,2274)
    And sdo_nn(cf.geom,r.geom,
                    'sdo_batch_size=10 distance=50 unit=U.S. Foot',1) =
  'TRUE';
  Commit;
```

Visualizing the CITY_FURNITURE_EVENT table for these streets' processed furniture objects reveals.

```
  Select cfe.road_id, cf.fid as furniture_id, cf.featType, cf.geom,
         cfe.from_measure, cfe.from_offset, cfe.to_measure,cfe.to_offset
    From city_furniture cf
         Inner Join
         city_furniture_event cfe
         On (cfe.cf_id = cf.fid);
  - Results
  -
```

ROAD_ID	CF_ID	FROM_MEASURE	FROM_OFFSET	T_MEASURE	TO_OFFSET
2274	536300	959.74	-22.26	(null)	(null)
2274	5460	1235.93	-21.99	(null)	(null)
2510	104	439.04	-25.52	(null)	(null)
2510	537000	1266.35	-21.72	(null)	(null)
2510	5507	1413.98	-22.44	(null)	(null)

(The bench objects are linestrings and so have length as against the light objects which are points. The processing of the processed the benches as single points which led to a single from-measure. They could be processed as linestring objects with the first and last coordinate being used to generate a from-measure and a to-measure. It is left up to the reader to implement this processing.)

Modifying a linestring's measures

The previous chapters introduced member functions that could be used with measured linestrings.

Member function	Description
ST_isMeasured	Reports if a geometry is measured or not.
ST_Lrs_Dim	Reports actual ordinate holding the measure value.
ST_GetVertex	Get nominated vertex in line (includes measure Values).
ST_StartVertex(	Gets geometry 's first vertex including measure ordinate if exists.
ST_EndVertex	Gets geometry's last vertex including measure ordinate if exists.
ST_InsertVertex	Inserts new vertex into geometry including measure ordinate.
ST_UpdateVertex	Updates vertices describing geometry including measure ordinate.
ST_DeleteVertex	Delete any vertex describing geometry.
ST_To2D	Remove measures from linestrings with SDO_GTypes of 3302/3306
ST_SquareBuffer	Offsets linestring and then produced square ended buffer.

In the previous section ST_Add_Measure was used to add a measure dimension to each vertex in the ROADS.GEOM linestrings. Additional editing of the measures may be needed.

For example, ST_Add_Measure was used because no measure records could be processed in time for the roads whose lamp posts needed to be processed. Since then, the measures have been made available. However, they only exist at the start and end of the roads. Yes, the measures could be removed (ST_To2D) and then added on again (ST_Add_Measure) but the following function is made available that works with existing measured linestrings.

[🖋 The SDO_LRS equivalent is SCALE_GEOM_SEGMENT]

```
Member Function
ST_Scale_Measures(p_start_measure in Number,
                  p_end_measure   in Number,
                  p_shift_measure in Number default 0.0 )
  Return T_Geometry Deterministic;
```

This function can redistribute measure Values between the supplied p_start_measure (start vertex) and p_end_measure (end vertex) by adjusting/scaling the measure Values of all coordinates. In addition, if p_shift_measure is not 0 (zero), the supplied value is added to each modified measure value performing a translation/shift of those Values.

As an example, the old/new start/end measure Values for road FID equals 2510 are as follows:

Old Start	Old End	New Start	New End
0	1679.671	1.1	1682.0

The new Values can be applied as follows. The query will only show the first and last four vertices in the linestring.

```
Select id as vertexId, t.x,t.y,round(t.z,3) as z
  From roads r,
       Table(t_geometry(r.geom,0.005)
              .ST_Scale_Measures(1.1,1682.0,0)
              .ST_Vertices()) t
  Where r.fid in (2510)
    And t.id in (1,2,3,4,19,20,21,22);

-- Results
--
VERTEXID          X           Y          Z
-------- ---------- ---------- --------
       1 6012515.99  2115785.2        1.1
       2 6012468.33 2116124.79    105.698
       3  6012420.6 2116464.91    210.459
       4  6012371.8 2116812.58    317.546
      19 6011858.74 2120478.86    1446.76
      20 6011811.01 2120819.17   1551.579
      21 6011767.72 2121149.38   1653.162
      22 6011794.43 2121240.07       1682

 8 rows Selected
```

For the rest of this chapter, the original length-based measurement of the roads will be used.

Examining properties of a measured linestring

Once a linestring is measured, a number of useful property investigation functions are possible. As indicated at the start of this chapter, these include:

```
/* Query first measure value in linestring */
Member Function ST_Start_Measure
        Return Number deterministic,

/* Query largest/last measure value in linestring */
Member Function ST_End_Measure (p_unit in Varchar2 default null)
        Return Number deterministic,

/* Find range of measured Values in linestring */
Member Function ST_Measure_Range (p_unit in Varchar2 default null)
        Return Number deterministic,

/* Convert measure value to a percentage of the whole range */
Member Function
ST_Measure_To_Percentage (p_measure IN NUMBER DEFAULT 0,
                            p_unit in Varchar2 default null)
        Return Number deterministic,

/* Convert percentage value (0-100) to a measure */
Member Function
ST_Percentage_To_Measure (p_percentage IN NUMBER DEFAULT 0,
                            p_unit in Varchar2 default null)
        Return Number deterministic,

/* Checks all vertices from beginning to end ensuring measure Values
    are always decreasing in value */
Member Function ST_Is_Measure_Decreasing
        Return Varchar2 Deterministic,

/* Checks all vertices from beginning to end ensuring measure Values
    are always increasing in value */
Member Function ST_Is_Measure_Increasing
        Return Varchar2 Deterministic,
```

Examining these properties of our new measured lines can be done as follows:

> The SDO_LRS equivalents for these functions are:
> ST_Start_Measure → Geom_Segment_Start_Measure;
> ST_End_Measure → Geom_Segment_End_Measure;
> ST_Measure_Range → Measure_Range;
> ST_Measure_To_Percentage → Measure_To_Percentage;
> ST_Percentage_To_Measure → Percentage_To_Measure;
> ST_is_Measure_Increasing → Is_Measure_Increasing;
> ST_is_Measure_Decreasing → Is_Measure_Decreasing

```
Select a.geom.ST_Start_Measure()                      as startM,
       round(a.geom.ST_End_Measure(),2)               as endM,
       round(a.geom.ST_Measure_Range(),2)             as rangeM,
       round(a.geom.ST_Measure_To_Percentage(1000),1) as M2Pcnt,
       round(a.geom.ST_Percentage_To_Measure(50),2)   as Pcnt2M,
       a.geom.ST_is_Measure_Increasing()              as isIncr,
       a.geom.ST_is_Measure_Decreasing()              as isDecr
  From (Select T_Geometry(r.geom,0.005) as geom
          From roads r
         Where r.fid in (2510,2274)
       ) a;

-- Results
--
STARTM    ENDM   RANGEM M2PCNT  PCNT2M ISINCR ISDECR
------ ------- ------- ------ ------- ------ ------
     0 1494.42 1494.42   66.9  747.21 TRUE   FALSE
     0 1679.67 1679.67   59.5  839.84 TRUE   FALSE
```

All these functions work on non-measured linestrings. Where one is supplied, all measures are based on the linestring's length property.

Reversing measures and linestring directions – ST_Reverse_Measures and ST_Reverse_Linestring

It is important to remember that any measured linestring is a graphic representation only of the actual road. A graphic linestring's direction, for example, is often used to denote physical characteristics such as direction of flow (sewer pipe) or direction of traffic (one-way street). It is possible that the measures of a road are based on physical monuments in the ground that were placed there at some time in the past. Since that placement, the direction of the road could have been changed as anyone familiar with the central business districts of most cities would know: roads that were once two-way are now one-way. In this situation it is quite possible that the measures as originally designed increase in a particular direction which is now opposite to that of the one-way traffic.

Since database spatial graphics often serve many purposes, a situation may arise where the linestring's graphic direction is opposite to the measures. Any processing of such linestrings need to cope with such a situation. Here the two ST_is_Measure_Increasing/ST_is_Measure_Decreasing functions are useful. But in addition, a function that allows for the measures of a linestring to be reverse independently of the graphic direction of the linestring, is of use.

```
Member Function ST_Reverse_Measure
        Return T_Geometry Deterministic
```

This function reverses a linestring's vertices' measures: the first measure becomes last, second becomes second last, and so on (circular arcs are honored). The following example takes the linestring for Sherman Rd, and adds measures to it using the linestring's length property. The same linestring has its measured reversed. The linestring's graphic direction (right to left) is left alone.

```
Select T_Geometry(rc.geom,0.005)
          .ST_Add_Measure(null,null,'unit=U.S. Foot')
          .ST_RoundOrdinates(2)
          .geom as normalGeom,
       T_Geometry(rc.geom,0.005)
          .ST_Add_Measure(null,null,'unit=U.S. Foot')
          .ST_Reverse_Measure()
          .ST_RoundOrdinates(2)
          .geom as reversedGeom
  From roads r
       Inner Join
       road_clines rc
       On (rc.road_id = r.fid)
 Where r.street_name = 'SHERMAN RD';
```

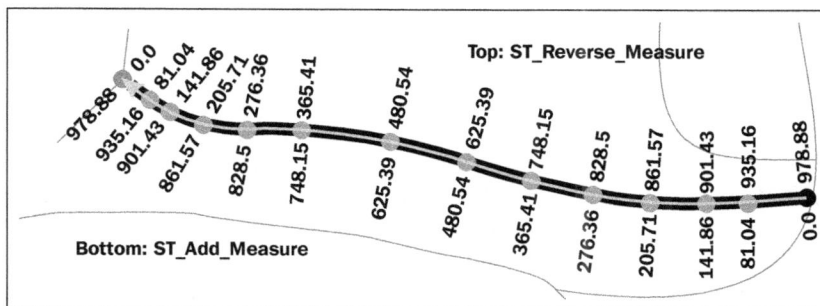

In addition, a function to reverse a complete linestring should accompany this function:

```
Member Function ST_Reverse_Linestring
        Return T_Geometry Deterministic;
```

This function reverses a linestring (or multi-linestring). Note that ST_Reverse_ Linestring reverses the XY direction of whole linestring, keeping the association between a vertex and its Z and/or W ordinates. The function honors circular arcs.

```
Select T_Geometry(rc.geom, 0.005)
          .ST_Add_Measure(null,null,'unit=U.S. Foot')
          .ST_Reverse_Linestring()
          .ST_RoundOrdinates(2)
          .geom as reversedGeom
   From roads r
        Inner Join
        road_clines rc
        On (rc.road_id = r.fid)
   Where r.street_name = 'SHERMAN RD';
```

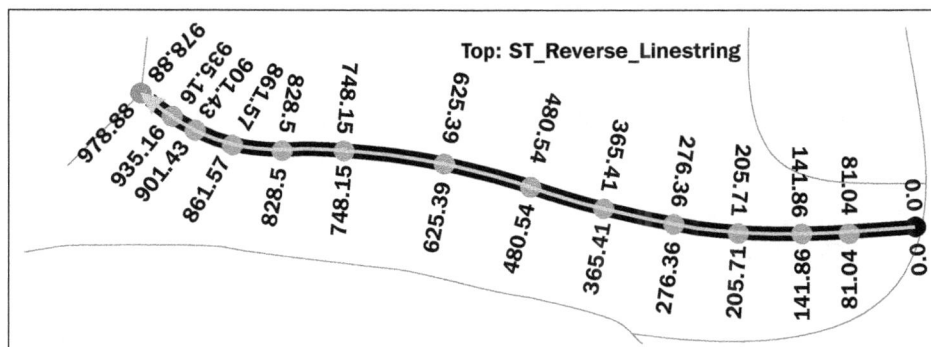

Note how the measures have remained in the same locations but the linestring's graphic direction is now left to right.

Calculating a linestring's centroid – ST_ Centroid

The ability to create a centroid point on a (multi)linestring is a common requirement when working with linear data. There are many reasons for this, from the need to label linear features in a map, through the need to conduct point-based statistical analysis of variable. As it is a common requirement, a function is presented that will compute a centroid (or centroids) of linestring (or multi-linestring) data based on measures or length.

```
Member Function ST_Centroid(p_option in Varchar2 := 'LARGEST',
                            p_unit   in Varchar2 default null)
         Return T_Geometry deterministic;
```

This function creates a single centroid if the linestring being operated on has a single part. The position of the centroid is either the mid-length point if the line is not measured, or the mid-measure position if measured. For a single linestring any supplied p_option value is ignored.

If the geometry is a multi-linestring a number of options are available.

- **LARGEST**: This returns centroid of largest (measure/length) linestring in multi-linestring (DEFAULT)

- **SMALLEST**: This returns centroid of smallest (measure/length) linestring in multi-linestring

- **MULTI**: This returns all centroid for all parts of multi-linestring as a single multi-point (x005 gType) geometry.

The centroid of each part of a multi-linestring is constructed using the same rules as for a single linestring. An example using a meaningless geometry demonstrates ST_Centroid's use on a measure multi-linestring:

> The equivalent in SDO_LRS is to use LOCATE_PT with GEOM_SEGMENT_START_MEASURE and MEASURE_RANGE.

```
With lrs_routes As (
  Select T_Geometry(
         SDO_GEOMETRY(3306,NULL,NULL,
           SDO_ELEM_INFO_ARRAY(1,4,3,1,2,1,7,2,2,13,2,1,25,2,1),
           SDO_ORDINATE_ARRAY(
             1.0,   0.0,   0.0,   2.0,   0.0,   1.0,
             3.0,   0.0,   2.0,   4.0,   1.0,   1.5,  5.0,  0.0,  5.14,
             5.0,  -0.5,   5.64,  3.0,  -0.5,   7.64,2.0,-0.48,8.64,
```

```
                      0.33,-1.17, 8.64, 1.57,-1.85,10.06,
                      2.64,-1.25,11.28, 3.68,-1.92,12.52))
                   ,0.005) as geom
      From dual
)
Select Case When mod(rownum,2) = 1
             Then t.which
             Else t.which||'_2D'
         End as Position,
       Case When mod(rownum,2) = 1
             Then a.geom.ST_Centroid(t.which)
                       .ST_RoundOrdinates(2)
                       .geom
             Else a.geom.ST_To2D()
                       .ST_Centroid(t.which)
                       .ST_RoundOrdinates(2)
                       .geom
         End as centroid
   From lrs_routes a,
        (Select level
           From dual
         Connect By Level < 3) b, /* Execute for M and no-M geom */
        (Select Case Level
                When 1 then 'SMALLEST'
                When 2 then 'LARGEST'
                When 3 then 'MULTI'
                End as which
           From dual
         Connect By Level < 4) t; /*Generate all position Types*/
   Order By 1;

-- Results
--
POSITION     CENTROID
----------   ----------------------------------------------------------
LARGEST      SDO_GEOMETRY(3001,NULL,
                SDO_POINT_Type(4.8,0.6,4.5),NULL,NULL)
LARGEST_2D   SDO_GEOMETRY(2001,NULL,
                SDO_POINT_Type(4.8,0.6,NULL),NULL,NULL)
MULTI        SDO_GEOMETRY(3005,NULL,NULL,
                SDO_ELEM_INFO_ARRAY(1,1,1,4,1,1),
                SDO_ORDINATE_ARRAY(4.8,0.6,4.5,2.08,-1.56,10.64))
MULTI_2D     SDO_GEOMETRY(2005,NULL,NULL,
                SDO_ELEM_INFO_ARRAY(1,1,1,3,1,1),
                SDO_ORDINATE_ARRAY(4.8,0.6,2.08,-1.56))
SMALLEST     SDO_GEOMETRY(3001,NULL,
                SDO_POINT_Type(2.08,-1.56,10.64),NULL,NULL)
SMALLEST_2D  SDO_GEOMETRY(2001,NULL,
                SDO_POINT_Type(2.08,-1.56,NULL),NULL,NULL)
```

The centroids generated from the measured multi-linestring are shown in the following screenshot. Note, that because the measures for the multi-linestring were generated from the same geometry the results are the same.

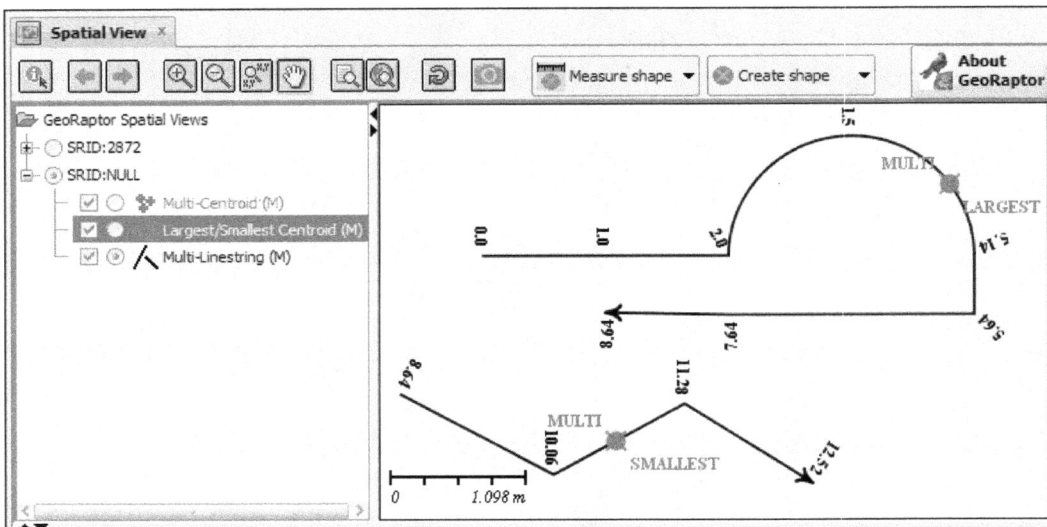

Creating a point at a known measure – ST_ Locate_Measure

ST_Locate_Measure (the SDO_LRS equivalent is Locate_Pt) creates a point at a known measure along a linestring. The position of the point can be offset a set distance to the left or right of the line (in the vertex direction from start to end) or on the line (no offset). This member function looks like this:

```
Member Function
ST_Locate_Measure(p_measure in Number,
                  p_offset  in Number,
                  p_unit    in Varchar2 default null)
       Return T_Geometry deterministic,
```

If the source geometry's is not measured, the function will interpret p_measure as a length from the beginning of the supplied linestring and will ignore any measure defined in the geometry. As such, the function will operate on a linestring geometry without measures.

The function can be tested against a measured linestring as follows. For each segment or vector within the linestring a random measure is generated from the Z measure start and end Values of that segment. The measure value is found and offset to the left and to the right.

```
Select f.rMeasure, f.offset,
       f.geom.ST_Locate_Measure(f.rMeasure,f.offset,null)
             .ST_RoundOrdinates(2)
             .geom.sdo_point as mPoint
  From (Select a.geom,
               Round(m.vector.startCoord.z +
                    ((m.vector.endCoord.z-m.vector.startCoord.z) /
                    dbms_random.value(1,10)),2)  as rMeasure,
               Case When o.offset = 1
                    Then -0.2 Else 0.2 End as offset
          From (Select level as offset
                  From dual
                Connect By Level <= 2) o,
                (Select T_Geometry(
                        SDO_GEOMETRY(3302,NULL,NULL,
                          SDO_ELEM_INFO_ARRAY(1,4,3, 1,2,1,
                                              7,2,2, 13,2,1),
                          SDO_ORDINATE_ARRAY(
                          1.0,0.0,0.0, 2.0,0.0,1.162,
                          3.0,0.0,2.324, 4.0,1.0,3.967,
                          5.0,0.0,5.61, 5.0,-0.5,6.191,
                          3.0,-0.5,8.514, 0.0,-0.5,12,
                          0.75,-2.0,13.677, 1.75,-1.25,14.927))
                        ,0.005) as geom
                  From dual) a,
               Table(a.geom.ST_ArcVectorize()) m
      ) f
Order By f.Offset, f.rmeasure;

-- Results
--
RMEASURE OFFSET MPOINT
-------- ------ --------------------------------
    0.15   -0.2 SDO_POINT_Type(1.13,0.2,0.15)
    1.41   -0.2 SDO_POINT_Type(2.21,0.2,1.41)
    2.74   -0.2 SDO_POINT_Type(2.89,0.46,2.74)
    5.69   -0.2 SDO_POINT_Type(5.2,-0.07,5.69)
    7.29   -0.2 SDO_POINT_Type(4.05,-0.7,7.29)
    8.95   -0.2 SDO_POINT_Type(2.62,-0.7,8.95)
   13.66   -0.2 SDO_POINT_Type(0.92,-1.9,13.66)
    14.2   -0.2 SDO_POINT_Type(1.05,-1.53,14.2)
    0.62    0.2 SDO_POINT_Type(1.53,-0.2,0.62)
    1.55    0.2 SDO_POINT_Type(2.33,-0.2,1.55)
    4.83    0.2 SDO_POINT_Type(4.59,0.54,4.83)
```

```
 5.68    0.2 SDO_POINT_Type(4.8,-0.06,5.68)
 6.48    0.2 SDO_POINT_Type(4.75,-0.3,6.48)
 9.62    0.2 SDO_POINT_Type(2.05,-0.3,9.62)
12.24    0.2 SDO_POINT_Type(-0.07,-0.8,12.24)
13.94    0.2 SDO_POINT_Type(1.08,-2,13.94)

16 rows Selected
```

It looks like the following screenshot when executed in SQL developer and rendered using `GeoRaptor`:

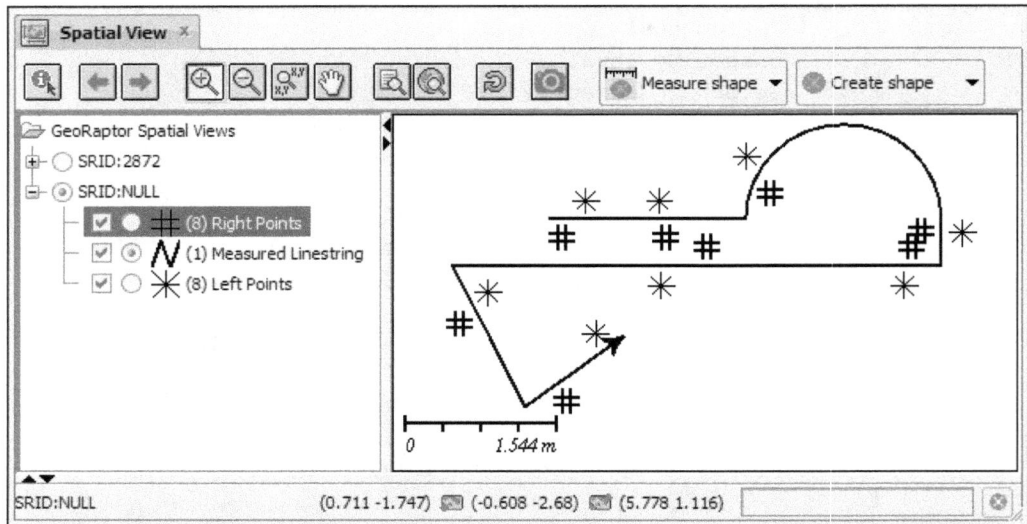

The hash (#) symbols are points generated via a left (-ve) offset of 0.2 units using measures extracted from each unique vertex in the linestring. The asterisks (*) are the corresponding right offset points at each vertex.

A test of this function against an unmeasured linestring using 10 randomly generated measures (lengths) can be executed as follows:

> The located measure is displayed in Well Known Text (WKT) format to save space.

```
Select f.rMeasure,f.offset,
       f.geom.ST_Locate_Measure(f.rMeasure,f.offset,null)
            .ST_RoundOrdinates(2)
            .ST_WKT() as mPoint
```

```
    From (Select a.geom, o.offset,
              round(dbms_random.value(1,
                                a.geom.ST_End_Measure()),3)
                   as rMeasure
          From (Select Case When MOD(LEVEL,2)=1
                        Then -0.2 Else 0.2 End
                     as offset
               From dual
              Connect By Level <= 10) o, -- Create 10 samples
              (Select T_Geometry(
                      SDO_GEOMETRY(2002,NULL,NULL,
                        SDO_ELEM_INFO_ARRAY(1,4,3, 1,2,1,
                                        5,2,2, 9,2,1),
                        SDO_ORDINATE_ARRAY(1.0,0.0, 2.0, 0.0,
                                      3.0,0.0, 4.0, 1.0,
                                      5.0,0.0, 5.0, -0.5,
                                      3.0,-0.5, 0.0, -0.5,
                                      0.75,-2.0,1.75,-1.25)
                      )
                   ,0.005) as geom
              From dual) a
        ) f
Order By f.Offset, f.rmeasure;

-- Results
-
RMEASURE OFFSET MPOINT
-------- ------ ----------------
   3.328   -0.2 POINT (3.71 1.16)
   6.617   -0.2 POINT (4.02 -0.7)
   7.379   -0.2 POINT (3.26 -0.7)
    8.64   -0.2 POINT (2.0 -0.7)
   9.236   -0.2 POINT (1.41 -0.7)
   1.385    0.2 POINT (2.39 -0.2)
   6.622    0.2 POINT (4.02 -0.3)
   9.579    0.2 POINT (1.06 -0.3)
  10.327    0.2 POINT (0.31 -0.3)
  12.057    0.2 POINT (0.45 -1.86)

10 rows Selected
```

It looks like the following when executed in SQL Developer and rendered using GeoRaptor. The asterisks (*) denote points created by randomly generating a measure between the start and end vertices measures and then left (-ve) offsetting it by 0.2 units. The asterisks (*) are randomly generated measures, right (+ve) offset by 0.2 unit points.

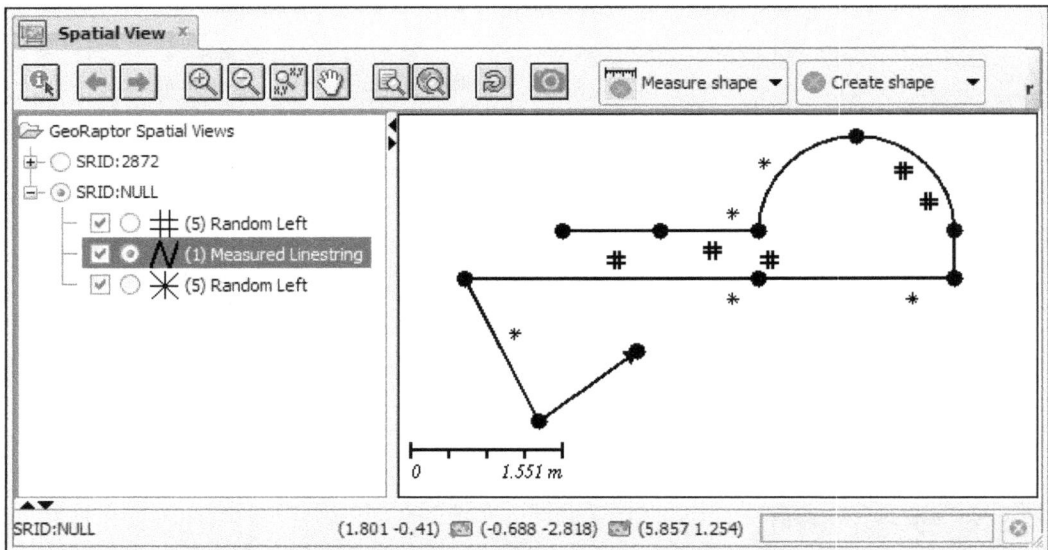

The asterisks (*) denote left (-ve) offset (-0.2 units) points at 5 random lengths from the start of the linestring. The asterisks (*) denote right (+ve) offset (0.2 units) at 5 random lengths from the start of the linestring.

A real-world example

It is often the case that linear asset management systems do not themselves include SDO_GEOMETRY columns describing their assets. The idea that spatial data should be managed by a specialist spatial system with descriptive asset data stored in another (with a "link" between the two) is common or pervasive. Such hybrid systems are problematic for a large number of reasons, a discussion of which is beyond the scope of this book. In these systems the asset data is managed via linear reference tables such as CITY_FURNITURE_EVENT with SDO_GEOMETRY data generated in consort with the "spatial system". Thus a common scenario is that the asset managers enter the linear reference data and then have the GIS "specialists" in the team generate the spatial representation (sometimes not even held within the same Oracle database and potentially in a non-Oracle form).

Another common scenario is where there are large numbers of existing records (paper or otherwise) held in a variety of systems that need to be imported into the database and spatial representations created.

The `ST_Locate_Measure` function can be used for taking point assets recorded by measures (not GPS coordinates) in tables like `CITY_FURNITURE_EVENT` and to create their `SDO_GEOMETRY` representation in the main `CITY_FURNITURE` table.

Assume that a bunch of new lamp posts exist for our two streets whose existing paper records have been imported into the `CITY_FURNITURE_EVENT` table.

```
Insert Into city_furniture_event(road_id,from_measure,from_offset)
                     Values (2274,1248.339,46);
Insert Into city_furniture_event(road_id,from_measure,from_offset)
                     Values (2274,622.841,63.58);
Insert Into city_furniture_event(road_id,from_measure,from_offset)
                     Values (2274,352.572,89.3);
Insert Into city_furniture_event(road_id,from_measure,from_offset)
                     Values (2510,266.517,-42.4);
Insert Into city_furniture_event(road_id,from_measure,from_offset)
                     Values (2510,503.353,-60.91);
Insert Into city_furniture_event(road_id,from_measure,from_offset)
                     Values (2510,864.125,37.77);
Insert Into city_furniture_event(road_id,from_measure,from_offset)
                     Values (2510,1619.157,-34.84);
```

The inserted data can be viewed using SQL Developer's table viewer as shown in the following screenshot:

	ROAD...	CF_ID	FROM_MEASURE	FROM_OFFSET	TO_MEASURE	TO_OFFSET
1	2274	(null)	352.572	89.3	(null)	(null)
2	2274	(null)	622.841	63.58	(null)	(null)
3	2274	(null)	1248.339	46	(null)	(null)
4	2510	(null)	266.517	-42.4	(null)	(null)
5	2510	(null)	503.353	-60.91	(null)	(null)
6	2510	(null)	864.125	37.77	(null)	(null)
7	2510	(null)	1619.157	-34.84	(null)	(null)

The CF_ID Foreign Key is null because no geometry objects for these records in the CITY_FURNITURE table have been created. From the new event data new lamp posts in the CITY_FURNITURE table can be created as follows:

```
Declare
  v_max_fid Number;
  Cursor c_furniture_events Is
  Select row_Number() over (order by 1) as fid,
         cfe.road_id,
         cfe.cf_id,
         cfe.from_measure,cfe.from_offset
    From city_furniture_event cfe
   Where cfe.road_id in (2510,2274)
     And cfe.cf_id Is null
   For Update Of cf_id;
  r_furniture_event c_furniture_events%ROWType;
Begin
  Execute Immediate
  'Alter Table city_furniture_event
      modify constraint CITY_FURNITURE_EVENT_CF_ID_FK disable';
  Select max(fid) as maxFid
    Into v_max_fid
    From city_furniture;
  Open c_furniture_events;
  Loop
     FETCH c_furniture_events Into r_furniture_event;
     If ( c_furniture_events%NOTFOUND ) Then
       EXIT;
     Else
       Insert Into city_furniture (fid,featType,geom)
       Select r_furniture_event.fid + v_max_fid,
              'LIGHT',
              T_Geometry(r.geom,0.005)
                 .ST_Locate_Measure(
                     r_furniture_event.from_measure,
                     r_furniture_event.from_offset,0,
                    'unit=U.S. Foot')
                 .ST_To2D().geom
        From roads r
       Where r.fid = r_furniture_event.road_id;
       Update city_furniture_event
          Set cf_id = r_furniture_event.fid + v_max_fid
        Where current of c_furniture_events;
     End If;
  End Loop;
  Close c_furniture_events;
  Commit;
  Execute Immediate
  'Alter Table city_furniture_event
```

```
        modify constraint CITY_FURNITURE_EVENT_CF_ID_FK enable';
End;
/
```

This result in the following records added to the CITY_FURNITURE table and the CF_ID of CITY_FURNITURE_EVENT being updated as shown in the SQL Developer table viewer:

The geom column is displayed in Well Known Text (WKT) format to save space.

	FID	FEATTYPE	GEOM
1	6971001	LIGHT	POINT (6012312.243410329 2120044.1407264485)
2	6971002	LIGHT	POINT (6012614.497614952 2118015.4196354947)
3	6971003	LIGHT	POINT (6012760.56566386 2117140.062593188)
4	6971004	LIGHT	POINT (6012352.475564571 2116645.219606871)
5	6971005	LIGHT	POINT (6012226.159680347 2117412.126727433)
6	6971006	LIGHT	POINT (6012155.961115988 2118597.461389879)
7	6971007	LIGHT	POINT (6011746.693296837 2121041.7401196454)
8	6971008	LIGHT	POINT (6012614.497669857 2118015.419272904)
9	6971009	LIGHT	POINT (6012760.565426593 2117140.064260801)
10	6971010	LIGHT	POINT (6012352.475702775 2116645.2186222556)
11	6971011	LIGHT	POINT (6012226.1596848285 2117412.126695503)
12	6971012	LIGHT	POINT (6012155.9612950515 2118597.4601128437)
13	6971013	LIGHT	POINT (6012312.243636251 2120044.139006675)
14	6971014	LIGHT	POINT (6011746.693239056 2121041.740560388)

CITY_FURNITURE

	ROAD...	CF_ID	FROM_MEASURE	FROM_OFFSET	TO_MEASURE	TO_OFFSET
1	2274	6971009	352.572	89.3	(null)	(null)
2	2274	6971008	622.841	63.58	(null)	(null)
3	2274	6971013	1248.339	46	(null)	(null)
4	2510	6971010	266.517	-42.4	(null)	(null)
5	2510	6971011	503.353	-60.91	(null)	(null)
6	2510	6971012	864.125	37.77	(null)	(null)
7	2510	6971014	1619.157	-34.84	(null)	(null)

CITY_FURNITURE_EVENT

Deriving the lamp post metric

Earlier in the chapter the idea that the San Francisco City Council passed a new by-law requiring that street lights must be placed at a minimum, every 150 feet. At that point the only data available were point data and some old linear measurements. A compliance metric for the two streets processed so far can be calculated.

The metric can be calculated by querying the CITY_FURNITURE_EVENT table to get the measures of the existing street lamps and then computing the difference in measures between them (and the start and end of each street). The new by-law defines its separation in feet, but it is safe to use the existing CITY_FURNITURE_EVENT records directly as they are defined in feet, created from the ROADS table's GEOM column whose measure Values were created by calculations based on length (feet).

The metric for the lamps can be generated via a query as shown in the following code:

```
Select f.road_id, round(avg(f.measure_diff),1) as measure_metric
  From (Select a.road_id,
            Lead(a.from_measure,1) over /* next measure */
               (partition by a.road_id
                   order by a.from_measure)
             - a.from_measure as measure_diff
        From (Select cfe.road_id,
                    cfe.from_measure
              From city_furniture_event cfe
             Where cfe.road_id in (2510,2274)
            Union All
            Select r.fid as road_id,
                   T_Geometry(r.geom,0.005).ST_Start_Measure()
                      as from_measure
              From roads r
             Where r.fid in (2510,2274)
            Union All
            Select r.fid as road_id,
                   T_Geometry(r.geom,0.005).ST_End_Measure()
                      as from_measure
              From roads r
             Where r.fid in (2510,2274)
            ) a
      ) f
 Group By f.road_id
 Order By f.road_id;

-- Results
--
```

```
ROAD_ID MEASURE_METRIC
------- --------------
   2274          249.1
   2510            210
```

Fairly obviously, more lamp posts are going to have to be placed within these two streets!

Having all the data the position of the missing lamps can now be calculated.

```
With postDetails As (
Select f.road_id,
       f.from_measure,
       f.measure_diff,
       ceil(f.measure_diff/150) -
       Case When from_measure = 0
               or from_measure+measure_diff = end_measure
            Then 1 /* last post before end */
            Else 2 /* Ignore first and last posts in range */
        End as newPosts,
       round(f.measure_diff / ceil(f.measure_diff/150),0)
         as PostSeparation
  From (Select a.road_id,
               a.from_measure,
               Lead(a.from_measure,1) over /* next measure */
                  (partition by a.road_id
                      order by a.from_measure)
                 - a.from_measure as measure_diff,
               last_value(a.end_measure) over
                  (partition by a.road_id)
                 as end_measure
          From (Select cfe.road_id,
                       cfe.from_measure,
                       cast(null as Number) as end_measure
                  From city_furniture_event cfe
                 Where cfe.road_id in (2510,2274)
                Union All
                Select r.fid as road_id,
                       T_Geometry(r.geom,0.005).ST_Start_Measure()
                         as from_measure,
                       cast(null as Number) as end_measure
                  From roads r
                 Where r.fid in (2510,2274)
                Union All
```

```
                   Select r.fid as road_id,
                       T_Geometry(r.geom,0.005).ST_End_Measure()
                           as from_measure,
                       T_Geometry(r.geom,0.005).ST_End_Measure()
                           as end_measure
                    From roads r
                    Where r.fid in (2510,2274)
                 ) a
            ) f
    Where f.measure_diff > 150
      And Ceil(f.measure_diff/150) Is not null
      And Ceil(f.measure_diff/150) > 0
  )
Aelect f.road_id, f.measure,f.offset,
       T_Geometry(r.geom,0.005)
          .ST_Locate_Measure(f.measure,f.offset,'unit=U.S. Foot')
          .ST_RoundOrdinates(2)
          .geom as lGeom
    From (Select pd.road_id as RDID,
             Round(pd.from_measure + (pd.newPosts +
                 (pd.postseparation * g.column_value)))
               As MEAS,
             Case When Mod(row_Number() Over
                         (partition by pd.road_id
                             order by pd.from_measure,
                                 g.column_value),2) = 0
                 Then 1 Else -1 End * offset
               as OFFST
          From (Select cfe.road_id,
                    Round(avg(abs(from_offset)),1) as offset
                 From city_furniture_event cfe
                 Group By cfe.road_id) a,
             postDetails pd,
             Table(book.generate_series(1,pd.newPosts,1)) g
          Where pd.newPosts > 0
            And a.road_id = pd.road_id
        ) f,
        roads r
  Where r.fid = f.road_id
  Order By f.road_id, f.measure;

-- Results
--
```

```
RDID MEAS OFFST LGEOM
---- ---- ----- ----------------------------------------
2274  177 -48.6 SDO_GEOMETRY(3001,2872,
                    SDO_POINT_Type(6012705.16,2116550.36,177),
                    NULL,NULL)
2274 1372  48.6 SDO_GEOMETRY(3001,2872,
                    SDO_POINT_Type(6012260.02,2120446.83,1372),
                    NULL,NULL)
2510  134 -35.1 SDO_GEOMETRY(3001,2872,
                    SDO_POINT_Type(6012420.13,2116215.69,134),
                    NULL,NULL)
2510  999  35.1 SDO_GEOMETRY(3001,2872,
                    SDO_POINT_Type(6012092.13,2119035.31,999),
                    NULL,NULL)
```

A subset of the lamps is shown in the following screenshot:

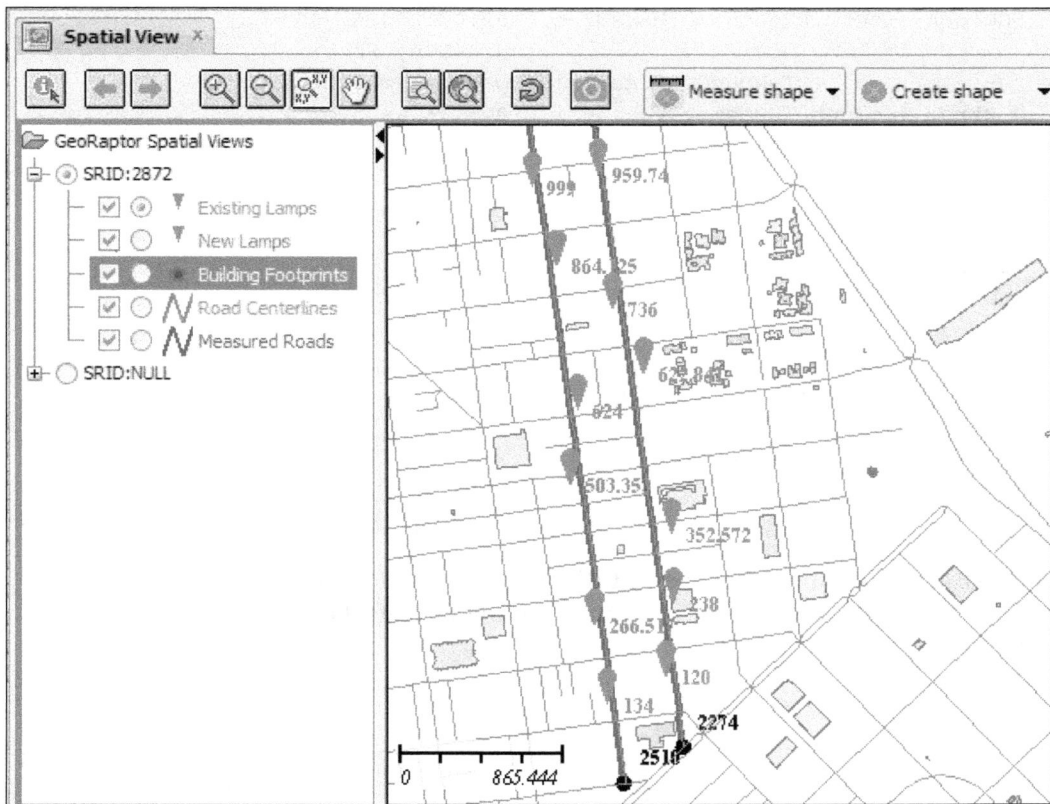

Selecting and offsetting a segment – ST_Locate_Measures

A common requirement when managing linear assets is the ability to define, dynamically, a segment along the linear asset by supplying a start and end measure. (the SDO_LRS equivalent is Offset_Geom_Segment).

```
Member Function
ST_Locate_Measures(p_start_measure In Number,
                   P_end_measure   In Number,
                   p_offset        In Number Default 0,
                   p_unit          In Varchar2 Default null)
        Return T_Geometry Deterministic;
```

The function can be used to Select and offset a segment of a measured linestring as follows:

```
Select a.geom.ST_Locate_Measures(1102.0,1200.0,-60,
                                 'unit=U.S. Foot')
          .ST_RoundOrdinates(2).geom as msegment
  From (Select T_Geometry(r.geom,0.005,3) as geom
          From roads r
         Where r.fid in (2510)
       ) a;

-- Results
--
MSEGMENT
----------------------------------------------------------------
SDO_GEOMETRY(3302,2872,NULL,
             SDO_ELEM_INFO_ARRAY(1,2,1),
             SDO_ORDINATE_ARRAY(6011951.95,2119357.13,1102.0,
                                6011939.08,2119451.33,1130.98,
                                6011908.41,2119675.69,1200.0))
```

This extracts a linestring between a fifth to two-fifths of the total measure of the original linestring and, via the p_offset parameter, offsets the extracted segment 60 feet to the left.

The following image, generated using SQL Developer and `GeoRaptor`, shows the extracted segment of the road center-line linestring as a directed line to the left of the road.

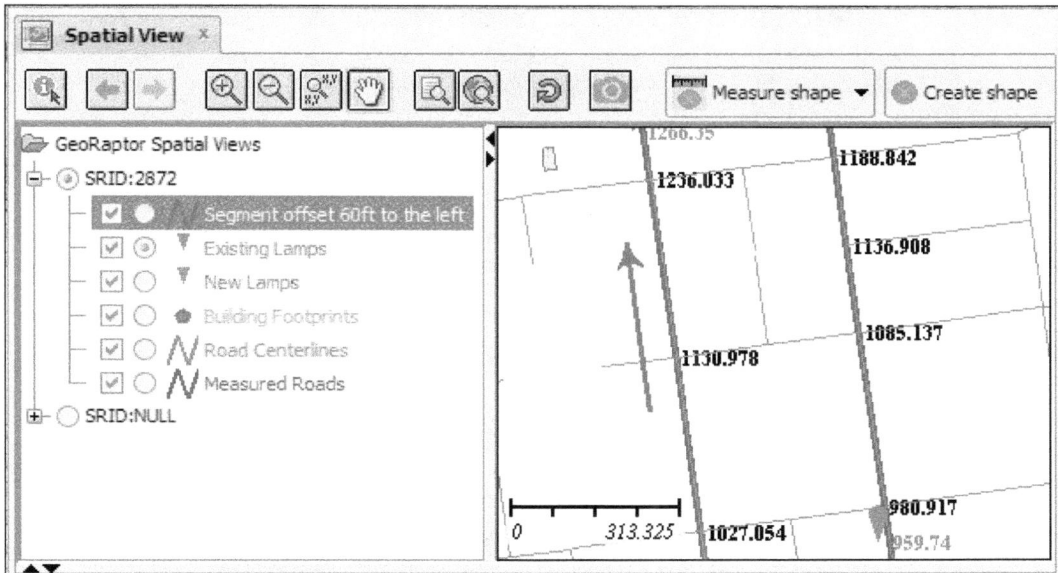

Mapping road surfaces

Real roads are not linestrings. They are composed of a vehicular pavement, plus kerb and guttering, and sidewalks. The vehicular pavement is normally not homogeneous being composed of patches of road surface made up of different materials such as concrete or asphalt. Road surface recording can be done using linear measures in a table such as the following code:

```
Create Table material (
  id              NUMBER  NOT NULL ,
  material_Type Varchar2 (100 BYTE) NOT NULL
);

Alter Table material
  Add Constraint material_pk Primray Key (id) ;

Comment On Table material
  IS 'Look up table. Domain of all pavement surface materials.';
Comment On Column material.id
    IS 'Unique Primary Key';
```

```
Comment On Column material.material_Type
    IS 'Textual description of the road surface material';

Insert Into material (id,material_Type) Values (1,'Gravel');
Insert Into material (id,material_Type) Values (2,'Asphalt');
Insert Into material (id,material_Type) Values (3,'Concrete');
Insert Into material (id,material_Type) Values (4,'Cobbles');
Commit;

Create Table road_surface (
  id           Integer not null,
  road_id      Number,
  material_id  Number not null,
  from_measure Number not null,
  to_measure   Number not null,
  offset       Number not null,
  width        Number not null,
  start_date   Date    not null,
  end_date     Date
);

Alter Table road_surface
  ADD Constraint road_surface_pk Primary Key ( id ) ;
Alter Table road_surface
  ADD Constraint road_surface_material_fk
      Foreign Key (material_id) REFERENCES material( id ) ;
Alter Table road_surface
  ADD Constraint road_surface_roads_fk
      Foreign Key (road_id) REFERENCES roads ( fid ) ;

Comment On Table road_surface
    IS 'Describes top surface of a road pavement.';
Comment On Column road_surface.id
    IS 'Unique Primary Key';
Comment On Column road_surface.road_id
    IS 'Holds id of measured road which it describes.';
Comment On Column road_surface.material_id
    IS 'Holds id of the road surface's material Type.';
Comment On Column road_surface.from_measure
```

```
      IS 'Measure that defines start of patch.';
Comment On Column road_surface.to_measure
      IS 'Measure that defines end of patch.';
Comment On Column road_surface.offset
      IS 'Offset to measured segment of patch.';
Comment On Column road_surface.width
      IS 'Full width of patch, 1/2 either side of offset measured
segment.';
Comment On Column road_surface.start_date
      IS 'Date surface patch was put in operation.';
Comment On Column road_surface.end_date
      IS 'Date surface patch was Replaced.';
```

Visually these changes look like this:

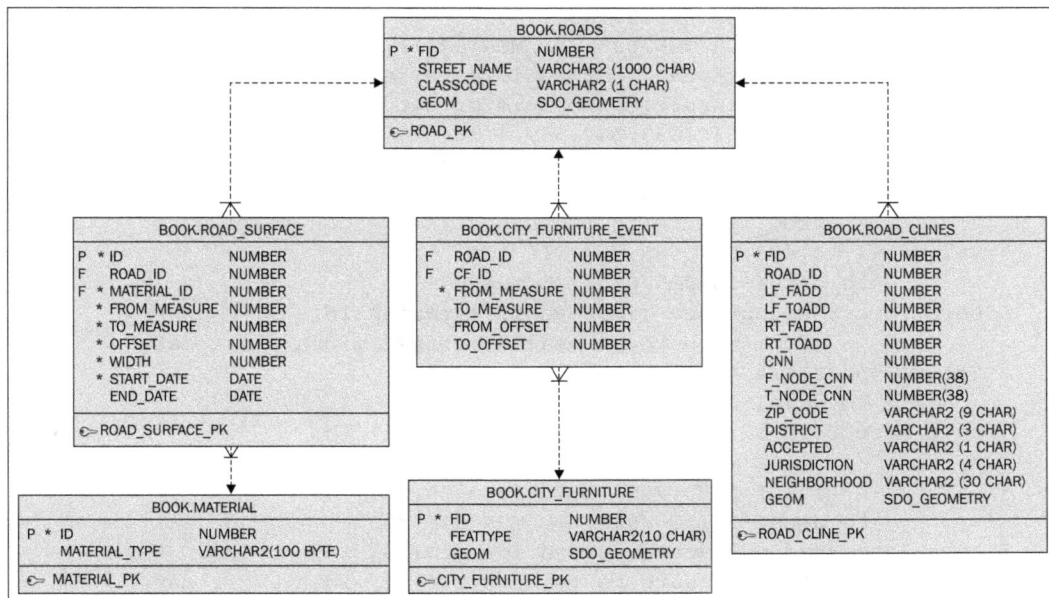

Over time, road surfaces are affected by wear and tear: their condition deteriorates as they develop faults such as cracks or potholes (chuckholes). The recording of these is not covered in this chapter, but these often use the same linear measure tables that records assets such as city furniture.

For the two streets that have been processed so far, some existing records from an older asset register can be loaded as follows:

[📝 Records are randomly generated.]

```
-- Inserting into ROAD_SURFACE
Set Define Off;
Insert Into road_surface (id,road_id,material_id,
            from_measure,to_measure,offset,width,start_date
) Values (1,2274,2,0,22.2,6,9,
        to_date('05/FEB/93','DD/MON/RR'));
Insert Into road_surface (id,road_id,material_id,
            from_measure,to_measure,offset,width,start_date
) Values (2,2274,1,22.2,512.033,6,10,
        to_date('02/JUN/03','DD/MON/RR'));
Insert Into road_surface (id,road_id,material_id,
            from_measure,to_measure,offset,width,start_date
) Values (3,2274,2,512.033,687.559,4,16,
        to_date('20/OCT/90','DD/MON/RR'));
Insert Into road_surface (id,road_id,material_id,
            from_measure,to_measure,offset,width,start_date
) Values (4,2274,3,687.559,918.916,1,9,
        to_date('10/JAN/01','DD/MON/RR'));
Insert Into road_surface (id,road_id,material_id,
            from_measure,to_measure,offset,width,start_date
) Values (5,2274,4,918.916,1190.797,12,16,
        to_date('02/MAY/06','DD/MON/RR'));
Insert Into road_surface (id,road_id,material_id,
            from_measure,to_measure,offset,width,start_date
) Values (6,2274,1,1190.797,1197.639,0,20,
        to_date('22/AUG/94','DD/MON/RR'));
Insert Into road_surface (id,road_id,material_id,
            from_measure,to_measure,offset,width,start_date
) Values (7,2274,4,1197.639,1283.798,16,8,
        to_date('19/OCT/09','DD/MON/RR'));
Insert Into road_surface (id,road_id,material_id,
            from_measure,to_measure,offset,width,start_date
) Values (8,2274,3,1283.798,1285.822,1,12,
        to_date('08/OCT/94','DD/MON/RR'));
```

```
Insert Into road_surface (id,road_id,material_id,
           from_measure,to_measure,offset,width,start_date
) Values (9,2274,1,1285.822,1358.659,16,8,
       to_date('09/AUG/04','DD/MON/RR'));
Insert Into road_surface (id,road_id,material_id,
           from_measure,to_measure,offset,width,start_date
) Values (10,2274,1,1358.659,1489.632,9,8,
       to_date('06/SEP/04','DD/MON/RR'));
Insert Into road_surface (id,road_id,material_id,
           from_measure,to_measure,offset,width,start_date
) Values (11,2274,4,1489.632,1494.421,1,9,
       to_date('25/SEP/08','DD/MON/RR'));
Commit;
```

To visualize these as polygon patches the following can be executed:

```
Select Case rs.material_id
       When 1 Then '255,0,0'
       When 2 Then '0,255,0'
       When 3 Then '0,0,255'
       When 4 Then '102,51,0'
       End as rgb,
       (Select material_Type
         From material m
        Where m.id = rs.material_id) as material,
       T_Geometry(r.geom,0.005)
            .ST_Locate_Measures(
                rs.from_measure,
                rs.to_measure,
                case when floor(dbms_random.value(-1,1))=0
                     then 1
                     Else -1
                  end*rs.offset,'unit=U.S. Foot')
            .ST_SquareBuffer(rs.width/2,0,'unit=U.S. Foot')
          as geom
 From road_surface rs
       Inner Join roads r
       On (r.fid = rs.road_id)
Where rs.road_id = 2274;
```

Visually it looks like this:

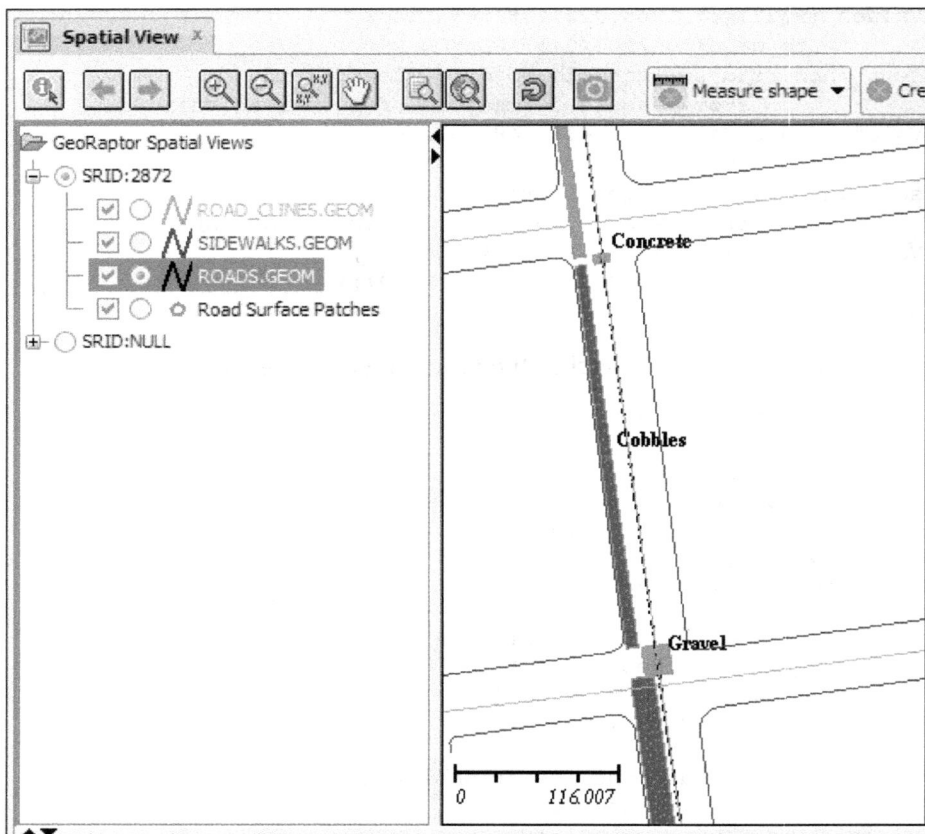

This sort of analysis can be applied to other data that is described linearly. This can include road speed zones, pavement fault data, kerb and gutter description. Linear processing does not need to be about roads or streets.

Summary

In this chapter you were introduced to a basic, yet complete, set of functional components that are required to carry out linear processing and analysis. In some situations, those functions could be used as "drop in" Replacements for existing Oracle functionality that is out of reach of an organization due to licensing and cost situations; such a decision should only be made after carefully weighing all options. While some of the examples were contrived, care was taken to create some "real-world" scenarios that showed how linear processing can be used to:

- Solve problems as generating metrics to test compliance with local council or city by-laws

- Represent, spatially, road surface, and condition data, stored in traditional linear referencing systems format

While the examples given were based on road or street data, linear referencing and processing can be used in other situations, such as the life sciences where linear plots or transects are created for the purpose of measuring trees or other fauna, or in the management of sewer, water, and electricity networks.

All the functions created in *Chapters 6, Implementing New Functions, Chapter 7, Editing, Transforming, and Constructing Geometries,* and this chapter were coded in pure PL/SQL. *Chapter 10, Integrating Java Technologies with Oracle Spatial,* will show how Java can be used to implement new functions based on existing open source code or custom programming.

9
Raster Analysis with GeoRaster

Spatial features can be represented in vector or raster format. So far we have discussed the vector related features of Oracle Spatial, and we introduce the raster related features called GeoRaster in this chapter. Traditional GISs propose to store raster data as BLOBs in the database. This approach might be sufficient if the raster data is only used as backdrop images in maps. But if any raster data processing and analysis is required, storing raster data as GeoRaster objects offers many features and advantages over storing this data just as BLOBs. Loading and storing any raster data inside a database simply for the purpose of storage or visualization provides limited utility. Storing raster data for use within a transactional system has engendered a view that one must see all data as part of a complete model; the data loaded must be seen in relation to all other data under the control of that model. The goal of this chapter is to demonstrate how to use raster data in conjunction with all the data in the database to answer questions that otherwise could not be answered in the database. The following topics are covered in this chapter to show how this goal can be achieved:

- Introduction to GeoRaster
- Loading and storing raster data inside a database
- Raster data for visualization applications
- Raster data for analytical applications
- Mapping between raster and vector space

Working with GeoRaster

GeoRaster can be used with data from any technology to capture or generate images, such as remote sensing, photogrammetry, and thematic mapping. It can be used in a wide variety of application areas including location-based services, geo-imagery archiving, environmental monitoring and assessment, geological engineering and exploration, natural resource management, defense, emergency response, telecommunications, transportation, and urban planning. In general, raster data covers a wide range of data types, such as digital images, geo-imagery (data collected by satellites), Digital Elevation Models (DEMs), Digital Terrain Models (DTMs), and gridded data (used in raster GIS). The GeoRaster feature in Oracle Spatial manages all of these diverse data types.

Raster data usually has some or all of the following elements:

- Cells or pixels
- Spatial domain (footprint)
- Spatial, temporal, and band reference information
- Cell attributes
- Other secondary metadata (such as the date of data acquisition)

GeoRaster defines a generic raster data model that is logically layered and is multidimensional. The core data in a raster is a multidimensional array or matrix of raster cells. Each cell is one element of the matrix, and its value is called the cell value, which is sampled at the center of the cell. If the GeoRaster object represents an image, a cell can also be called a pixel with only one value (in GeoRaster, the terms cell and pixel are interchangeable). The matrix has a number of dimensions, a cell depth, and a size for each dimension. The cell depth is the data size of the value of each cell. The cell depth defines the range of all cell values, and it applies to each single cell, not to an array of cells. For example, an RGB image in raster terms can be considered as a three dimensional matrix. Along the third dimension, it has three layers, one each for red, green, and blue bands of the image, and each pixel along the rows and columns stores the byte value that is used to represent the red, green, and blue color. If the image can represent true color used in most modern monitors, then each pixel has a size of 1 byte that is, 24-bits to represent a color (one byte each for R, G, and B). In GeoRaster terms, such an image has three bands with a pixel size of 1 byte for each band. GeoRaster has a very generic data model to support several different sizes and types for the pixels, such as 1BIT, 2BIT, 4BIT, 8BIT_U, 8BIT_S, 16BIT_U, 16BIT_S, 32BIT_U, 32BIT_S, 32BIT_REAL, or 64BIT_REAL.

A digital elevation model typically has one band where each pixel in the raster denotes the elevation value at that particular location. Since the elevation values can be any real number, the pixels in the raster will be of type `real` and the size can be 32 or 64-bits depending on the resolution of the DEM. If the raster is georeferenced, it also has a geometric footprint that can be used to spatially search for all rasters corresponding to a spatial location. A hyperspectral image typically has hundreds of bands that store information to derive a continuous spectrum for each image cell. Hundreds of readings for each pixel in hyperspectral images are used to recognize and map surface materials such as soil types of vegetation and mineral composition in soil. GeoRaster has a single database type called `SDO_GEORASTER` to store, index, query, and analyze all of these types of raster data.

Each raster can also have a spatial reference system associated with it if the raster is geo referenced. In such cases, there are two types of coordinates associated with the raster: the coordinates of each pixel in the raster matrix (the column, row, and band value for each pixel) and the coordinates on the Earth that these pixels represent. The pixel coordinates in the raster are in the cell coordinate space and the corresponding points these pixels represent on the Earth or any other coordinate system are in the model or the ground coordinate system. The spatial reference system associated with each raster gives a mapping from the cell coordinate space
to the model coordinate space and vice versa.

GeoRaster physical storage

GeoRaster data consists of a multidimensional matrix of cells and metadata. The cell data is the raw raster data and the metadata stores information such as the number of bands, spatial reference system, number of pixels in each band, pixel type, and size. In the GeoRaster data model, this metadata is stored as an XML document according to the GeoRaster XML metadata schema. The multidimensional matrix of cells is blocked into small subsets for GeoRaster objects. Each block is stored in a secondary table as a binary large object (BLOB), and a geometry object (of type `SDO_GEOMETRY`) is used to define the precise extent of the block. This multidimensional matrix of pixel values has to be linearized for storage on disk, and GeoRaster supports several methods of linearization of these pixels. Linearization here refers to the concept of taking two (or three) dimensional data and laying them out on a disk as a linear sequence of bytes; several methods exist for doing this transformation. In raster space, these are called interleaving methods in raster terminology, and the most common methods are BSQ, BIL, and BIP. BSQ is band-sequential (where each band is linearized separately), BIL is band-interleaved by line, and BIP is band-interleaved by pixel.

One can also combine these different techniques so that the first set of bands of a raster uses one type of interleaving and the second set of bands uses a different type of interleaving.

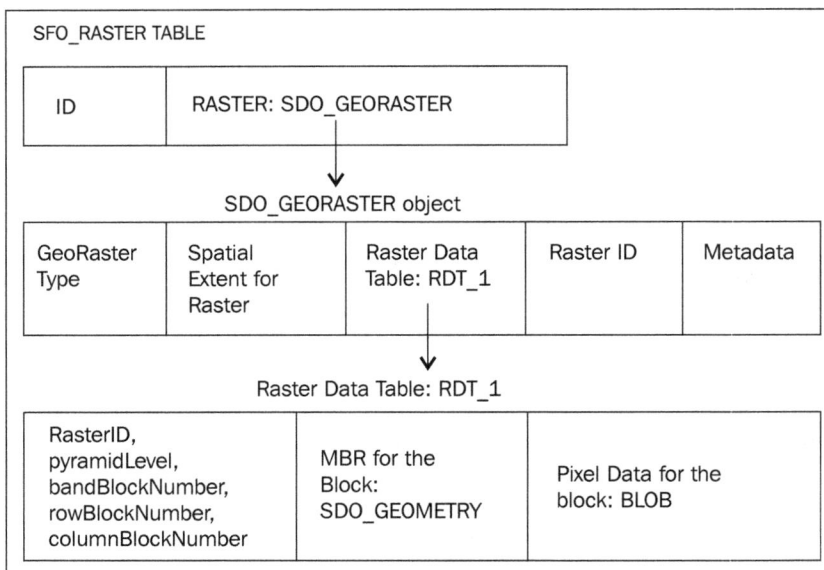

```
SFO_RASTER TABLE

  ┌──────────┬──────────────────────────────────────────┐
  │   ID     │        RASTER: SDO_GEORASTER              │
  └──────────┴──────────────────────────────────────────┘
                            │
                            ▼
              SDO_GEORASTER object

  ┌───────────┬───────────┬───────────┬───────────┬──────────┐
  │ GeoRaster │ Spatial   │ Raster Data│ Raster ID │ Metadata │
  │ Type      │ Extent for│ Table: RDT_1│          │          │
  │           │ Raster    │           │           │          │
  └───────────┴───────────┴───────────┴───────────┴──────────┘
                            │
                            ▼
              Raster Data Table: RDT_1

  ┌──────────────────┬──────────────┬──────────────────┐
  │ RasterID,        │ MBR for the  │ Pixel Data for the│
  │ pyramidLevel,    │ Block:       │ block: BLOB      │
  │ bandBlockNumber, │ SDO_GEOMETRY │                  │
  │ rowBlockNumber,  │              │                  │
  │ columnBlockNumber│              │                  │
  └──────────────────┴──────────────┴──────────────────┘
```

Pyramiding of raster data

The concept of pyramids for raster data can be explained with the help of map scales common in GIS. The resolution of raster data denotes the area on the ground covered by each raster pixel. Since raster is a discretized representation of the model space, a raster cell represents a rectangular area on the ground. Often, the rectangle can be a square, but this is not always the case. The resolution of the raster is also similar to the scale of the map. Consider a raster with a resolution of 10 x 10 meters. If this raster is to be displayed on a map where each pixel in the map is 10 x 10 meters, the raster cell values can be directly used in the map. If the map scale is smaller than the resolution of the raster cells (for example, a map where each pixel is 40 x 40 meters), then the raster cell values have to be combined to generate cells that cover more area on the ground than the original raster data. This process of generating new cell values for pixels is called pyramiding.

GeoRaster storage allows pyramiding of the pixel data that enables the raster data to be represented at different resolutions. GeoRaster provides several native methods for generating pyramids for any type of raster data. We illustrate this concept of pyramiding with a simple example using the following figure. The figure shows a raster with 12 x 16 pixels, where each pixel represents a particular land use type. Let us construct the next level in the pyramid by generating a new pixel for four pixels in the original raster such that the next pyramid level will have 6 x 8 pixels in the raster. We generate a new pixel by looking at the corresponding four pixels from the original raster and assigning a value for the new pixel. We use a simple majority rule for assigning the values for new pixels, for example, if two or more pixels out of the four candidate pixels are red, the new pixel gets the value of red. If there is a tie, we can use priority values for these colors to choose the value for the new pixel. Let us assign priorities so that the values are ordered as red, pink, yellow, green, and light green. So, in the four candidate pixels, each having red, pink, yellow, and green values, the new pixel will get the value of red using this simple algorithm.

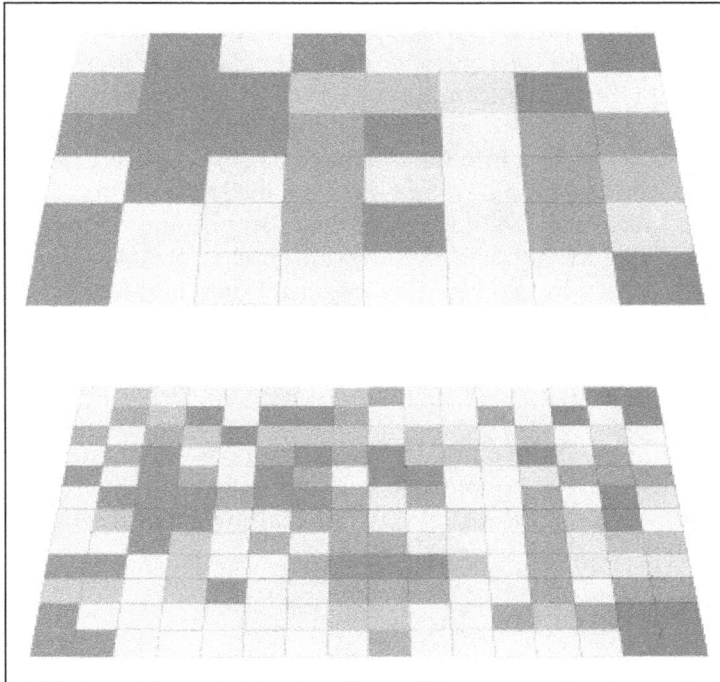

In practice, more complex algorithms are used to generate pixel values in the pyramids, and users are encouraged to read the GeoRaster users guide if interested in learning about these algorithms. The pyramids are numbered from 0...N, where 0 is the highest resolution raster. All the examples in this chapter use the level 0 pyramid.

Georeferencing

Georeferencing establishes the relationship between cell coordinates of GeoRaster data and real-world ground coordinates (or some local coordinates). Georeferencing assigns ground coordinates to cell coordinates and cell coordinates to ground coordinates. Georeferencing can be accomplished by providing an appropriate mathematical formula, enough **ground control point (GCP)** coordinates, or rigorous model data from the remote sensing system. Georeferencing does not change the GeoRaster cell data or other metadata except as needed to facilitate the transformation of coordinates between the cell coordinate system and the model coordinate system. This topic itself is very complex, and the readers are urged to read the GeoRaster documentation to get a better understanding of this topic. Since most of this information is automatically set up by the raster loading tools, we skip detailed discussion of this topic in this book.

Loading data into GeoRaster

Raster data is usually produced in files. There are two main components to raster data: the raw pixel data and the corresponding metadata that includes the source information, spatial reference information, and so on. One of the most difficult tasks for developers planning to use GeoRaster is to find the right approach for loading data on the file system into the database. This process is error prone and time consuming, if the right tools are not used for the loading process. **Geospatial Data Abstraction Library (GDAL)** is one of the most popular open source tools for handling raster data in several different well-known formats. GDAL also has a GeoRaster driver that can be used for loading raster data into GeoRaster. GDAL is mostly used via a command-line interface. But you need to be an experienced GDAL user to take advantage of all the command-line options provided by the tool. Oracle now provides a GUI-based tool on top of GDAL that is easy to use. This can be downloaded from `http://www.oracle.com/technetwork/indexes/samplecode/spatial-1433316.html`. The included ZIP file has the GDAL distribution of Linux 64-bit and Windows 32 and 64 platforms. It also has the GUI wizard along with the user guide for the ETL tool. The rest of the data loading examples in this chapter are described using this GDAL ETL tool.

Using GDAL to load raster data

The examples in this section are described using the Linux version of the GDAL tool obtained from the previous link. Windows users should adapt these steps accordingly. The directory paths in this example are with respect to the directory used to unzip the GDAL distribution from the earlier OTN page. GDAL requires an OCI client or the DB server installed on the machine used to run the GDAL commands. It requires the setup of LD_LIBRARY_PATH to point to the OCI libraries for it to work correctly. It also requires the GDAL_DRIVER_PATH environment variable to point to the plugins directory of GDAL, where the GeoRaster driver is located. This directory is located under gdal/lib/gdalplugins. Users can look at the startGeoRasterETL.sh file to see how these two variables are set up. If the GDAL tool is running on a machine that is different from the machine where the DB is installed, it requires a TNSNames entry to be created so that the DB can be accessed from the client machine. The examples in this chapter use the GDAL command-line tools as these tools give the greatest flexibility for loading raster data into GeoRaster.

GDAL provides a command to get information about the data stored in the filesystem. We used sample data from a couple of the websites for building the examples in this chapter. A color image of San Francisco bay area is available from http://www.terracolor.net/sample_imagery.html. Another good source for the land cover images is available at http://www.mrlc.gov/nlcd06_data.php. Users can also get sample data sets from the USGS website that provides several types of raster data for the Unites States. The national MapViewer site, http://viewer.nationalmap.gov/viewer, is also a good place to get DEM and other data for the United States.

Once some of this sample data is copied onto the local machine, the gdalinfo command can be used to find more details about each of the data sets. For our examples, we use a land cover image, a DEM, and a color image. The gdalinfo command gives some basic information about the raster along with the coordinate system used for the raster and any other georeferencing information. For example, this command gives the following information for the San Francisco color image. It is important to notice the coordinate reference system used for the image along with the pixel size and the actual number of pixels in the images. This particular image has 7204 x 7204 pixels and it is in the WGS 84 coordinate system:

```
./gdal/bin/gdalinfo tc_sanfrancisco_us_geotiff/TerraColor_
SanFrancisco_US_15m.tif
Driver: GTiff/GeoTIFF
Files: tc_sanfrancisco_us_geotiff/TerraColor_SanFrancisco_US_15m.tif
       tc_sanfrancisco_us_geotiff/TerraColor_SanFrancisco_US_15m.tfw
Size Is 7204, 7204
Coordinate System Is:
```

```
GEOGCS["WGS 84",
    DATUM["WGS_1984",
        SPHEROID["WGS 84",6378137,298.257223563,
            AUTHORITY["EPSG","7030"]],
        AUTHORITY["EPSG","6326"]],
    PRIMEM["Greenwich",0],
    UNIT["degree",0.0174532925199433],
    AUTHORITY["EPSG","4326"]]
Origin = (-122.809804036593036,38.285436652047451)
Pixel Size = (0.000138800000000,-0.000138800000000)
Metadata:
  AREA_OR_POINT=Area
Image Structure Metadata:
  INTERLEAVE=PIXEL
Corner Coordinates:
Upper Left   (-122.8098040,   38.2854367)
Lower Left   (-122.8098040,   37.2855215)
Upper Right  (-121.8098888,   38.2854367)
Lower Right  (-121.8098888,   37.2855215)
Center       (-122.3098464,   37.7854791)
Band 1 Block=7204x1 Type=Byte, ColorInterp=Red
Band 2 Block=7204x1 Type=Byte, ColorInterp=Green
Band 3 Block=7204x1 Type=Byte, ColorInterp=Blue
```

GDAL has options to create tables with SDO_GEORASTER columns, but it is not recommended. It is better to create these tables directly in the DB as it gives the complete flexibility of setting up all the storage and other parameters as these might not be supported through the GDAL driver. For our example, we create a table called SDO_RASTER as follows:

```
Create Table CITY_RASTER
(
  ID NUMBER, RASTER SDO_GEORASTER
);

Create Table RDT_1 OF MDSYS.SDO_RASTER
(
  Primary Key (RASTERID, PYRAMIDLEVEL, BANDBLOCKNUMBER,
  ROWBLOCKNUMBER, COLUMNBLOCKNUMBER)
)
Lob(RASTERBLOCK) Store As (NOCACHE NOLOGGING);
```

The GeoRaster storage model also requires the RDT tables to store the actual raw pixel data, so we create a table (called `RDT_1`) as well as part of the initial step. Once these two tables are created, we can load a GeoTiff-formatted file using the following commands:

```
gdal/bin/gdal_translate -of georaster TerraColor_SanFrancisco_US_15m.
tif georaster:book/book@11gr2,city_raster,raster -co
Insert="(ID, RASTER)  Values(1, SDO_GEOR.INIT('RDT_1', 10))"
```

The GDAL driver requires the specification of `georaster` as the output format along with the input file name. The `INSERT` option specifies that the driver should insert the new row in the existing GeoRaster table and is expecting two column values: a first column of type integer for `ID` and a second column of type `SDO_GEORASTER` for storing the raster data. The `SDO_GEOR.INIT()` function returns an object of type `SDO_GEORASTER` and is invoked with the syntax as shown in the example. The `SDO_GEOR.INIT()` function initializes the GeoRaster storage and links the RDT table to the new GeoRaster object that is created. Users also need to specify the `ID` used for identifying the GeoRaster object in the RDT table. It is recommended that this ID match the ID stored in the first column of the table to make it easy to do the association between raster blocks and the GeoRaster objects.

In the following code example, we load a DEM file into GeoRaster. The DEM file is stored in flt format. This raster is loaded with an ID of 16:

```
gdal/bin/gdal_translate -of georaster floatn41w124_1.flt
georaster:book/book@11gr2,city_raster,raster
-co Insert="(ID, RASTER)  Values(2, SDO_GEOR.INIT('RDT_1', 16))"
```

Finally, the last file is a land cover raster in GeoTiff format, and we load it into the DB with an ID of 15:

```
gdal/bin/gdal_translate -of georaster NLCD2001_LC_N36W120_v2.tif
georaster:scott/tiger@11gr2,city_raster,raster
-co Insert="(ID, RASTER)  Values(2, SDO_GEOR.INIT('RDT_1', 15))"
```

Loading multiple files into a single raster object

It is very common to start with a set of raster files and load the data from all the files into a single GeoRaster object. Due to various size limits with filesystems, very large rasters are usually shipped in smaller files. But, when this data is loaded into the DB, data from all the relevant files should be mosaicked and loaded into a single GeoRaster object. There are two options for loading multiple files into a single GeoRaster object. GeoRaster provides a powerful mosaicking functionality that can take several raster objects and mosaic them into a single raster object. One disadvantage with this method is the space overhead. Raster data has to be loaded into an individual GeoRaster object first before they can be mosaicked and combined into a single GeoRaster object. The advantages are the advanced mosaicking capability that allows the overlapping of raster data. This method even supports raster data with gaps in between them. The second option is to use the GDAL driver and mosaic the data while loading it into GeoRaster. The main disadvantage of this approach is the limitations of the GDAL driver for dealing with overlapping raster data. But if the input data is fairly regular, the GDAL driver is the best option. In the following code example, we show how to load multiple files into GeoRaster with an ID of 20:

```
-- create a virtual mosaic list of files
gdal/bin/gdalbuildvrt MY.VRT *.flt -overwrite

-- load the virtual mosaic into the DB
gdal/bin/gdal_translate -of georaster MY.VRT georaster:scott/
tiger@11gr2,city_raster,raster
-co Insert="(ID, RASTER)  Values(20, SDO_GEOR.INIT('RDT_1', 20))"
```

Verification of data after the load

After the data is loaded into GeoRaster, it is very important to make sure all of the information is transferred into the database accurately. For example, if the georeferencing information is not accurate, the queries to retrieve pieces of the raster based on other spatial objects will return wrong results. GeoRaster provides a very useful UI tool that can be used by DBAs and other database users to verify the raster data loading is done accurately. This tool is shipped as part of the sample schema and demos that come with the Oracle DB distribution. Under the $ORACLE_HOME/md/demo/georaster.java directory, users can find the thick client Java tool to visualize raster data. The tool can be invoked using the startGeoRasterViewer.sh script. Once the tool is started, under the menu option **Rasters**, there are choices for loading the data from the DB or the filesystem. Pick the DB option here and follow the workflow to create a connection to the DB and retrieve the list of rasters from the database. Once a raster with the given ID is picked, it will be displayed in the window. If the view is switched to the metadata tab, the raster metadata will be displayed. This should be checked to make sure it matches the metadata data listed by running the gdalinfo command on the corresponding file.

Working with GeoRaster

Once the raster data is loaded into a GeoRaster object, the raster data is available for use by other applications. There are two main types of applications for raster data: analytical applications that use the cell values in raster data to solve a spatial or non-spatial business problem, and visualization applications that use the raster data as background images for other vector maps.

Coordinating system transformations of GeoRaster

In the sample data, we loaded rasters that are in different coordinate systems. The land cover image has an SRID of 32775 (equal-area projection) while the DEM data has an SRID of 4269 (Geodetic). The rest of the data used for our San Francisco examples has an SRID 2872. So, we first show how to transform this raster into the same coordinate system as the rest of our sample data. The following code example shows the PL/SQL code required to transform the DEM data (that is loaded with an ID of 16) and create a new raster with an ID of 30. Note that different rasters with different coordinate systems can be stored in the same GeoRaster table. So, we store the new transformed raster object in the same CITY_RASTERS table:

```
Declare
    gr1 sdo_georaster;
    gr2 sdo_georaster;
Begin
  Select raster Into gr1 From CITY_RASTER Where id=16;
  Insert Into CITY_RASTER Values (30, SDO_GEOR.init('rdt_1', 30))
  RETURNING raster Into gr2;
   -- target SRID Is 2872
   -- use the same interleaving as the source raster
  sdo_geor.Reproject(gr1, 0,  SDO_NUMBER_ARRAY(0, 0, 3611, 3611),
                 null, null, 'blocking=true,
             blocksize=(256,256,1),  interleaving=BSQ', 2872, gr2);
  Update CITY_RASTER SET raster=gr2 Where id=30;
  Commit;
End;
```

Visualization applications for GeoRaster

A common use of raster data is for background images of vector maps. When the data is stored in GeoRaster, users can easily retrieve portions of the raster corresponding to an area of interest and generate an image with a specified format. Since the raster data itself does not have to be image type, it is necessary to convert general raster data types to one that can be displayed as an image. For this purpose, GeoRaster supports conversion to BMP, TIFF, PNG, and ESRI world file formats. There are certain size limits while using the GeoRaster functions for this format conversion, and so we describe an alternate method for exporting the raster data using the GDAL tool.

First, we describe the native GeoRaster function for exporting the data to any one of the earlier mentioned formats. Since this procedure runs in the database JavaVM, it requires write privileges to be granted to MDSYS and the user who is executing the export procedure. For this, connect as a SYSDBA user and grant the following privileges. Here, we are assuming the output file will be called landcover.bmp:

```
connect sys as sysdba

EXEC dbms_java.grant_permission( 'MDSYS', 'SYS:java.
io.FilePermission', '/private/Book/Schema/Chapter9/landcover.bmp',
'write' );
EXEC dbms_java.grant_permission( 'BOOK', 'SYS:java.io.FilePermission',
'/private/Book/Schema/Chapter9/landcover.bmp', 'write' );
```

Once the required permissions are granted, we can use the GeoRaster native SDO_GEOR.exportTo function to create a BMP image of a subset of rasters. This function takes a GeoRaster object as an input along with the required output file format. In addition, users can also supply an area of interest that is used to crop the input raster. In the following code example, the area of interest is specified in the cell space:

```
Declare
  geor SDO_GEORASTER;
  fileName Varchar2(1024);

Begin

Select raster Into geor From CITY_RASTER Where id = 15;

-- Export a subset to a file with a world file.
fileName := '/private/Book/Schema/Chapter9/landcover.png';
sdo_geor.exportTo(geor, 'cropArea=(0,0,500,500)',
  'PNG', 'file', fileName);

End;
```

The native export function has a limit of 67 Mbytes for the output file generated. For many visualization applications, this limit does not cause any problems, as it is a good idea to get smaller images if these images are used as background images for maps. In case large image files are required, users can use the open source GDAL tool to export larger subsets of rasters to file formats. The other advantage of GDAL is that it supports many more formats than the native GeoRaster export function. In the following code example, we show how to use the GDAL tool for exporting subsets of GeoRaster to files:

```
-- first set the GDAL_DATA variable to point gdal/data directory
gdal/bin/gdal_translate -of gtiff geor:book/book@11gr2,city_
raster,raster,id=30 output.tif
```

Another most common way to use GeoRaster data in visualization applications is via **Web Mapping Service (WMS)**. Oracle Application Server MapViewer provides WMS capability, and when the raster data is stored in GeoRaster, the MapViewer can be configured to provide WMS functionality on top of the raster data stored in the database.

Analytical applications for GeoRaster

Rasters can be analyzed at pixel level to solve many common problems in GIS, for example, a DEM can be used to find the average or the maximum elevation value in a given land parcel. Similarly, a land cover raster can be analyzed to find the area of a piece of land covered by water or vegetation. For all of these types of applications, the raster cells have to be retrieved from the GeoRaster object corresponding to the given area of interest. GeoRaster provides a powerful function called SDO_GEOR.getRasterSubset used to retrieve all the cells corresponding to a given area of interest.

Since the raster cell values are stored as a BLOB, it is important to understand how to read these cell values out of the Oracle BLOB storage and cast them to standard types such as numbers and bytes. We first describe a simple function used to read cell values from the raster BLOB and return the result as a number type based on the cell depth of the raster. Note that the pixels in the raster can have any of the following types: 1BIT, 2BIT, 4BIT, 8BIT_U, 8BIT_S, 16BIT_U, 16BIT_S, 32BIT_U, 32BIT_S, 32BIT_REAL, or 64BIT_REAL. We keep the following code example simple and cover only some of these types, but the procedure can be easily extended to read other types as well:

```
Create Or Replace Function getValue(cell_depth Number, buf  raw, idx
Number, is_float Number)
Return Number Is
r1 raw(1);
r2 raw(2);
```

```
r4 raw(4);
r8 raw(8);
val Number;
Begin
  If cell_depth = 1 Then
     r1 := utl_raw.substr(buf, (idx-1)*cell_depth+1, cell_depth);
     val := utl_raw.cast_to_binary_integer(r1);
  ElsIf cell_depth = 2 Then
     r2 := utl_raw.substr(buf, (idx-1)*cell_depth+1, cell_depth);
     val := utl_raw.cast_to_binary_integer(r2);
  ElsIf cell_depth = 4 Then
     r4 := utl_raw.substr(buf, (idx-1)*cell_depth+1, cell_depth);
     -- assuming float Is 4 bytes
     If is_float = 0 Then
       val := utl_raw.cast_to_binary_integer(r4);
     Else
       val := utl_raw.cast_to_binary_float(r4);
     End If;
  ElsIf cell_depth = 8 Then
     r8 := utl_raw.substr(buf, (idx-1)*cell_depth+1, cell_depth);
     val := utl_raw.cast_to_binary_double(r8);
  End If;
  Return val;
End;
```

This function takes a raw buffer as its main input along with the cell depth in bytes of the raster data. An index into the raw buffer is also required so that the next set of bytes can be read starting from that index. Finally, the last parameter specifies whether to read the value as a floating number or as an integer type. Note that the byte types of pixels are read as integer values without the loss of any data, and the applications using this data can make appropriate adjustments to cast the integer value to whatever type is used in the application for these pixel values. For each cell depth, a corresponding number of bytes is read from the raw buffer and cast into an Oracle number type. If it is a floating-point type, then the data is read in as a binary float value. This function can be easily extended to cover other pixel types supported by GeoRaster. Note that a similar function can be written in Java or C to read the pixel data from the raster BLOBs.

Analyzing DEM data

Using this utility function, we can now create a function to read all the pixel values corresponding to a given area of interest from the GeoRaster object. We use the `getRasterSubset` function as the main component of this function to retrieve all the pixels corresponding to the given area of interest. This area of interest can be specified in the model space (in the geographically referenced space) or in the pixel space. Which type is used for the area of interest depends on the application. The following function finds the maximum value associated with a pixel corresponding to the given area of interest:

```
-- gr: the georaster object
-- plevel: pyramid level
-- bno: band Number
-- win_c: window of interest in the cell space
-- win_g: window of interest in the cell model space
Create Or Replace Function getMaxCellValue
        (gr sdo_georaster, plevel Number, bno Number,
                win_c sdo_Number_array, win_g SDO_GEOMETRY)
Return Number As
c_type Varchar2(80);
is_float Number := 0;
cell_depth Number;
parm Varchar(200);
lb blob;
buf raw(32767);
amt0 Integer;
amt Integer;
offset Integer;
len Integer;
maxv Number := null;
val Number;
Begin
  -- first figure out the cell type from the metadata
  c_type := gr.metadata.extract('/georasterMetadata/rasterInfo/
cellDepth/text()',
  'xmlns=http://xmlns.oracle.com/spatial/georaster').getStringVal();
  If c_type = '32BIT_REAL' Then
    is_float := 1;
  End If;
  cell_depth := sdo_geor.getCellDepth(gr);
  If cell_depth < 8 Then
  -- If celldepth<8bit, get the cell values as 8bit integers
  cell_depth := 8;
```

```
      parm := 'celldepth=8bit_u';
   End If;
   parm := parm || ' compression=none';

   -- create a temporary LOB to store the result of the raster subset
   dbms_lob.createTemporary(lb, true);
   If (win_g Is null) Then
      sdo_geor.getRasterSubset(gr,plevel,win_c,to_char(bno),lb,parm);
   ElsIf (win_c Is null) Then
      sdo_geor.getRasterSubset(gr,plevel,win_g,to_char(bno),lb,parm);
   End If;
   len := dbms_lob.getlength(lb);
   -- cell Depth in Bytes
   cell_depth := cell_depth / 8;

   -- make sure to read all the bytes of a cell value at one run
   amt := floor(32767 / cell_depth) * cell_depth;
   amt0 := amt;

   offset := 1;
   While offset <= len Loop
      dbms_lob.read(lb, amt, offset, buf);
      For i In 1..amt/cell_depth Loop
         val := getValue(cell_depth, buf, i, is_float);
         If maxv Is null or maxv < val Then
            maxv := val;
         End If;
      End Loop;
      offset := offset+amt;
      amt := amt0;
   End Loop;
   dbms_lob.freeTemporary(lb);
   Return maxv;
End;
```

This function also takes a pyramid level and a band number in addition to the area of interest parameter. Since this area of interest can be either in model space or cell space, this function uses the corresponding getRasterSubset call depending on which type of area of interest is used. One of the main features of the method getRasterSubset is that it is very flexible and can subset the rasters in any dimension. So, users can retrieve the cells from any pyramid level or band number. For DEM type data, there is only one band, so it is not applicable for elevation data. This function finds the maximum cell value corresponding to an area of interest, but it can be easily modified to find the average of the cell values or any other type of aggregate measure. It can also be modified to return all the cell values as an array of numbers.

With this function, we can now analyze our DEM data to find the maximum cell values for a given area of interest.

```
-- find the max value for the given area of interest in cell space
Select getMaxCellValue(raster,0,0,sdo_Number_array(0,0,511,511),null)
From city_raster Where id=16;

-- find the max cell value in the entire raster
Select getMaxCellValue(raster,0,0,null, null)
From city_raster Where id=16;

-- find the max value for the given area of interest in model space
Select getMaxCellValue(raster,0,1, null, SDO_GEOMETRY(2003,4269,null,
  sdo_elem_info_array(1, 1003, 3),
  sdo_ordinate_array(-122.8, 40.7, -123.6, 40.8)))
From city_raster Where id=16;
```

Next, we show a more complex procedure used to calculate the slope of each pixel in the DEM. For this, we first need the mathematics required to compute a slope given the elevation values for each pixel. There are several methods for finding the slope, and we use the simple method that takes eight neighboring pixels of each pixel to compute the slope. In the following example, the slope for a pixel, e, is calculated using the elevation values of its neighboring pixels. Since each pixel can have up to eight neighbors, every slope computation needs to retrieve eight elevation values from the DEM.

a	b	C
d	e	F
g	h	I

```
dz/dx = [(a+2d+g)-(c+2f+i)]/[8*x_resolution]

dz/dy = [(g+2h+i)-(a+2b+c)]/[8*y_resolution]

slope = rise/run = SQRT[(dz/dx)^2 + (dz/dy)^2]
```

This slope calculation is done for every pixel in the DEM, so we can just scan the DEM in the pixel space, compute the slope values, and write them back to the DB to create a new raster object.

We first create a set of convenience types used in the following code examples. We need a matrix type data structure to temporarily store the pixel elevation values, and a matrix-like structure will also make it easier to compute the slope values based on the above formula.

```
-- first create a varray of Number type to store 1024 Numbers
Create Or Replace Type dem_array_1 As varray(1024) of Number;

-- then we create a row type object to store a row of the matrix
Create Or Replace Type dem_row_obj As object (dem_row dem_array_1);

-- finally the type to store a 2D array of Numbers
Create Or Replace Type dem_array As varray(1024) of dem_row_obj;
```

Since the DEM can be very large, it is a good idea to retrieve subsets of the DEM and perform the slope calculations. We can think of a moving window of 1024 x 1024 pixels used to subset the DEM. Once each set of 1024 x 1024 pixels are processed, they can be stored or sent to some other program for further analysis.

```
Create Or Replace Function slopeArray
(gr sdo_georaster, plevel Number, bno Number, win_c sdo_Number_array)
Return DEM_ARRAY DETERMINISTIC As
c_type Varchar2(80);
is_float Number := 0;
cell_depth Number;
parm varchar(200);
lb blob;
buf raw(32767);
amt0 Integer ;
amt Integer ;
offset Integer ;
len Integer ;
val Number;
result dem_array;
d_r_obj dem_row_obj;
idx1 Number;
idx2 Number;
Begin
  is_float := 1;
  cell_depth := sdo_geor.getCellDepth(gr);
  parm := parm || ' compression=none';

  dbms_lob.createTemporary(lb, true);
  sdo_geor.getRasterSubset(gr,plevel,win_c,to_char(bno),lb,parm);
  len := dbms_lob.getlength(lb);
```

```
-- cell Depth in Bytes
cell_depth := cell_depth / 8;

-- make sure to read all the bytes of a cell value at one run
amt := floor(32767 / cell_depth) * cell_depth;
amt0 := amt;

offset := 1;
idx1 := 1;
idx2 := 1;
result := dem_array();
result.extend(1024);
While offset <= len Loop
  dbms_lob.read(lb, amt, offset, buf);
  For i in 1..amt/cell_depth Loop
    val := getValue(cell_depth, buf, i, is_float);
    If (idx1 = 1025) Then
        idx1 := 1;
        idx2 := idx2 + 1;
    End If;
    If (idx1 = 1) Then
      d_r_obj := dem_row_obj(dem_array_1());
      result(idx2) := d_r_obj;
      result(idx2).dem_row.extend(1024);
    End If;
    result(idx2).dem_row(idx1) := val;
    idx1 := idx1 + 1;
  End Loop;
  offset := offset+amt;
  amt := amt0;
End Loop;
dbms_lob.freeTemporary(lb);
Return result;
End;
```

It is important to note the creation of the output matrix based on the result of the raster subset operation. Note that the raster data is stored in the BSQ format for this DEM data. Since there is only one band of data, we do not focus on how to deal with band inter-leaving in this example. Furthermore, each line of the data is stored in sequence, followed by the next line of the data, and so on. Since we are doing the subset operation on a 1024 x 1024 size window, we first read the first 1024 values of the result. These become the first of the resulting matrix. The next 1024 values become the second row of the matrix, and so on. Therefore, this procedure can be used to construct elevation values for 1024 x 1024 pixels at a time. We then need another procedure that can loop through the whole raster and construct the slope value of each pixel. The procedure to construct the actual slope is not described here, but given a matrix of elevation values as the input, it can be easily developed. The following code example shows how to repeatedly call this preceding routine to retrieve elevation values for 1024 x 1024 blocks:

```
Declare
result dem_array;
idx1 Number;
idx2 Number;
row_s Number;
col_s Number;
row_c Number;
col_c Number;
Begin

  row_s := 3611;
  col_s := 3611;
  row_c := 0;
  col_c := 0;
  While row_c < row_s Loop
    While col_c < col_s Loop

    Select      slopeArray(raster,0,0,
          sdo_Number_array(row_c,col_c,1023+row_c,1023+col_c) )
          Into result
    From  city_raster Where id=16;

    For idx2 in 1 .. 1024 Loop
      For idx1 in 1 .. 1024 Loop
        dbms_output.put_line(to_char(result(idx2).dem_row(idx1)));
      End Loop;
    End Loop;
```

```
      col_c := col_c + 1024;
    End Loop;
    col_c := 0;
    row_c := row_c + 1024;
  End Loop;
End;
```

Analyzing land cover data

Land cover data is an example of thematic raster data that is produced using multi-spectral images. Land cover data has discrete sets of values assigned to different pixels in the raster. These values denote a specific characteristic for the land covered by each pixel. In the example of land cover data we have, the classification for pixel data is as follows:

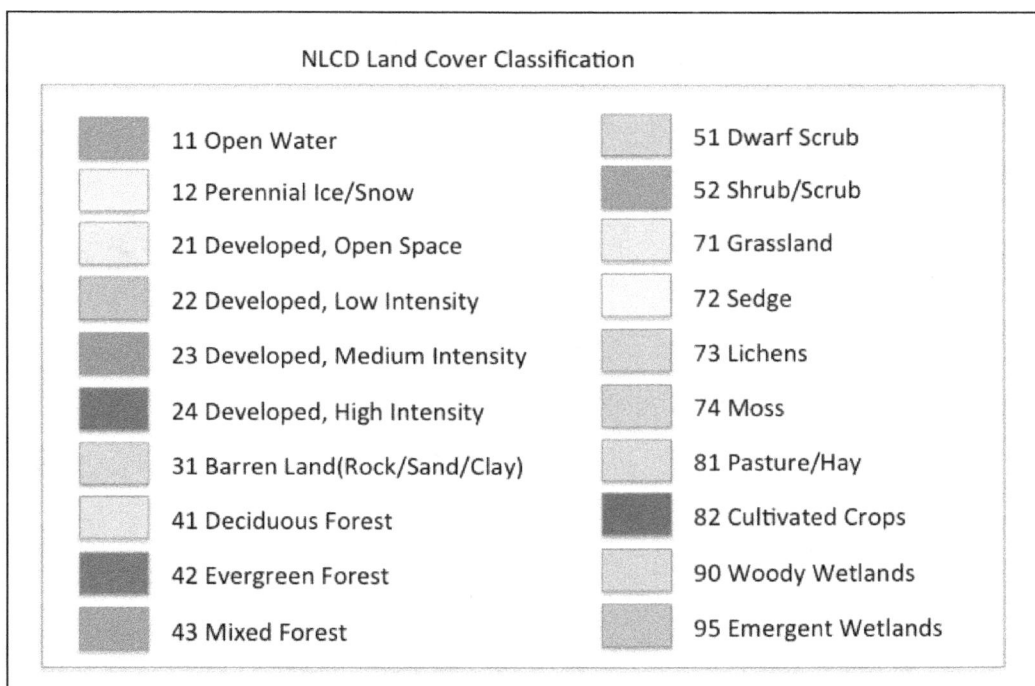

NLCD Land Cover Classification

11 Open Water
12 Perennial Ice/Snow
21 Developed, Open Space
22 Developed, Low Intensity
23 Developed, Medium Intensity
24 Developed, High Intensity
31 Barren Land(Rock/Sand/Clay)
41 Deciduous Forest
42 Evergreen Forest
43 Mixed Forest

51 Dwarf Scrub
52 Shrub/Scrub
71 Grassland
72 Sedge
73 Lichens
74 Moss
81 Pasture/Hay
82 Cultivated Crops
90 Woody Wetlands
95 Emergent Wetlands

With this data classification, it is often desirable to compute the average area covered by a specific land cover for a given area of interest. For example, a city planner might want to calculate how much area of a new proposed development is covered by open water. This can be done using `getRasterSubset` as the core function from GeoRaster. We cover two types of cases in our example. In the first case, the pixels are counted based on the exact value used as a search parameter, that is, one can find how much of the given area of interest is covered by land of type 21 (developed, open space). In the second case, a general category of the land cover is calculated, that is, one can find out how much of the given area of interest is covered by developed land (type 21, 22, 23, 24).

```
Create Or Replace Function computeLandCover
(gr sdo_georaster, plevel number, bno number, win_c sdo_number_array,
    land_c_type number, groups number default 0)
Return Number As
c_type Varchar2(80);
is_float Number := 0;
cell_depth Number;
parm varchar(200);
lb blob;
buf raw(32767);
amt0 Integer ;
amt Integer ;
offset Integer ;
len Integer ;
percent Number := null;
val Number;
total_cells Number := 0;
lc_type Number := 0;
Begin

  cell_depth := 8;
  parm := 'celldepth=8bit_u';
  parm := parm || ' compression=none';

  dbms_lob.createTemporary(lb, true);
  sdo_geor.getRasterSubset(gr,plevel,win_c,to_char(bno),lb,parm);
  len := dbms_lob.getlength(lb);
  -- cell Depth in Bytes
  cell_depth := cell_depth / 8;

  -- make sure to read all the bytes of a cell value at one run
  amt := floor(32767 / cell_depth) * cell_depth;
  amt0 := amt;
```

```
offset := 1;
While offset <= len Loop
  dbms_lob.read(lb, amt, offset, buf);
  For i In 1..amt/cell_depth Loop
    val := getValue(cell_depth, buf, i, is_float);
    total_cells := total_cells + 1;
    If (groups = 0) Then
      If ( val Is not null and  val = land_c_type) Then
        lc_type := lc_type + 1;
      End If;
    Else
      If ( val Is not null and  land_c_type = 10*(floor(val/10)))
      Then

        lc_type := lc_type + 1;
      End If;
    End If;
  End Loop;
  offset := offset+amt;
  amt  := amt0;
End Loop;
dbms_lob.freeTemporary(lb);

Return 100*(lc_type/total_cells);

End;
```

This function takes the land use type along with a parameter to specify whether to look for a group of land use types. This function can then be used to find the land use type for any given area of interest. Note that the function is defined in cell space, but can be easily extended to work with model space as well.

```
-- look for Open Water
Select computeLandCover(raster,0,0,sdo_number_array(0,0,511,511), 11)
From city_raster Where id=15;

-- look for Developed Open space
Select computeLandCover(raster,0,0,sdo_number_
array(4700,6300,5300,7000), 21)
From city_raster Where id=15;

-- look for any develop land; the groups parameter is used here
Select computeLandCover(raster,0,0,sdo_number_
array(4700,6300,5300,7000), 20,1)
From city_raster Where id=15;
```

Mapping from cell space to model space

When the raster data is processed as an array of numbers, integer addressing using row and column numbers is sufficient in most applications. However, the raster data is a discretized representation of a continuous space (the geographic space), and a one-to-one mapping of coordinates between the cell space and the model space is required. Since the cell space is discrete, a pixel will cover a certain area in the corresponding geographic space. This area covered by each pixel depends on the coordinate system used in the model space and the resolution of the raster cells. For very high-resolution rasters, each pixel can cover less than a square meter on the ground, while low-resolution rasters usually cover tens of square meters on the ground. The GeoRaster metadata explicitly stores this information and, using the GeoRasterViewer, this information can be checked under the metadata tab.

It is also important to understand how the cell coordinates map to the model coordinates. In the model coordinate space, the lower-left corner is considered the origin of the space. In the cell space, the top-left corner is considered the origin of the space. It is important to note that the Y coordinates flip while going from model space to cell space; that is, the minimum Y in the model space maps to the maximum Y in the cell space, and the maximum Y in the model space maps to the minimum Y in the cell space. Most of the functions provided by GeoRaster do not require the end users to understand this mapping. However, there are some functions that can return data in cell space, for example, some of the subset operations described in this chapter used the clip window in pixel space that requires the users to understand how to map a polygon from the model space to the cell space.

Converting raster cells to rectangles in model space

Using the concepts of resolution and georeferencing, we will show how to construct rectangles in model space using cell values. This type of mapping is useful in applications where the exact overlap of a set of raster cells with a polygon is important. We have used the land cover raster in the previous example, to compute the area covered by open water for a given land parcel. In that example we assumed if a land parcel overlaps a raster cell, the whole raster cell is completely covered by the land parcel. But, in practice, this is not always true, as each cell covers a certain area on the ground and a land parcel might only cover part of the area represented by the raster cell. The land cover raster has a resolution of 30 meters by 30 meters. If a typical land parcel covers a few tens of square meters, the error will be very high in our calculations if partially covered raster cells are assumed to be completely covered by the land parcel.

A solution for this problem is to find the actual extent of each raster cell that overlaps a given land parcel polygon and find out how much area of each raster cell is actually covered by the land parcel polygon. Then, depending on the percentage overlap, appropriate values can be assigned to the land parcel polygon. As we have already seen in previous examples, we know how to identify all the raster cells that are associated with a polygon when the polygon is specified in cell space.

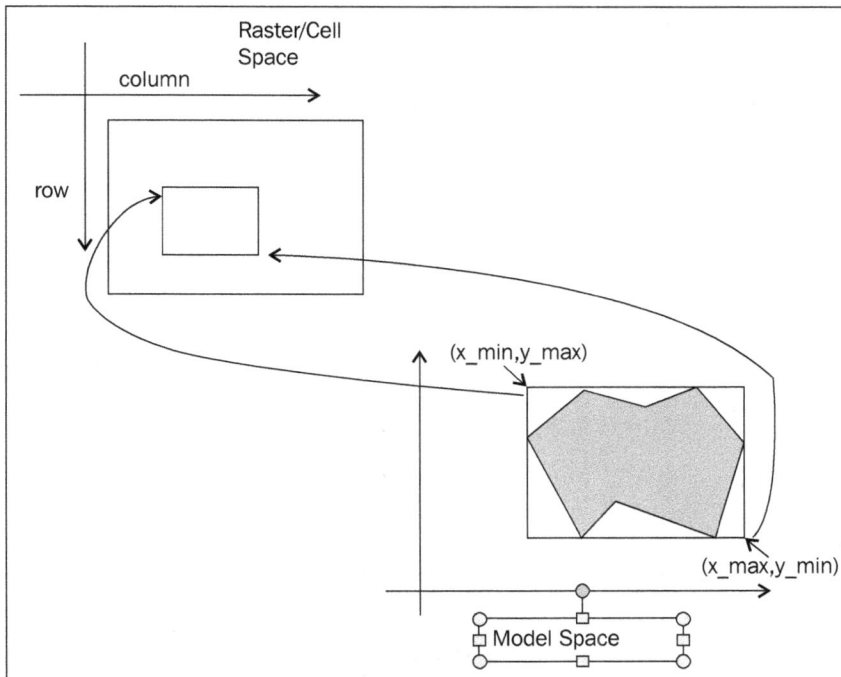

In the following code example, we show how to switch between cell space and model space. We use the functions provided by GeoRaster to develop a new method that returns the SDO_GEOMETRY for the cell given a coordinate in cell space. We then use that function to generate geometry values corresponding to all the cells covering a given area of interest:

```
-- this function takes a row, col value in cell space and returns the
-- the corresponding geometry in the cell model space
Create Or Replace Function cell_geometry (row_in number, col_in
number, raster sdo_georaster)
Return SDO_GEOMETRY Is
raster_mbr SDO_GEOMETRY;
spatial_res sdo_number_array;
```

```
x_res Number;
y_res Number;
x_offset Number;
y_offset Number;
x_divs Number;
y_divs Number;
cell_x_y sdo_number_array;
Begin

    raster_mbr := sdo_geom.sdo_mbr (raster.spatialextent);
    spatial_res := sdo_geor.getSpatialResolutions(raster);

    x_res := spatial_res(1);
    y_res := spatial_res(2);
    x_offset := raster_mbr.sdo_ordinates(1) + col_in*x_res;
    y_offset := raster_mbr.sdo_ordinates(4) - row_in*y_res;

    -- switch the Y values since the min/max along Y is flipped
    Return SDO_GEOMETRY(2003, raster_mbr.sdo_srid, null,
      sdo_elem_info_array(1, 1003, 3),
      sdo_ordinate_array(x_offset,y_offset-y_res, x_offset+x_res,y_
offset));

End;
```

GeoRaster provides a function to get the cell coordinates for a given point geometry value. This function works as follows:

```
Select sdo_geor.getCellCoordinate(raster, 0, SDO_GEOMETRY(2001,32775,
  sdo_point_type(-2214093,2026020,null), null,null)) coord
From city_raster Where id=2;

COORD
------------------------------------------------------------------
SDO_NUMBER_ARRAY(2526, 5525)

-- now use this returned value in our function to get cell geometry
Select cell_geometry(2526, 5525, raster) cell_g
From city_raster
Where id = 2;
```

```
CELL_G
-----------------------------------------------------------------
SDO_GEOMETRY(2003, 32775, NULL, SDO_ELEM_INFO_ARRAY(1, 1003, 3), SDO_
ORDINATE_ARRAY(-2214105, 2026005, -2214075, 2026035))

-- check that function works as expected
-- the input point should ANYINTERACT with this cell geometry
Select sdo_geom.relate(SDO_GEOMETRY(2001,32775,
  sdo_point_type(-2214093,2026020,null), null, null), 'determine',
  cell_geometry(2526, 5525, raster), 0.05) relation
From city_raster
Where id = 2;


RELATION
-----------------------------------------------------------------
INSIDE
```

Now we can use this function to generate the cell geometry values for any given polygon geometry. Since we will also know the cell values in the cell coordinate space, we can get the raster values corresponding to those cells and do the appropriate adjustment based on the area of intersection between the cell geometry and the given polygon geometry. The following code example assumes the input polygon geometry is in the same spatial reference system as the raster object:

```
Create Or Replace Function computeLandCoverExact( gr sdo_georaster,
       plevel number, bno number, win_g SDO_GEOMETRY, tol number,
  land_c_type number) Return Number Is
col_m Number;
row_m Number;
mbr SDO_GEOMETRY;
row_col_min sdo_number_array;
row_col_max sdo_number_array;
x_min Number;
y_min Number;
x_max Number;
y_max Number;
cell_value Number;
cell_geom SDO_GEOMETRY;
int_geom SDO_GEOMETRY;
total_cells Number;
cells_w_type Number;
weight Number;
Begin
  -- find the MBR of the input geometry for reference
```

```
mbr := sdo_geom.sdo_mbr(win_g);
x_min := mbr.sdo_ordinates(1);
y_min := mbr.sdo_ordinates(2);
x_max := mbr.sdo_ordinates(3);
y_max := mbr.sdo_ordinates(4);
-- cell coordinate of upper left corner
row_col_min := sdo_geor.getCellCoordinate(gr, 0, SDO_GEOMETRY(2001,
    mbr.sdo_srid, sdo_point_type(x_min,y_max,null),null,null));
-- cell coordinate of bottom right
row_col_max := sdo_geor.getCellCoordinate(gr, 0, SDO_GEOMETRY(2001,
    mbr.sdo_srid, sdo_point_type(x_max,y_min,null),null,null));

-- iterate over all cells to identify cells that really intersect
total_cells := 0;
cells_w_type := 0;
For row_m in row_col_min(1) .. row_col_max(1) Loop
  For col_m in row_col_min(2) .. row_col_max(2) Loop
      cell_geom := cell_geometry(row_m, col_m, gr);
      int_geom := sdo_geom.sdo_intersection(cell_geom, win_g, tol);
      -- if the current cell has interaction with win_g do more
      If (int_geom Is not NULL) Then
        total_cells := total_cells + 1;
        cell_value := sdo_geor.getCellValue(gr, 0, row_m, col_m, 0);
        If (cell_value = land_c_type) Then
          weight := sdo_geom.sdo_area(int_geom,tol)/
                         sdo_geom.sdo_area(cell_geom,tol);
          cells_w_type := cells_w_type + weight;
        End If;
      End If;
  End Loop;
End Loop;
Return (cells_w_type*100)/total_cells;
End;
```

Note that the input geometry's MBR is used to find the rectangle of interest in the cell space. Once we know the upper and lower bounds in the cell space corresponding to the input geometry, we can iterate over all the cells and compute their corresponding geometry in the model space. With this geometry in the model space, we can use the Oracle Spatial intersection function to compute the area of overlap between the cell geometry and the input geometry. If the cell geometry is fully covered by the input geometry, the weight for that cell will be equal to 1. If a cell is only partially covered by the input geometry, the weight is calculated as a percentage of the area of the cell covered by the input geometry. Then, we sum over all the cells and find the final percentage of land cover type in the given input geometry. This function can be used as follows:

```
Select computeLandCoverExact(raster, 0, 0, SDO_
GEOMETRY(2003,32775,null, sdo_elem_info_array(1, 1003, 1),
  sdo_ordinate_array(-2217449,2026800, -2216129,2025210,
-2213909,2025090, -2213009,2027250, -2215679,2027820,
-2217449,2026800)), 0.05, 21) weighted_coverage
From city_raster Where id = 2;


WEIGHTED_COVERAGE
--------------------------------------------------

12.6008587
```

This example illustrates how to combine the raster and vector analysis functions to solve common raster analysis problems. Using PL/SQL, it is very efficient to do this type of raster processing in the database, since this eliminates the need for a specialized raster client application to analyze the raster data. And because the processing is done close to the data storage, this also reduces the overhead of transferring large amounts of raster data over the network to client applications.

Summary

In this chapter, we described Oracle Spatial's raster data management feature called GeoRaster. We learned different methods to load raster data from files into the DB. The examples explained the public domain GDAL tool for loading raster data into the database. We also learned different applications for raster data, and how different types of raster data can be used in land management applications. These examples explained how to combine the vector and raster functions available in the database to solve many problems encountered in raster applications. GeoRaster features are used to illustrate the ease of doing raster analysis in the database, and the advantages of storing raster and vector data in the same data store are explained using the examples.

In the next chapter, we will discuss Java Stored Procedures as another way of extending Oracle Spatial capabilities. We'll also learn how to integrate existing Java code into the database using Java Stored Procedure concepts.

10
Integrating Java Technologies with Oracle Spatial

Chapter 6, Implementing New Functions, Chapter 7, Editing, Transforming, and Constructing Geometries, and *Chapter 8, Using and Imitating Linear Referencing Functions,* outlined how to embrace and extend the standard functionality available with Oracle Locator and Spatial using PL/SQL. PL/SQL is a programming language that is native to Oracle. Oracle also supports the creation of Java stored procedures. This chapter explores the application of Java to spatial processing involving the SDO_GEOMETRY type.

In this chapter, we will cover the following topics:

- Why Java and Oracle Spatial?
- Available Java spatial technologies
- Matching requirements to source code project
- Strengths and limitations of using Java
- How to download, modify, compile, and install external libraries
- Calling an external method
- Converting an SDO_GEOMETRY object to a Java object
- Exposing **JTS Topology Suite** functionality:
    - One-sided buffers
    - Snapping geometries
- Building polygons from lines
- Performance of Java-based **SQL** processing

Why Java and Oracle Spatial?

PL/SQL language can be used to develop extensions with Oracle Spatial. But PL/SQL is not a common, cross-platform development language for most programmers outside the database world. Java is a good choice for such a situation. In addition, there already exist a lot of spatial algorithms that are written in Java and licensed for use as free and open source software.

Java is very much a cross-platform technology that Oracle has invested in heavily, and it runs on all platforms on which the Oracle database runs. It can be used to develop applications that can be deployed in any architectural tier (client, middle, or data). This chapter, however, looks specifically at data-tier deployment of functionality inside the **Java Virtual Machine (JVM)** embedded in the Oracle database.

It is the case that many existing Oracle Spatial functions provided by the Locator and Spatial products are implemented in Java. Therefore, what is promoted in this chapter is not out of the ordinary; it is following a well-established and accepted practice of using Java inside Oracle database!

Developing for deploying inside an Oracle database comes with a requirement that source code and compilation of custom classes is released and version-compatible. Each Oracle database release supports one and only one JVM version. The following table outlines the Java and Oracle versions to date:

Oracle database version	Java version
8*i* (8.1.5)/8*i* (8.1.6 or later)	1.1.6/1.2
9*i*Rx	1.3.x
10*g*Rx	1.4.x
11*g*Rx	1.5.x

Developing in Java does not negate the need to know a little PL/SQL! When one has developed a Java class plus method and deployed it, or one wishes to access an existing Java method that is deployed with the JVM itself, these class methods can only be accessed via SQL, via a PL/SQL function declaration. For example, to access the actual Java version deployed with a particular Oracle database, one can access a Java system property indicated by a specified key as follows:

```
Create Or Replace Function getJVMProperty(p_property In varchar2)
Return varchar2 Is
Language Java Name
'Java.lang.System.getProperty(Java.lang.String)
return Java.lang.String';
/
```

```
show errors
-- Result
--
No Errors.
-- Get Java Runtime Environment version
--
Select getJVMProperty('Java.version') as JVersion
  From dual;
-- Result
--
JVERSION
--------
1.5.0_10
```

Some valid `p_property` values include: `Java.io.tmpdir` (default temp file path); `os.name` (operating system name), and `file.separator` (file separator, for example, '/' on Linux/Unix).

Java stored procedures are complementary rather than competitive

Some of the functions created in this chapter, look like they implement already existing functionality (for example, overlay, buffer, or centroid generation). It is not fair to say that this is done to compete with Oracle itself. It is not an "either/ or" proposition; it is "both/and". It is better to consider each implementation on its merits, and to look at it from the perspective of what it offers that is complementary to what Oracle itself has provided.

Some functions are there because of licensing issues. While it is always better to purchase the vendor's full product, when one needs such functionality, it is not always possible. Just as Oracle sets pricing to suit its needs, customers are also allowed to make such choices as well, especially since deploying alternate solutions is supported by the Oracle database via its open development and deployment framework. But here, customers need to be careful. If development, deployment, and support costs are fully and transparently accounted for; a decision to develop and deploy "competitive" functions simply may not be cost effective.

Some of the presented functions appear to compete with existing functions, for example, centroid generation until one looks at the results. Algorithmic differences can provide a better solution to a business or functional requirement than the existing one. So, having two functions to attack a problem may be better than just one. In the case of buffering, the **JTS** solution offers complementary or extended functionality that is not in the base Locator version. Finally, some functions like `ST_PolygonBuilder` are simply not available in Locator or Spatial.

Disclaimer

This chapter will concentrate on developing and deploying specific spatial functions via Java Stored Procedures(JSP), which are coded in Java (as opposed to PL/SQL), stored and executed by the database JVM in the database memory space, but accessed via PL/SQL. It will neither be a thorough tutorial on Java development, nor on all the different and alternate methods for developing, compiling, and deploying JSPs. The relevant Oracle database documentation is thorough and complete, and the reader is directed to read this documentation to gain a more complete and detailed understanding. However, it will include those steps that the author uses when deploying and accessing spatial functions, deployed as JSPs.

Sourcing available Java spatial technologies

There are a number of open source spatial technologies and projects that are a potential source of implemented algorithms, which could be deployed within the Oracle database's JVM, to complement or extend the existing Oracle Spatial software.

The following are some of the main ones that are available:

deegree	deegree is an open source software for spatial data infrastructures and the geospatial web.	**JTS Topology Suite**	The JTS Topology Suite is an API for modeling and manipulating two-dimensional linear geometry. It provides numerous geometric predicates and functions. JTS conforms to the **Open GIS Consortium Simple Features Access– Part 2: SQL Option, Version 1.1 and 1.2 (OGC SFA)**
GeoServer	GeoServer is an open source software server written in Java that allows users to share and edit geospatial data.	**JASPA**	Java Spatial is a spatial extension for relational database systems that implements the OpenGIS Simple Features for SQL and partially the SQL/MM standard.
GeoTools	GeoTools is an open source Java library that provides tools for geospatial data.	**HatBox**	HatBox is a user space spatial extension for the H2 and Derby Java databases.

| JUMP | The Unified Mapping Platform (JUMP) is a GUI-based application for viewing and processing spatial data. Spawned OpenJUMP and other projects. | Proj4j | Proj4J is a Java library to transform point coordinates from one geographic coordinate system to another, including datum transformations. |
| uDig | The goal of uDig is to provide a complete Java solution for desktop GIS data access, editing, and viewing. | Sexante | Sexante is a spatial data analysis library and a powerful geoprocessing framework. |

Whether any one project's source code can be used in a specific Oracle database version, depends on the compliance of the code with the correct Java version (refer the previous table on Oracle databases and Java versions).

Basic requirements for Java processing

To build a new Java-based functionality that can complement what is already available through the Oracle SDO_GEOMETRY data type; an understanding of what is needed in that technology is required before deciding, which existing open source projects to use. The following things are the basic requirements:

- A common geometry type hierarchy
- Conversion methods
- A source of quality existing spatial algorithms

Common geometry type hierarchy

To access existing open source Java functionality, we would prefer that functionality to be based on a common method of representing a Geometry object. Otherwise, we could waste a lot of time and effort converting between representations, when we could be doing more useful work!

The ISO TC211/OGC SFA 1.x standards provide a geometry type hierarchy that is common to many of the existing open source Java spatial projects as possible. The following diagram shows this type hierarchy:

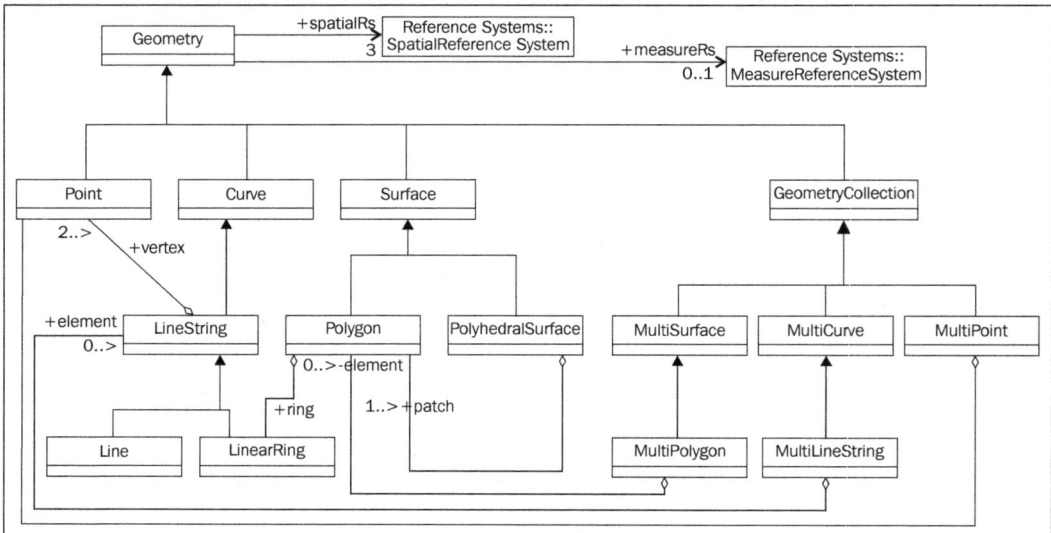

What adoption of this standard means in practice is shown in the following example. The encoding of a single polygon with a ring within the Oracle database of SDO_ GEOMETRY object, complies with the OGC SFA 1.x standards. Thus, conversion of the polygon to a Java representation is greatly simplified, if the target Java encoding complies with the same standard.

While the SDO_GEOMETRY object is itself singly inherited (roughly equivalent to Geometry in the preceding hierarchy diagram), each geometry subclass is correctly described in that object. So, while there is no SDO_POLYGON object, polygons are correctly encoded within the SDO_GEOMETRY object.

The compliance of Oracle Spatial and Locator enabled databases for versions 9 through 11 determines the Java technology that can be used. For example, if an open source Java project is not SFA 1.1 compliant, then its use could be problematic Refer the following table for a better understanding:

Product name	OGC Spec
Oracle Locator 11*g*, Release 1 (11.1.0.7)	SFA(TF) 1.1 (compliant 2009-09-14)
Oracle Locator, 10*g* Release 1 (10.1.0.4)	SFA(TF) 1.1 (compliant 2005-07-26)
Oracle Locator, 10*g* Release 2 (10.2.0.1)	SFA(TF) 1.1 (compliant 2005-11-01)
Oracle Spatial, 9*i* (9.0.1) and Release 2 (9.2.0)	SFA(NG) 1.1 (compliant 2002-09-30)

Choosing a geometry type hierarchy implementation

Two projects offer suitable geometry type hierarchies and associated **Application Programming Interfaces (API)** that are geared for independent use. These are as follows:

- JTS Topology Suite
- deegree

Candidate 1 – the JTS Topology Suite

JTS is not a full implementation of the TC211/OGC standards, and is not certified against either (it has never been submitted for certification). However, its representation and claimed SFA based approach has been tried and tested in a lot of products, and is accepted as being a faithful and useful implementation. It does (like most open source spatial projects) have the following additional limitations:

- Does not use current standard class names in certain situations (for example, `MultiSurface`).

- The `Coordinate` class that describes the vertices of geometry has only X, Y, and Z ordinates.

- There is no support for measures (namely, `MDSYS.VERTEX_TYPE`'s W attribute).

- JTS is 2.5D in that each `Coordinate` can have an elevation or Z value, but there is no structure for representing a solid (2008/2009 `SDO_GTYPE`).

- Circular arcs are not supported, thus classes such as `Curve` and `MultiCurve` do not exist. As a result all `SDO_GEOMETRY` objects that contain circular arcs should be "stroked" or densified via `SDO_GEOM.Sdo_Arc_Densify` before calling JTS `Geometry` based external methods.

- Geometry stores a **Spatial Reference System ID (SRID)** but does not use it; a *"coordinate system is infinite, planar and Euclidean (i.e. rectilinear and obeying the standard Euclidean distance metric). [...] JTSJTS does not specify any particular units for coordinates and geometries. Instead, the units are implicitly defined by the input data provided."*

Many of these limitations will be addressed over time.

JTS Topology Suite is ideal for deployment in Oracle database 10*g*R2 and above, as it is written using Java 1.4.2 (10*g*R2) and runs in all newer versions (for example, 11*g*R2 is 1.5).

In addition, JTS is used in a large number of open source spatial projects and has a strong commercial, application, and user focus. Also, in addition to deegree, GeoServer, GeoTools, uDig, Sexante, JUMP (OpenJUMP), JASPA, and HatBox, many other implementations and ports of the source code exist.

In summary, JTS is a mature, respected, and widely adopted application programming interface that is well aligned to the existing Oracle technology. It provides a highly suitable platform for developing complementary functionality. All of its limitations are well known, and each is scheduled to be addressed in future releases.

Candidate 2 – deegree

The geometry type hierarchy provided by deegree (deegree [...] successfully graduated from the OSGeo incubation process in February 2010 `http://wiki.deegree.org/deegreeWiki/OSGeoIncubation`) is a complete implementation of the TC211/OGC standards. It not only provides support for circular arcs and solids, it also supports Bezier, BSpline, and Cubic representations (Oracle Spatial 12*c* has introduced support for non-uniform rational B-spline - NURBS - curve geometries). It also provides some support for solids. Its limitations include the following:

- deegree currently only supports a Z (or third ordinate) it does not support a measure (that is, greater than three ordinates)

- The processing of circular arcs, and so on appear to be done by linearization to vertex-connected geometry objects

- deegree requires Java 1.6 and above limiting its potential usefulness for deployment inside current Oracle database versions

deegree is a well-supported project. However, its academic focus and the lack of support for its fundamental geometry type library in other open source projects makes its choice as a suitable library for this chapter, less enticing than JTS.

Alternate candidates

Two projects have implemented a complete geometry type system for other databases: JASPA for PostgreSQL and HatBox for the H2 and Derby Java databases.

While JASPA (at time of writing) is not production ready software, it does use JTS for its geometry type hierarchy and programming API. While based on PostgreSQL, it does provide an implementation template, and source for understanding how another project has implemented spatial data processing within an object relational database. JASPA requires Java 1.5 and above for compilation and deployment.

Similarly, JTS is a critical library for Hatbox, and while many of its implementation details are specific to H2 and Derby, it also provides an implementation template, and example source code, for understanding how to implement spatial processing within a database. HatBox requires Java 1.5 and above for compilation and deployment. While they have limitations, both projects should be kept in mind when developing functionality for Oracle, especially when based on JTS.

Deciding between JTS Topology Suite and deegree

While JTS appears to be behind deegree in its semantic breadth and implementation of its geometry type hierarchy, it is Java-compliant across a larger number of Oracle databases, and has greater extension into and adopted by the wider open source community's projects of choice. The limitations that have been identified are all scheduled to be addressed in later versions of JTS. As such, it will be adopted as the source for a geometry type hierarchy and implementation for this chapter.

Converting between SDO_GEOMETRY and Java geometry

Now that we have settled on a representation, we need stable and supported methods for converting SDO_GEOMETRY to and from the geometry types, this representation provides. There are two that are of specific use with Oracle and JTS; these are as follows:

- JTS's own com.vividsolutions.JTS.io.oracle package (c.f., OraReader/OraWriter)
- GeoTools' org.geotools.data.oracle.sdo package

The fairly obvious benefit of the first is that it is a part of the JTS project requiring no additional effort to extract and compile source code to build and deploy a suitable JAR file.

The GeoTools' implementation is excellent and relatively easy to extract specific elements from its very large code base (over 8,000 classes). All GeoTools data source packages are a part of its approach to organizing and packaging data sources for use by its other classes. For example, GeoTools SDO functionality is deployed within the jdbc-oracle plugin, which includes functionality for accessing other data types, such as blobs and clobs through the use of Data Stores. While this code is reasonably packaged, the data access context needs to be kept in mind.

Considering product release cycles

GeoTools and any other Java project that we may use are all subject to source code release cycles that are different to each other; code synchronization becomes another issue to manage in this situation. Wherever possible, your project or code choice should try to keep such dependencies at a minimum. For this reason, and the tight integration in a single project, the JTS reader/writer classes will be used by the Java code in this chapter.

Sourcing existing spatial algorithms

Each of the previously mentioned projects provides a source for existing spatial algorithms that can complement the functionality available in any one version of the Oracle database, or to that provided by our chosen core library, JTS.

For example, JTS provides the ability to import a number of sources of geospatial data, such as WKT, but it does not provide an ability to import/export KML or GeoJSon. JTS's implementation of GML for import/export is behind projects that demand implementations of more current specifications. As such other projects need to be examined, if this functionality is required for your Oracle database.

Useful algorithms

The following table lists a limited set of the algorithms and functionality that could be potentially useful companion functions to "plug" existing Oracle Locator/Spatial functionality (until available in future releases):

Algorithm/function	Oracle database version
Intersection/Union/Difference/XOR	All versions but only for licensed Spatial users
Relationships	All versions but only for licensed Spatial users.
Insert/Update/Delete vertices	None (custom)
One-sided and stylized buffers	None
Geometry Snapping	None
Building polygons from lines	None

Algorithm/function	Oracle database version
Line Noding or Merging	None
Delaunay Triangulation	11*g*R2 (SDO_GEOM.SDO_TRIANGULATE) for which a Spatial license is required
Voronoi	None

As we peruse this list, we note that JTS itself provides quite a number of these functions already. However, it does not, for example, provide an ability to insert, update, or delete the individual vertices of an existing Geometry. We need to look elsewhere. JASPA provides such functionality, and being already based on JTS means such functionality is a primary source for a possible implementation.

Such analysis needs to be carried out for each new function that is required. That is, as you meet each requirement, you will be forced to make decisions about a technology source that will meet your needs in the most effective and timely manner, while minimizing complexity both now and in the future; welcome to the world of software development!

Downloading the JTS source code

To download the latest version of JTS Topology Suite, either visit its sourceforge home page or use subversion as follows:

- svn co https://JTS-topo-suite.svn.sourceforge.net/svnroot/JTS-topo-suite JTS-topo-suite

Or for the latest released version (at the time of writing was 1.13) visit the following location:

- svn co https://JTS-topo-suite.svn.sourceforge.net/svnroot/JTS-topo-suite/tags/Version_1.13

Or the current trunk can be found at the following location:

- svn co https://JTS-topo-suite.svn.sourceforge.net/svnroot/JTS-topo-suite/trunk

Normally, you would build your solutions based on an existing release, such as 1.13, as the trunk is where developmental code exists prior to the next release; it is not designed or guaranteed to be of production quality and may contain bugs.

Once you have decided which release you require, download the version you want to a specific directory/folder on your computer.

Modifying JASPA and JTS

JASPA supports 4D coordinates, whereas JTS (as at 1.14) does not. So how can 4D data be interchanged?

The SQLMM and OGC **Well Known Text (WKT)** standards do not specify how to handle more than two ordinates (this will change). In addition they do not include the **Spatial Reference System Identifier (SRID)** of the geometry. PostGIS was the first to introduce the **Extended Well Known Text (EWKT)** and **Binary (EWKB)** interchange formats which handle 3D and 4D data and SRIDs. An example of converting an EWKT point with multiple dimensions can be seen from the following PostGIS SQL:

```
Select ST_Y(ST_GeomFromEWKT(
       'SRID=4326;POINT(-44.3 60.1)')) as ord
 union all
Select ST_Z(ST_GeomFromEWKT(
       'SRID=4326;POINT(-44.3 60.1 3.4)')) as ord
 union all
Select ST_M(ST_GeomFromEWKT(
       'SRID=4326;POINT(-44.3 60.1 3.4 8.9)')) as
ord;

ord
double precision
----------------
60.1
3.4
8.9
```

Thus, to exchange such data we have to rely on the EWKT/B nonstandardized forms until such time as the relevant standards specify what to do.

To support JASPA's EWKT text conversion functions with JTS, the changes that JASPA has made to the JTS's base `Coordinate.Java` class must also be made to the core JTS class. Not only does `Coordinate` need changing, but also three other classes need to be changed: `CoordinateArraySequence`, `CoordinateArraySequenceFactory` and `OraWriter`. In addition, JASPA's Java is Java 1.5-based. For 11g deployment this is acceptable, but for 10g deployment the code needs to be made 1.4 compliant. This is relatively simple to do as the main issue is JASPA's use of generics as in:

```
public static List<Geometry> geomArray2geomList (Geometry[] geometry)
```

The preceding code can be changed to the following:

```
public static List geomArray2geomList (Geometry[] geometry)
```

For this book, the following JASPA source classes were modified:

- `es.upv.jaspa/Core.Java`
- `es.upv.jaspa/GeomProperties.Java`
- `es.upv.jaspa.editors/GeometryEditor.Java`
- `es.upv.jaspa.editors/JTSSegmentizeEditor.Java`
- `es.upv.jaspa.editors/JTSSnapToGridEditor.Java`
- `es.upv.jaspa.io/WKT2JTS.Java`

And, the following JTS source classes were modified to support the 4D SDO_ GEOMETRYs, for example, SDO_GTYPE 4002 and 4402 that JASPA support:

- `com.vividsolutions.JTS.geom/Coordinate.Java`
- `com.vividsolutions.JTS.geom.impl/CoordinateArraySequence.Java`
- `com.vividsolutions.JTS.geom.impl/CoordinateArraySequenceFactory.Java`
- `com.vividsolutions.JTS.io.oracle/OraWriter.Java`

The changes are imperfect as the `OraReader`/`OraWriter` classes do not support the identification of the position of a measure, such as when `OraReading` the following SDO_GEOMETRY object:

```
SDO_GEOMETRY(4302,8307,NULL,
            SDO_ELEM_INFO_ARRAY(1,2,1),
            SDO_ORDINATE_ARRAY(0,0,2,3,50,50,100,200))
```

The preceding, when converted from its JTS Geometry object back to an SDO_GEOMETRY object by the modified `OraWriter` class, results in the following code:

```
SDO_GEOMETRY(4002,8307,NULL,
            SDO_ELEM_INFO_ARRAY(1,2,1),
            SDO_ORDINATE_ARRAY(0,0,2,3,50,50,100,200))
```

Once JTS moves to fully support measured coordinates (mooted) these changes will no longer be required.

Compiling and building a JTS .jar file

The JTS source code includes an `Apache Ant build.xml` file. (**Another Neat Tool (ANT)** is a simple, platform independent, tool for compiling and packaging/building jar files, and applications from Java source code). This `build.xml` file implements all that is needed for building all possible outputs from the JTS project: installable JAR files, documentation, test data, and classes. It is quite comprehensive.

However, for deployment to an Oracle database, what is needed is a more suitable JAR file that:

- Compiles the `Java` classes
- Merges the `OraReader/OraWriter` classes in the separate `JTSio` directory into the same `com.vividsolutions.JTS.io.oracle` package
- Optionally adds the source code to the JAR file in case both the source and the classes are to be distributed in a manner that keeps them together

A replacement `build.xml` file has been created that implements these requirements. It is not documented here as it is included in the source code that accompanies this book. This `build.xml` file has in its props target the following reference:

```
<property name="ora.lib" value="c:/oracle/product/10.2.0/db_1/jdbc/
lib" />
```

This needs to be modified to suit your installation.

The `build.xml` file has two main build targets with a single, wrapper target for the most common of the two):

```
<target name="ora-JTS-source"
        depends="props, JTS-jar, copy-source"
        description="Cleans, compiles, builds jar with source">
  <jar destfile="${pre-zip.JTS-jar}">
    <fileset dir="${build.classes}" includes="**/*.class"/>
    <fileset dir="${pre-zip}/src"
          includes="**/*.Java"
          excludes="**/JTStest/ **/JTSplugin/ **/JTSexample/" />
  </jar>
</target>

<target name="ora-JTS-nosource"
        depends="props, JTS-jar, copy-source"
        description="Cleans, compiles, builds jar">
  <jar destfile="${pre-zip.JTS-jar}">
      <fileset dir="${build.classes}" includes="**/*.class"/>
  </jar>
</target>

<!-- Wrapper target -->
<target name="ora-JTS" depends="ora-JTS-nosource"
        description="Cleans, compiles, builds jar">
</target>
```

To build a JAR file that does not contain the source Java files, execute the following code:

```
ant -buildfile build.xml ora-JTS
```

This creates the following JAR file:

```
{JTS_HOME}\Version_1.13\build\dist\lib\JTS-1.13.jar
```

Installing the JTS .jar file

In the introduction to Java and Oracle Spatial we loaded an actual Java class' source code into the database and created its PL/SQL access function (getJVMProperty) in one step. This approach is fine for simple one-off, independent functions that are entirely standalone or only use functionality in the actual Oracle JVM.

When developing a collection of new functions based on an external source code project like JTS, one cannot load the source for the new function, and all the sources of the required project in the same way. It is more efficient to load the dependent project's JAR files containing its compiled classes into the databases in one activity, and then build the functionality, which one needs afterwards.

> As is the case for all Oracle database activities, the creating or loading of Java source files, classes, or JAR files requires specific database permissions. To be able to load Java classes the following permission, expressed as a grant is required for the schema that will own the code:
>
> ```
> GRANT JAVAUSERPRIV TO <Schema>;
> loadJava -force -oci -stdout -verbose -user book/
> book@GISDB -resolve -grant PUBLIC -f {JTS_HOME}\
> Version_1.13\build\dist\lib\JTS-1.13.jar > JTS-1.13.
> log
> ```

The JTS-1.13.log file should contain something like the following:

```
arguments: '-user' 'codesys/***@GISDB' '-force' '-oci' '-stdout'
'-verbose' '-resolve' '-grant' 'PUBLIC' '-f' 'c:\Projects\JTS\
Version_1.13\build\dist\lib\JTS-1.13.jar'
creating : resource META-INF/MANIFEST.MF
loading  : resource META-INF/MANIFEST.MF
creating : class com/vividsolutions/JTS/JTSVersion
loading  : class com/vividsolutions/JTS/JTSVersion
(Every class will create and load unless there is a problem such as
a reference to a jar file that is not already available to the Oracle
database JVM)
```

```
.... . . . .
Classes Loaded: 525
Resources Loaded: 1
Sources Loaded: 0
Published Interfaces: 0
Classes generated: 0
Classes skipped: 0
Synonyms Created: 0
Errors: 0
```

Checking the installation

Once the JTS JAR file is loaded; there are a number of ways to check the success of
that load. The first is to query the Oracle metadata views, and the other is to actually
try and use the new JTS functionality in a PL/SQL function.

Checking by querying Oracle's Java metadata tables

There are a number of metadata views relating to Java within the database.
Those that are of interest are as follows:

```
Select view_name
  From all_views
 Where view_name Like 'USER_JAVA_CLASSES'
    Or view_name Like 'USER_JAVA_METHODS'
 Order By 1;
-- Results
--
VIEW_NAME
-------------------------
USER_JAVA_CLASSES
USER_JAVA_METHODS

 2 rows selected
```

The following `Select` statement shows those classes relating to the conversion
of an Oracle `SDO_GEOMETRY` object to a JTS geometry:

```
Select Replace(name,'com/vividsolutions/','') as name,
       accessibility,
       Replace(outer,'com/vividsolutions/','') as outer
  From user_Java_classes
 Where name Like 'com/vividsolutions/JTS/io/oracle%'
 Order By 1;
```

```
-- Results
--
NAME                           ACCESSIBILITY OUTER
------------------------------ ------------- -------
JTS/io/oracle/OraGeom          PUBLIC        -
JTS/io/oracle/OraGeom$ETYPE    (null)        -
JTS/io/oracle/OraGeom$GEOM_TYPE (null)       -
JTS/io/oracle/OraGeom$INTERP   (null)        -
JTS/io/oracle/OraReader        PUBLIC        -
JTS/io/oracle/OraUtil          PUBLIC        -
JTS/io/oracle/OraWriter        PUBLIC        -

 7 rows selected

-- What are the public methods for OraReader
-- with its arguments & return types?
--
Select m.method_name,
       Replace(Case When m.base_type <> 'void'
                    or m.base_type is null
                 Then m.RETURN_CLASS
                 Else m.base_type
              End,'com/vividsolutions/','') as return_type,
       Replace(Case When a.base_type is null
                 Then a.argument_class
                 Else a.base_type
              End,'com/vividsolutions/','') as arg_type
  From user_Java_methods m
       Inner Join USER_JAVA_ARGUMENTS a
       On (a.name = m.name and a.method_index = m.method_index)
 Where m.name = 'com/vividsolutions/JTS/io/oracle/OraReader'
   And m.accessibility = 'PUBLIC'
 Order BY m.method_name, a.argument_position;

-- Results
--
METHOD_NAME RETURN_TYPE                    ARG_TYPE
----------- ------------------------------ ----------------------
<init>      void                           JTS/geom/GeometryFactory
read        JTS/geom/Geometry oracle/sql/STRUCT            (NULL)
setDimension void                                          int
```

Checking by trying to actually execute something

The JTS library comes with a Java class called `JTSVersion`. If the JAR file has been loaded then it should firstly exist in `USER_JAVA_CLASSES`, and secondly we should be able to execute one of its methods to get the loaded version. This can be done as follows:

```
-- 1. First create a Java class to query the main JTSVersion
--    class as there is not public static method we can use.
--
Create Or Replace AND COMPILE JAVA SOURCE NAMED JTSCheck
As
public class JTSCheck {
   public static String getJTSVersion() {
      return com.vividsolutions.JTS.JTSVersion
             .CURRENT_VERSION.toString();
   }
}
/

-- 2. Now create a PL/SQL function to execute
--    JTSCheck.getJTSVersion().
--
Create Or Replace Function getJTSVersion
Return Varchar2 Is
Language Java Name
'JTSCheck.getJTSVersion() return Java.lang.String';
/

-- 3. Execute getJTSVersion():
--    Returning 1.13 means our load is successful.
--
Select getJTSVersion() as JTSVersion
  From dual;

-- Result
--
JTSVERSION
----------
1.13.0
```

We can see that from an analysis of our load log, Java metadata views, and executing our `JTSCheck.getJTSVersion` function, that we have successfully loaded the JTS library. We are now ready to start using it to build new functionality for use with Oracle `SDO_GEOMETRY` object.

Creating Java Stored Procedures

We saw with `JTSVersion.getJTSVersion` and `getJVMProperty` how easy it is to write a new PL/SQL function that executes a Java method within the JVM. These Java classes and methods are called **Java Stored Procedures**. As we have seen, creating an interface between PL/SQL and Java is easy. But it is important to note that:

- Java Stored Procedures are by default executed with invokers rights
- PL/SQL procedures are by default executed with definer's rights

When writing functions using Java classes and methods, one must comply with the following rules:

- No constructor method is needed
- Variables and methods must be declared [b] `static` [/b]
- Use the default database connection (no user ID/password required: uses session connection)
- Declare output variables as arrays [where output is not singular]
- Console output from `System.out.println` statements will be written to trace files in the Oracle `UDUMP` destination directory [get friendly with your DBA!]

Our first Java function – buffering a single geometry

As we have seen previously, in order to be able to use an external geospatial Java-based algorithm or method, we first have to be able to convert between an `SDO_GEOMETRY` object and a JTS `Geometry` object (our chosen OGC SFA 1.x compliant geometry type hierarchy representation).

We will demonstrate how to do this as our first of many functions, by building a Java class that will convert an `SDO_GEOMETRY` object to a JTS `Geometry` object, construct a stylized buffer from a selection of parameters, and convert the result back to `SDO_GEOMETRY` for passing back to the calling Oracle session.

To implement our first useful Java Stored Procedure, we will:

- Create a class to hold our code

- Include three private methods:
  - ○ One that gets the default connection between the JVM and the database session
  - ○ Another computes a JTS `PrecisionModel` scale from an input ordinate precision value
  - ○ A method that returns the `SDO_SRID` value of an `SDO_GEOMETRY` object

- Add a method called `ST_Buffer()` that will:
  - ○ Convert an `SDO_GEOMETRY` object into an JTS `Geometry` object
  - ○ Buffer the converted `Geometry` object using the desired JTS processing
  - ○ Convert the buffered polygon back to an `SDO_GEOMETRY` object
  - ○ Return the result to the calling Oracle session

Before we present the (long) Java source that we will load, let's look at the key elements in the execution as this will give us a simple "recipe" for using JTS functionality.

JTS implements its buffering via the `com.vividsolutions.JTS.operation.buffer.BufferOp` class. This class can create both positive and negative buffer distances. The class has a number of constructors, but the main one that provides the most flexibility is as follows:

```
static Geometry bufferOp(Geometry g,
                         double distance,
                         BufferParameters params)
/*Computes the buffer for a geometry for a given buffer distance and
accuracy of approximation */
```

The `BufferParameters` class provides a collection of properties that control a buffer's construction. These include the following:

```
void setEndCapStyle(int endCapStyle)
    // Specifies the end cap style of the generated buffer.

void setJoinStyle(int joinStyle)
    // Sets the join style for outside (reflex) corners
    // between line segments.

void setMitreLimit(double mitreLimit)
    // Sets the limit on the mitre ratio used for very
    // sharp corners.
```

```
void setQuadrantSegments(int quadSegs)
    // Sets the number of line segments used to approximate
    // an angle fillet.
    // Controls the number of linear segments needed to
    // approximate any curves that may be created when buffering
    // (JTS does not support circular arcs in Geometries)

void setSingleSided(boolean isSingleSided)
    // Sets whether the computed buffer should be single-sided.
```

Three different options control the type of end cap:

```
PrecisionModel  pm = new PrecisionModel(
                          getPrecisionScale(_precision));
GeometryFactory gf = new GeometryFactory(pm,SRID);
OraReader       or = new OraReader(gf);
Geometry        geo = or.read(_geom);
```

Three different styles are provided to control the type of mitre at each corner between two line segments:

```
public static final int JOIN_ROUND = 1;
public static final int JOIN_MITRE = 2;
public static final int JOIN_BEVEL = 3;
```

The default for Quadrant Segments is as follows:

```
public static final int DEFAULT_QUADRANT_SEGMENTS = 8;
```

Converting SDO_GEOMETRY TO JTS Geometry

Conversion of an SDO_GEOMETRY object to a JTS Geometry object in our method (as shown in the following code) is carried out using the following four lines:

```
PrecisionModel  pm = new PrecisionModel(getPrecisionScale(_
precision));
GeometryFactory gf = new GeometryFactory(pm,SRID);
OraReader       or = new OraReader(gf);
Geometry        geo = or.read(_geom);
```

The creation of a JTS Geometry object requires a JTS PrecisionModel object:

```
PrecisionModel  pm = new PrecisionModel(
                          getPrecisionScale(_precision));
```

The chosen method here is to use a FIXED precision model that represents a model with a fixed number of decimal places. This is done by the caller providing the number of decimal digits of precision for JTS processing via the _precision parameter. The _precision value is converted to the equivalent JTS scale value via a simple formula:

*[If a user] specifies 3 decimal places of precision, [...] a scale factor of 1000 [is used].
To specify -3 decimal places of precision (that is, rounding to the nearest 1000), [...]
a scale factor of 0.001 [is used].*

This conversion is implemented in a private method called `getPrecisionScale`.
(A full double precision floating point model is JTS's default – this is not
implemented for this chapter, and is left to an exercise for the reader to investigate).

After creating a suitable `PrecisionModel`, a `GeometryFactory` object can then be
created that uses this `PrecisionModel`, when creating `Geometry` objects:

```
GeometryFactory gf = new GeometryFactory(pm,SRID);
```

A `GeometryFactory` is needed because the JTS conversion class `OraReader` uses
a `PrecisionModel`, when converting. A `GeometryFactory` object supplies a set of
utility methods for building `Geometry` objects from lists of coordinates enabling
the `OraReader` class to work with a single factory object to create possibly multiple
`Geometry` types (LineString, Point, and Polygon) without having to call each
individual `Geometry` type's constructor directly.

```
OraReader      or = new OraReader(gf);
```

In addition note that the `GeometryFactory` object has been supplied with the `SDO_
GEOMETRY` object's SRID (extracted via a private method called `getSRID`). This ensures
that any `Geometry` object created by the factory will be tagged with the correct SRID.

In respect of the processing that `GeometryFactory` provides, it is important to
note that:

*[...] the factory constructor methods do not change the input coordinates in any
way. In particular, they are not rounded to the supplied* `PrecisionModel`. *It is
assumed that input coordinates meet the given precision.*

This "do not change" behavior is exactly the behavior of Oracle Spatial; all
statements of tolerance in the `DIMINFO` metadata entry, or any function that
processes an `SDO_GEOMETRY` object are used only when comparing ordinates. The
tolerance value is not used to round ordinates to the implicit precision stated in
the tolerance value. As shown in *Chapter 7, Editing, Transforming, and Constructing
Geometries*, the `T_Geometry` member function `ST_RoundOrdinates`, so JTS is able
to round a Geometry's coordinates (XY only) via use of a method called `reduce()`
in a class called `GeometryPrecisionReducer`. While not presented in this chapter,
an implementation: `ST_CoordinateRounder` is available in the source code that
accompanies this book.

The final step in creating a `Geometry` object occurs when the OraReader's `read` method is called on the passed-in `SDO_GEOMETRY` object. This method determines the correct `Geometry` object to create, and may include the conversion of Oracle specific representations, such as an optimized rectangle polygon to a normal JTS Polygon:

```
Geometry geo = or.read(_geom);
```

Once this object is created, the relevant buffer parameters are determined, and then the buffer is created by executing the appropriate `Geometry` buffer method. The resulting polygon geometry is converted back to an `SDO_GEOMETRY` object via use of a `JTS` `OraWriter` object. The `getConnection` call returns the default connection that exists between the calling database session and the JVM:

```
BufferParameters bufParam =
    new BufferParameters(_quadrantSegments,
                         _endCapStyle,
                         _joinStyle,
                         BufferParameters.DEFAULT_MITRE_LIMIT);
// Single side only for Linestrings.
If ( geo instanceof LineString ||
     geo instanceof MultiLineString ) {
    bufParam.setSingleSided(_singleSided==0?false:true);
}
Geometry buffer = BufferOp.bufferOp(geo,_distance,bufParam);
OraWriter    ow = new OraWriter(getConnection());
```

Implementing ST_Buffer

Putting all this together we can construct the following Java source with a main method of `ST_Buffer` as follows:

```
Drop JAVA SOURCE "Buffer";
SET DEFINE OFF;
Create Or Replace And Compile Java Source Named "Buffer" As
package com.packt.spatial;

import com.vividsolutions.JTS.geom.Geometry;
import com.vividsolutions.JTS.geom.GeometryFactory;
import com.vividsolutions.JTS.geom.LineString;
import com.vividsolutions.JTS.geom.MultiLineString;
import com.vividsolutions.JTS.geom.PrecisionModel;
import com.vividsolutions.JTS.io.oracle.OraReader;
import com.vividsolutions.JTS.io.oracle.OraWriter;
```

```java
import com.vividsolutions.JTS.operation.buffer.BufferOp;
import com.vividsolutions.JTS.operation.buffer.BufferParameters;
import Java.sql.SQLException;

import oracle.jdbc.driver.OracleConnection;
import oracle.jdbc.pool.OracleDataSource;

import oracle.sql.Datum;
import oracle.sql.NUMBER;
import oracle.sql.STRUCT;

public class Buffer
{
    private static final int SRID_NULL = -1;

    protected static OracleConnection g_connection;

    public static OracleConnection getConnection()
    throws SQLException
    {
        /* See Source Supplied with this Book */
    }

    protected static double precisionModelScale = Math.pow(10,3);

    public static double getPrecisionScale(int _numDecPlaces)
    {
        /* See source supplied with this book */
    }

    public static int getSRID(STRUCT _st,
                              int    _nullValue)
    {
        /* See source supplied with this book */
    }

    public static STRUCT ST_Buffer(STRUCT _geom,
                                   double _distance,
                                   int    _precision,
                                   int    _endCapStyle,
                                   int    _joinStyle,
                                   int    _quadrantSegs,
```

```
                                     int     _singleSided)
      throws SQLException
      {
           /* See source supplied with this book */
      }
}
/
-- Results
--
anonymous block completed
```

We can check the success of this in a number of different ways:

```
Select count(*) as numLines
  From user_source
 Where type = 'JAVA SOURCE'
   And name like '%Buffer%';

-- Results
--
NUMLINES
--------
     201

Select name, Source, accessibility
  From user_Java_classes
 Where name like 'com/packt/spatial/Buffer%';

-- Results
--
NAME                      SOURCE ACCESSIBILITY
------------------------- ------ -------------
com/packt/spatial/Buffer Buffer PUBLIC

Select object_name, object_type, status
  From user_objects
  Where object_type like '%JAVA%'
    And object_name like 'com/packt/spatial/Buffer%';

-- Results
--
OBJECT_NAME               OBJECT_TYPE STATUS
------------------------- ----------- -------
com/packt/spatial/Buffer JAVA CLASS  VALID
```

But the ultimate test is to show that it actually works! To do this the ST_Buffer method must be capable of being called from an Oracle session. To do this, a PL/SQL function wrapper is created as follows:

```
Create Or Replace
   Function JTS_Buffer(p_geom        in mdsys.SDO_GEOMETRY,
                       p_distance    in Number,
                       p_precision   in Number,
                       p_endCapStyle in Number,
                       p_joinStyle   in Number,
                       p_quadSegs    in Number,
                       p_singleSide  in Number )
      Return mdsys.SDO_GEOMETRY
         As language Java name 'com.packt.spatial.JTS.ST_Buffer(oracle.
sql.STRUCT,double,int,int,int,int,int) return oracle.sql.STRUCT';
/
show errors
-- Result
--
FUNCTION JTS_BUFFER compiled
No Errors.
```

Note the As language Java name clause, and the matching of the PL/SQL parameters' data types, and their equivalent Java data types. Also, note that an SDO_GEOMETRY object has oracle.sql.STRUCT data type.

A restriction on a Java Stored Procedure PL/SQL function is that it cannot have default parameters. If we tried to do so (for example, p_precision in number default 3) we would get the following error:

```
Errors: check compiler log
1/10     PLS-00255: CALL Specification parameters cannot have default
values
0/0      PL/SQL: Compilation unit analysis terminated
```

Therefore, we need a second function, if we want to have default argument values:

```
Create Or Replace Function
   Function ST_Buffer(p_geom        in SDO_GEOMETRY,
                      p_distance    in Number,
                      p_precision   in Number   default 3,
                      p_endCapStyle in Number   default 1
                                                /*CAP_ROUND*/,
                      p_joinStyle   in Number   default 1
                                                /*JOIN_ROUND*/,
                      p_quadSegs    in Number   default 8
                                                /*QUADRANT_SEGMENTS*/,
```

```
                         p_singleSide  in Integer default 0)
    ReturnSDO_GEOMETRY
  As
  Begin
    If ( p_geom is null ) Then
       return null;
    End If;
    Return JTS_Buffer(p_geom,
                      p_distance,
                      p_precision,
                      p_endCapStyle,
                      p_joinStyle,
                      p_quadSegs,
                      case when ABS(p_singleSide)=0 then 0 Else 1
end);
  End ST_Buffer;
/
show errors

-- Result
--
FUNCTION ST_BUFFER compiled
No Errors.
```

Note that the possible values for the end cap/join styles must match the constants declared in the `BufferParameters.Java` class those and presented earlier in this chapter (for example, public static final int `CAP_ROUND` = 1; for end cap).

Executing ST_Buffer

Now that we have a compiled `Java` class and PL/SQL call functions, we call our new `Java` method in the same way as calling any other PL/SQL function:

```
Select SPEX.ST_Buffer(NULL,100.0,1) as geom
  From dual;
-- Result
--
GEOM
------
(NULL)
Prompt *1. 15m Buffer with _Round_ End Cap and Join Style*
Select SPEX.ST_Buffer(
           SDO_GEOMETRY('LINESTRING(20 1,50 50,100 0,150 50)'),
           15 /*DISTANCE*/,
            2 /*PRECISION*/,
            1 /*CAP_ROUND*/,
```

```
                    1 /*JOIN_ROUND*/,
                    8 /*QUADRANT_SEGMENTS*/,
                    0 /*FULL BUFFER*/) as buf
        From dual;
```

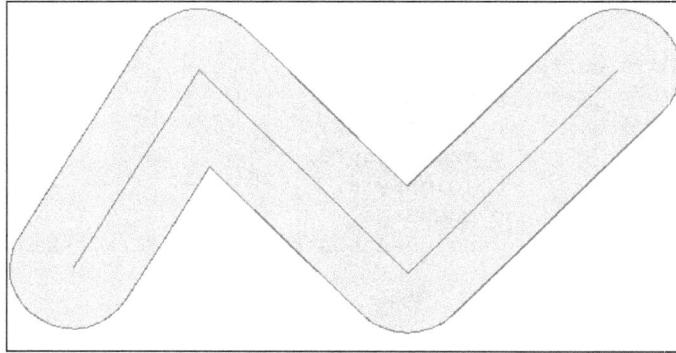

```
    Prompt *2. 15m Buffer with _SQUARE_ End Cap and ROUND Join Style*
    Select SPEX.ST_Buffer(
              SDO_GEOMETRY('LINESTRING(20 1,50 50,100 0,150 50)'),
                    15 /*DISTANCE*/,
                     2 /*PRECISION*/,
                     3 /*CAP_SQUARE*/,
                     1 /*JOIN_ROUND*/,
                     8 /*QUADRANT_SEGMENTS*/,
                     0 /*FULL BUFFER*/) as buf
        From dual;
```

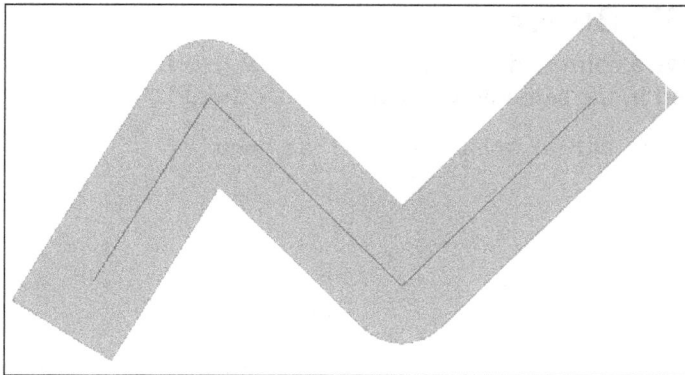

```
    Prompt *3. 15m Buffer with _BUTT_ End Cap and _ROUND_ Join Style*
    Select SPEX.ST_Buffer(
              SDO_GEOMETRY('LINESTRING(20 1,50 50,100 0,150 50)'),
                    15 /*DISTANCE*/,
```

```
                    2 /*PRECISION*/,
                    2 /*CAP_BUTT*/,
                    1 /*JOIN_ROUND*/,
                    8 /*QUADRANT_SEGMENTS*/,
                    0 /*FULL BUFFER*/) as buf
    From dual;
```

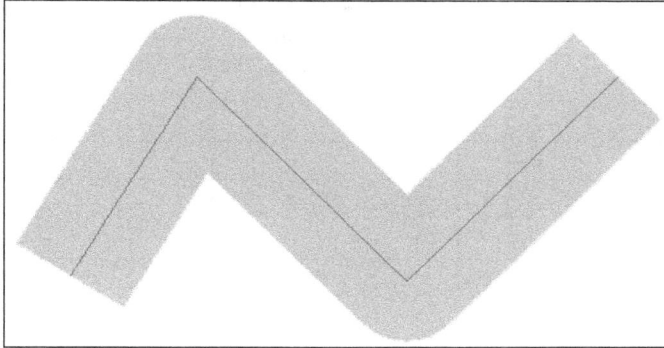

```
Prompt *4. 15m Buffer with _BUTT_ End Cap and _MITRE_ Join Style*
Select SPEX.ST_Buffer(
            SDO_GEOMETRY('LINESTRING(20 1,50 50,100 0,150 50)'),
                    15 /*DISTANCE*/,
                    2 /*PRECISION*/,
                    2 /*CAP_BUTT*/,
                    2 /*JOIN_MITRE*/,
                    8 /*QUADRANT_SEGMENTS*/,
                    0 /*FULL BUFFER*/) as buf
    From dual;
```

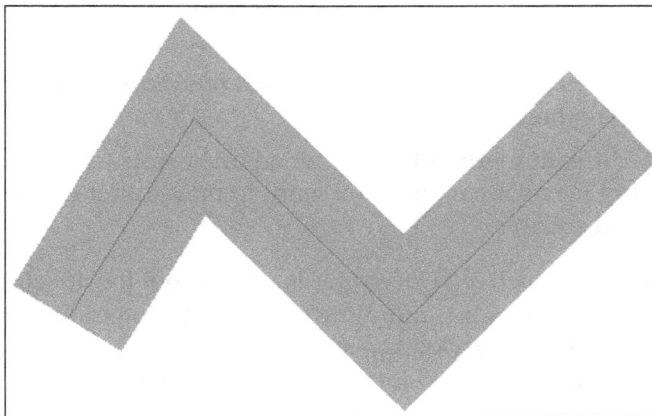

```
Prompt *5. 15m Buffer with _BUTT_ End Cap and _BEVEL_ Join Style*
Select SPEX.ST_Buffer(
              SDO_GEOMETRY('LINESTRING(20 1,50 50,100 0,150 50)'),
                   15 /*DISTANCE*/,
                    2 /*PRECISION*/,
                    2 /*CAP_BUTT*/,
                    3 /*JOIN_BEVEL*/,
                    8 /*QUADRANT_SEGMENTS*/,
                    0 /*FULL BUFFER*/) as buf
       From dual;
```

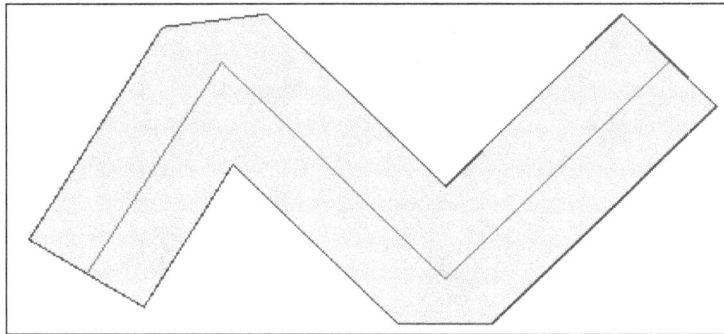

Packaging – source code versus .jar file

The creation of a Java class via the preceding method (creating the Java source and compiling it using CREATE AND COMPILE JAVA SOURCE NAMED ...) is fine for simple, one-off, implementation, but such an approach has a number of limitations:

- Many utility methods would be better off being moved to other classes, for example, getConnection, getPrecisionScale, getSRID

- Resolving errors and dependent class/JAR references is difficult the more complex a class becomes

- It is a very difficult method to scale when one wants to add more methods to an existing Java source, or if you want to build other classes that rely on earlier loaded sources

A GUI development tool such as JDeveloper is a solid tool for developing small, simple Java classes, to larger scale and more complex packages of Java classes, and compiling and testing, before packaging into suitable JAR files (ours will be called J6365EN.jar), before trying to load into the Oracle database.

In our first example earlier, we used a single standalone class `Buffer` and two PL/SQL functions to demonstrate how to create and call JTS's buffer functionality. From now on we will use a single Java class called `JTS` for all JTS related implementation, in the `com.packt.spatial` package in which to place most of our new Java classes.

In addition, in order to organize, collect, and manage variables, functions (allowing overloading), and procedures as a single unit, a single PL/SQL package called SPEX (`SpatialEXtend`) will be created. The SPEX package elements for `ST_Buffer` are:

```
Create Or Replace Package SPEX
AUTHID CURRENT_USER
As
  -- For Buffer, some constants
  --
  CAP_ROUND  CONSTANT Number := 1;
  CAP_BUTT   CONSTANT Number := 2;
  CAP_SQUARE CONSTANT Number := 3;

  JOIN_ROUND CONSTANT Number := 1;
  JOIN_MITRE CONSTANT Number := 2;
  JOIN_BEVEL CONSTANT Number := 3;

  QUADRANT_SEGMENTS CONSTANT Number := 8;

  /**
  * ST_Buffer
  * Buffer a geometry using variety of parameters including
  * single siding.Allows for executing using normal defaults.
  */
  Function ST_Buffer(
            p_geom       in SDO_GEOMETRY,
            p_distance   in Number,
            p_precision  in Number,
            p_endCapStyle in Number default book.SPEX.CAP_ROUND,
            p_joinStyle  in Number default book.SPEX.JOIN_ROUND,
            p_quadSegs   in Number default
                                book.SPEX.QUADRANT_SEGMENTS,
            p_singleSide in Integer default 0)
    Return SDO_GEOMETRY
          Deterministic;

End SPEX;
/
SHOW ERRORS

-- SPEX PACKAGE BODY not shown for brevity's sake.
```

Compiling and loading

Even though, at the moment all our application code that we are deploying is in a single Java class, this implementation differs from the buffer one, in that a lot of the common methods (such as `getSRID` and `getConnection`) have been placed into "helper" classes (Tools and SDO). As such, once compiled, all the classes are loaded into Oracle via a JAR file. The following steps are used.

Firstly, drop any existing classes with the same names as those in the JAR file:

```
dropJava -verbose -force -oci -stdout -user book/book@GISDB J6365EN.
jar > J6365EN_drop.log
```

Secondly, the new JAR file is loaded:

```
loadJava -force -oci -stdout -verbose -user book/book@GISDB -resolve
-grant PUBLIC -f J6365EN.jar > J6365EN_load.log
```

Finally, the PL/SQL package used to access the functionality is rebuilt:

```
sqlplus book/book@GISDB @ SPEX_Package.sql > SPEX_Package_load.log
BOOK
```

Functions for processing two geometries

In the previous example, we constructed a Java function that processed a single `Geometry` object to buffer a geometry using a number of styles. In this section, we will introduce a number of functions that will process two converted SDO_ GEOMETRY objects.

Spatially (Topological) comparing two geometries – ST_Relate

Oracle Spatial contains a special licensed function called `SDO_GEOM.RELATE` that examines two SDO_GEOMETRY objects to determine their spatial relationship.

This functionality is something that is often required when processing geometry objects and is often something Locator users look for.

The chosen PL/SQL call interface mimics the Oracle equivalent:

```
/**
 * ST_Relate
 * Implements a license free version of SDO_GEOM.RELATE.
 * @note Supports JTS named topological relationships and
```

```
*        not Oracle specific keywords like OVERLAPBDYDISJOINT
* @param p_geom1  : SDO_GEOMETRY : Geometry compared to second
* @param p_mask   : varchar2     : Mask containing DETERMINE,
*                                  ANYINTERACT or a list of comma
*                                  separated topological relationships
* @param p_geom2: SDO_GEOMETRY : Geometry compared to first.
* @param p_precision : Number  : Decimal places of precision
* @return String      : Result of processing
*/
Function ST_Relate(p_geom1    in SDO_GEOMETRY,
                   p_mask     in Varchar2,
                   p_geom2    in SDO_GEOMETRY,
                   p_precision in Number)
    Return varchar2 Deterministic;
```

Java implementation

All comparison operations in JTS are executed through the `RelateOp` class that implements a static method called `relate`:

```
public static IntersectionMatrix relate(Geometry a, Geometry b)
```

Computes the `IntersectionMatrix` for the spatial relationship between two Geometrys, using the default (OGC SFA) Boundary Node Rule

The returned `IntersectionMatrix` "Models a Dimensionally Extended Nine-Intersection Model (DE-9IM) matrix. DE-9IM matrices (such as "212FF1FF2") specify the topological relationship between two Geometrys."

To make the JTS function operation somewhat like its Oracle equivalent takes a little more Java coding than we have seen so far. Here is the relevant part of the ST_ Relate Java method in the JTS class:

```
public static String ST_Relate(STRUCT _geom1,
                               String _mask,
                               STRUCT _geom2,
                               int    _precision)
throws SQLException
{
    // Check parameters
    . . .
    String returnString = "";
    try
    {
      // Get valid connection
```

```
    . . .
            // Extract SRIDs from SDO_GEOMETRYs make sure same
            . . .
            // Convert Geometries
            . . .
            // Check converted geometries are valid
            . . .
            // Now get relationship mask
            IntersectionMatrix im = RelateOp.relate(geo1,geo2);
            // Process relationship mask
            int dimGeo1 = geo1.getDimension();
            int dimGeo2 = geo2.getDimension();
            If ( im.isEquals(dimGeo1,dimGeo2)) {
                returnString = "EQUAL";
            } Else If ( im == null ) {
                returnString = "UNKNOWN";
            } Else {
                ArrayList al = new ArrayList();
                If (im.isContains())              al.add("CONTAINS");
                If (im.isCoveredBy())             al.add("COVEREDBY");
                If (im.isCovers())                al.add("COVERS");
                If (im.isCrosses(dimGeo1,dimGeo2)) al.add("CROSS");
                If (im.isDisjoint())              al.add("DISJOINT");
                If (im.isIntersects())            al.add("INTERSECTS");
                If (im.isOverlaps(dimGeo1,dimGeo2)) al.add("OVERLAP");
                If (im.isTouches( dimGeo1,dimGeo2)) al.add("TOUCH");
                If (im.isWithin())                al.add("WITHIN");
                // Now compare to user mask
                //
                If ( mask.equalsIgnoreCase("ANYINTERACT") ) {
                  // If the ANYINTERACT keyword is passed in mask,
                  // TRUE is returned if two geometries not disjoint.
                  //
                  return al.size()==0
                          ?"UNKNOWN"
                          :(al.contains("DISJOINT")?"FALSE":"TRUE");
                } Else If ( mask.equalsIgnoreCase("DETERMINE") ) {
                  // If the DETERMINE keyword is passed in mask, return
                  // the one relationship keyword that best matches.
                  //
                  Iterator iter = al.iterator();
                  returnString = "";
                  while (iter.hasNext()) {
```

```
                    returnString += (String)iter.next() +",";
            }
            // remove unwanted end ","
            returnString = returnString
                        .substring(0,returnString.length()-1);
        } Else {
            // If a mask listing one or more relationships is
            // passed in, the function returns the name of the
            // relationship if it is true for the pair of
            // geometries. If all relationships are false,
            // the procedure returns FALSE.
            //
            StringTokenizer st =
                    new StringTokenizer(mask.toUpperCase(),",");
            String token = "";
            returnString = "";
            while ( st.hasMoreTokens() ) {
                token = st.nextToken();
                If ( al.contains(token) )
                    returnString += token + ",";
            }
            If ( returnString.length()==0 ) {
                // Passed in relationships do not exist
                returnString = "FALSE";
            } Else {
                // remove unwanted end ","
                returnString = returnString.substring(0,
                                    returnString.length()-1);
            }
        }
    }
} catch(SQLException sqle) {
    System.err.println(sqle.getMessage());
    returnString = "UNKNOWN";
} catch (Exception e) {
    returnString = "UNKNOWN";
}
return returnString;
}
```

Use of this function is demonstrated in the next section, where the overlay operations are tested.

Replicating existing licensed overlay functions – ST_Union

The following four overlay functions are not available to Locator users:

- SDO_GEOM.SDO_DIFFERENCE
- SDO_GEOM.SDO_INTERSECTION
- SDO_GEOM.SDO_UNION
- SDO_GEOM.SDO_XOR

JTS, however, includes the critical overlay functionality that enables a license free implementation of these SDO_GEOM package functions. Each of these functions can be created fairly easily as the following PL/SQL function signatures of our new functions are (extracted from SPEX PL/SQL package) shown as follows:

```
Function ST_Difference(p_geom1    in SDO_GEOMETRY,
                       p_geom2    in SDO_GEOMETRY,
                       p_precision in Number)
   Return SDO_GEOMETRY
       As language Java name
         'com.packt.spatial.JTS.ST_Difference(
oracle.sql.STRUCT,oracle.sql.STRUCT,int)
return oracle.sql.STRUCT';

Function ST_Intersection(p_geom1    in SDO_GEOMETRY,
                         p_geom2    in SDO_GEOMETRY,
                         p_precision in Number)
   Return SDO_GEOMETRY
       As language Java name
         'com.packt.spatial.JTS.ST_Intersection(
oracle.sql.STRUCT,oracle.sql.STRUCT,int)
return oracle.sql.STRUCT';

Function ST_Union(p_geom1    in SDO_GEOMETRY,
                  p_geom2    in SDO_GEOMETRY,
                  p_precision in Number)
   Return SDO_GEOMETRY
       As language Java name
         'com.packt.spatial.JTS.ST_Union(
oracle.sql.STRUCT,oracle.sql.STRUCT,int)
return oracle.sql.STRUCT';

Function ST_Xor(p_geom1    in SDO_GEOMETRY,
```

```
                p_geom2      in SDO_GEOMETRY,
                p_precision in Number)
    Return SDO_GEOMETRY
        As language Java name
            'com.packt.spatial.JTS.ST_Xor(
oracle.sql.STRUCT,oracle.sql.STRUCT,int)
return oracle.sql.STRUCT';
```

Java implementation

All overlay operations in JTS are executed through the `OverlayOp` class:

```
public class OverlayOp extends GeometryGraphOperation
// Computes the overlay of two Geometrys.
// The overlay can be used to determine any boolean combination
// of the geometries.
```

The critical method is:

```
Static Geometry overlayOp(Geometry geom0,
                          Geometry geom1,
                               int opCode)
// Computes an overlay operation for the given geometry arguments.
```

The spatial operations (`opCodes`) supported by this class are:

```
public static final int INTERSECTION  = 1;
public static final int UNION         = 2;
public static final int DIFFERENCE    = 3;
public static final int SYMDIFFERENCE = 4;
```

Four wrapper methods are now added to the JTS class for each of our overlay types; each calls a private common method that carries out the actual `overlay.JTS`.

Only one wrapper method is shown in the following example, with sections of the actual private overlay method removed for brevity's sake:

```
public static STRUCT ST_Xor(STRUCT _geom1,
                            STRUCT _geom2,
                            int    _precision)
throws SQLException
{
    Tools.setPrecisionScale(_precision);
    return Overlay(_geom1, _geom2, OverlayOp.SYMDIFFERENCE);
}

/**
 * Overlay
 * @param _geom1      : STRUCT : First geometry subject to overlay
```

```java
 * @param _geom2      : STRUCT : Second geometry subject to overlay
 * @param _operationType : int : Overlay operation eg INTERSECTION */
private static STRUCT Overlay(STRUCT _geom1,
                             STRUCT _geom2,
                             int    _operationType)
 throws SQLException
{
    // Check geometry parameters are not null
    STRUCT resultSDOGeom = null;
    try
    {
        // Get valid connection
        . . . .
        // Extract and Check SRIDs from SDO_GEOMETYs
        int SRID = SDO_Geometry.getSRID(_geom1,0);
        If ( SRID != SDO_Geometry.getSRID(_geom2,0) ) {
            throw new
                    SQLException("SDO_Geometry SRIDs not equal");
        }
        // Convert Geometries
        PrecisionModel  pm =
            new PrecisionModel(Tools.getPrecisionScale());
        GeometryFactory gf = new GeometryFactory(pm,SRID);
        OraReader       or = new OraReader(gf);
        Geometry        geo1 = or.read(_geom1);
        Geometry        geo2 = or.read(_geom2);
        // Check converted geometries are valid
         . . .
        // Now do the required overlay
        try {
            Geometry resultGeom = null;
            resultSDOGeom = OverlayOp.overlayOp(geo1,
                                                geo2,
                                                _operationType);
            OraWriter  ow = new OraWriter(conn);
            resultSDOGeom = ow.write(resultGeom);
        } catch (Exception e) {
            return null;
        }
    } catch(SQLException sqle) {
        throw new SQLException(sqle.getMessage());
    }
    return resultSTRUCT;
}
```

Testing the implementation

The following tests are presented as follows:

```
With testGeoms As (
Select SDO_GEOMETRY('POLYGON((1 1,10 1,10 10,1 10,1 1))',NULL) g1,
       SDO_GEOMETRY('POLYGON((5 5,15 5,15 15,5 15,5 5))',NULL) g2
  From dual
)
Select 'UNION' as gtype,
       book.SPEX.ST_Union(g1,g2,1) as rGeom
  From testGeoms a Union All
Select 'INTERSECTION' as gtype,
       book.SPEX.ST_intersection(g1,g2,1) as rGeom
  From testGeoms a Union All
Select 'XOR' as gtype,
       book.SPEX.ST_XOr(g1,g2,1) as rGeom
  From testGeoms a Union All
Select 'DIFFERENCE' as gtype,
       book.SPEX.ST_Difference(g1,g2,1) as rGeom
  From testGeoms;
```

Test polygons	Overlay results
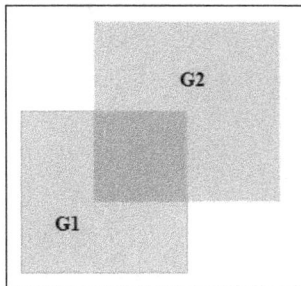	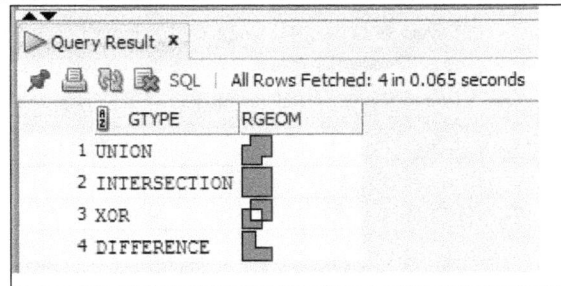
	Results shown in the rowset using GeoRaptor's display geometry as thumbnail capability.

The JTS overlay functions support the overlaying combinations of points, lines, and polygons.

We can now, finally, compare the results of Oracle's SDO_GEOM.RELATE, and our JTS implementation using the previous geometries. Note that the results are essentially the same:

```
With geoms As (
Select rownum as id, geom
  From (
Select SDO_GEOMETRY('LINESTRING (548766.398 3956415.329, 548866.753
  3956341.844, 548845.366 3956342.941)',32639) geom
```

```
      From Dual Union All
Select SDO_GEOMETRY('POINT (548766.398 3956415.329)',32639) geom
      From Dual Union All
Select SDO_GEOMETRY('LINESTRING (548938.421 3956363.864, 548823.852
      3956379.758, 548818.01 3956381.297, 548812.139 3956382.844,
      548683.715 3956400.404)',32639) geom
      From Dual Union All
Select SDO_GEOMETRY('LINESTRING (548766.398 3956415.329, 548866.753
      3956341.844, 548845.366 3956342.941)',32639) geom
      From Dual Union All
Select SDO_GEOMETRY('POLYGON ((548862.366 3956401.619, 548793.269
      3956409.845, 548785.043 3956369.812, 548850.302 3956361.587,
      548862.366 3956401.619))',32639) geom
      From Dual Union All
Select SDO_GEOMETRY('POINT (548711.3 3956106.6)',32639) geom
      From Dual Union All
Select SDO_GEOMETRY('POLYGON ((548710.0 3956105.0, 548715.0 3956105.0,
      548720.0 3956110.0, 548710.0 3956110.0, 548710.0 3956105.0))',32639)
geom
      From Dual
      )
)
Select Distinct
        mdsys.SDO_GEOM.Relate(a.geom,'DETERMINE',b.geom,0.05)
          as OraRel,
        book.SPEX.ST_Relate(a.geom,'DETERMINE',b.geom,2)
          as JTSRel
    From geoms a,
        geoms b
  Order By 1;

-- Results
--
ORAREL                JTSREL
---------------       --------------------------
CONTAINS              CONTAINS,COVERS,INTERSECTS
DISJOINT              DISJOINT
EQUAL                 EQUAL
INSIDE                COVEREDBY,INTERSECTS,WITHIN
OVERLAPBDYDISJOINT    CROSS,INTERSECTS
TOUCH                 COVEREDBY,INTERSECTS,TOUCH
TOUCH                 COVERS,INTERSECTS,TOUCH
```

Notice that the JTS code can return slightly different keywords to those which Oracle's RELATE returns. It is left up to the reader to modify the provided code to implement comparisons that return Oracle specific keywords like OVERLAPBDYDISJOINT, and so on.

Snapping geometries together

Oracle Spatial has no snapping functions. However, JTS has such capability which will now be explained and demonstrated.

PL/SQL implementation

The following three PL/SQL functions expose the three types of snapping that JTS provides between two geometries:

```
/**
 * ST_Snap
 * Snaps both geometries to each other with both being able to move.
 * Returns compound SDO_GEOMETRY ie x004
 */
Function ST_Snap(p_geom1        in mdsys.SDO_GEOMETRY,
                 p_geom2        in mdsys.SDO_GEOMETRY,
                 p_snapTolerance in Number,
                 p_precision    in Number)
   Return mdsys.SDO_GEOMETRY
       As language Java name  'com.packt.spatial.JTS.ST_Snap(oracle.
sql.STRUCT,oracle.sql.STRUCT,double,int) return oracle.sql.STRUCT';

/**
 * ST_SnapTo
 * Snaps the vertices in the component LineStrings of the source
geometry to
 * the vertices of the given snap geometry.
 */
Function ST_SnapTo(p_geom1        in mdsys.SDO_GEOMETRY,
                   p_snapGeom     in mdsys.SDO_GEOMETRY,
                   p_snapTolerance in Number,
                   p_precision    in Number)
   Return mdsys.SDO_GEOMETRY
       As language Java name       'com.packt.spatial.JTS.ST_
SnapTo(oracle.sql.STRUCT,oracle.sql.STRUCT,double,int) return oracle.
sql.STRUCT';

/**
 * ST_SnapToSelf
 * Snaps the vertices in the source Geometry's LineStrings to
themselves.
 **/
```

```
Function ST_SnapToSelf(p_geom          in mdsys.SDO_GEOMETRY,
                       p_snapTolerance in Number,
                       p_precision     in Number)
     Return mdsys.SDO_GEOMETRY
        As language Java name
'com.packt.spatial.JTS.ST_SnapToSelf(oracle.sql.STRUCT,double,int)
return oracle.sql.STRUCT';
```

Implementing the snapping methods

The three snapping functions all have a public method in the JTS class. Each of these methods call a common private method in the JTS class called _snapper.

```
private static STRUCT _snapper(STRUCT _geom1,
                               STRUCT _geom2,
                               double _snapTolerance,
                               int    _precision,
                               int    _snapType)
     throws SQLException
```

_snapper implements the geometry snapping by using the JTS GeometrySnapper class constructor, which is "told" what to do by each of the three public snapping methods using the required constant value from the following three:

```
private static final int SNAP       = 1;
private static final int SNAPTO     = 2;
private static final int SNAPTOSELF = 3;
```

For example, ST_SnapToSelf uses _snapper and the SNAPTOSELF constant as follows:

```
public static STRUCT ST_SnapToSelf(STRUCT _geom1,
                                   double _snapTolerance,
                                   int    _precision)
     throws SQLException
{
  return _snapper(_geom1,
                  null,
                  _snapTolerance,
                  _precision,
                  JTS.SNAPTOSELF);
}
```

The _snapper method uses the snapping methods provided by the GeometrySnapper JTS class which are:

```
public GeometrySnapper(Geometry srcGeom)
        // Constructor that creates a new snapper to act
        // on the provided geometry
public static Geometry[] snap(Geometry g0,
```

```
                    Geometry g1,
                    double snapTolerance)
    // Snaps two geometries together with a given
    // tolerance. Where snapTolerance is the desired
    // snapping tolerance eg snap two Geometrys together
    // by 0.5 meters.

public Geometry snapTo(Geometry snapGeom,
                    double snapTolerance)
    // Where snapTolerance is the desired snapping tolerance
    // eg snap two Geometrys together by 0.5 meters.

public Geometry snapToSelf(double snapTolerance,
                    boolean cleanResult)
    // Setting the parameter cleanResult to true will
    // force the result to be valid.
```

The common _snapper method executes the snapping via the following code:

```
try {
    Geometry resultGeom = null;
    GeometrySnapper gs = null;
    // Don't need to create a new Snapper when
    //   JTS.SNAP as method is static
    If ( _snapType!= JTS.SNAP ) {
        gs = new GeometrySnapper(geo1);
    }
    switch (_snapType) {
      case JTS.SNAPTOSELF :
          resultGeom = gs.snapToSelf(_snapTolerance,
                                    true /*cleanResult*/);
          break;
      case JTS.SNAPTO :
          resultGeom = gs.snapTo(geo2, _snapTolerance);
          break;
      case JTS.SNAP :
          resultGeom = gf.createGeometryCollection(
                            GeometrySnapper.snap(geo1,
                                            geo2,
                                        _snapTolerance));
          break;
    }
    OraWriter ow = new OraWriter(Tools.getConnection());
    resultSDOGeom = ow.write(resultGeom);
} catch (Exception e) {
    return null;
}
```

Testing the implementation

The use of the geometry snapping functions can be seen in the following examples. The first example snaps one linestring to another:

```
Select book.SPEX.ST_SnapTo(
        SDO_GEOMETRY(
        'LINESTRING(0.5 0.5,9.5 10.5,19.5 -0.5,30.5 9.5)'),
        SDO_GEOMETRY(
        'LINESTRING(0.0 0.0,10.0 10.0,20.0 0.0,30.0 10.0)'),
        1.0,3).get_wkt() as SnappedLine1
   From dual;

-- Results
--
SNAPPEDLINE1
----------------------------------------------------
LINESTRING (0.0 0.0, 10.0 10.0, 20.0 0.0, 30.0 10.0)
```

Note that in this example, the result is the same as the first.

The next example snaps a point to an area:

```
Select book.SPEX.ST_SnapTo(
        SDO_GEOMETRY('POINT(-7.091 1.347)'),
        SDO_GEOMETRY('POLYGON((-8.369 14.803,-8.191 8.673,
                              -8.072 0.4,5.737 0.4,
                              5.142 14.922, -8.369 14.803
            ))'),
        2.0,3).get_wkt()
        as snappedPoint
   From dual;

-- Results
--
SNAPPEDPOINT
-----------
POINT (-8.072 0.4)
```

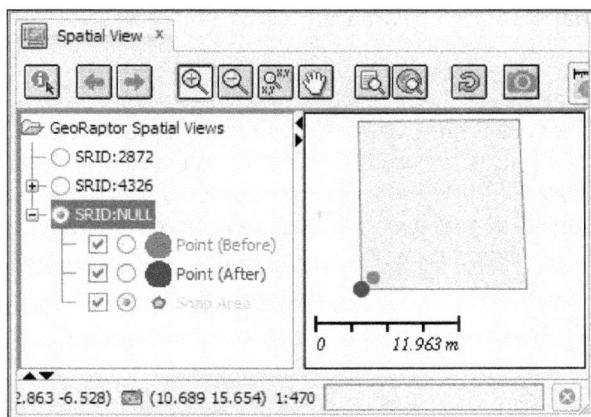

The next example snaps a line to an area:

```
Select book.SPEX.ST_SnapTo(
       SDO_GEOMETRY(
         'LINESTRING(-8.339 -1.553,-8.682 8.496,-8.476 16.728)'),
       SDO_GEOMETRY(
         'POLYGON ((-8.369 14.803,-8.191  8.673,-8.072 0.4,
                    5.737  0.4,   5.142 14.922,-8.369 14.803 ))'),
       0.75,3).get_wkt()
       as snappedLine
  From dual;

-- Results
--
SNAPPEDLINE
----------------------------------------------------
LINESTRING (-8.339 -1.553,-8.072 0.4,-8.191 8.673,
            -8.369 14.803,-8.476 16.728)
```

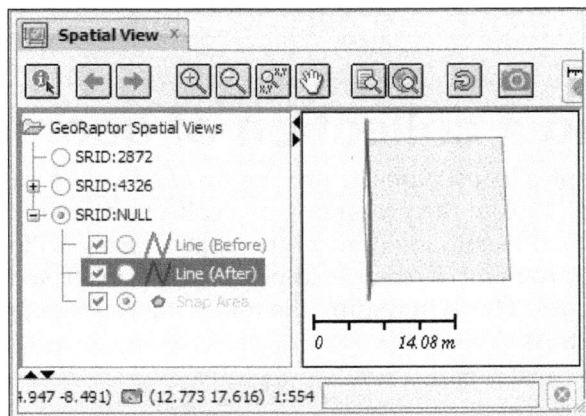

The final example snaps an area to an area:

```
Select book.SPEX.ST_SnapTo(
    SDO_GEOMETRY(
        'POLYGON((-24.089  0.348,  -8.339 0.553,  -8.682 8.496,
                  -8.476 14.728, -24.02 14.522, -24.089 0.348))'),
    SDO_GEOMETRY(
        'POLYGON((-8.369 14.803, -8.191  8.673, -8.072  0.4,
                  5.737  0.4,   5.142 14.922, -8.369 14.803))'),
    0.75,3).get_wkt()
  as snappedPoly
 From dual;

-- Results
--
SNAPPEDPOLY
------------------------------------------------------
  POLYGON ((-24.089  0.348, -8.072 0.4,   -8.191 8.673,
            -8.369 14.803,-24.02 14.522,-24.089 0.348))
```

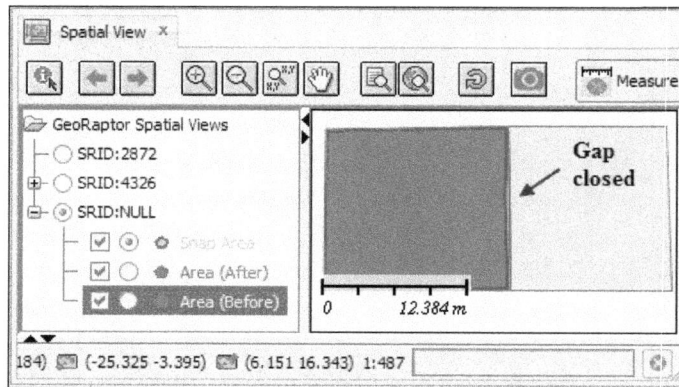

Processing a collection of geometries

The preceding examples focused on the processing of one, then two geometries. In this section we will look at the processing of a collection (more than two) of SDO_ GEOMETRY objects. The presentation of this type of processing will take place around solving a real world problem; creating land parcel polygons from their boundaries described by linestrings. This will require the following types of processing of collections of geometries:

- Forming nodes between a collection of linestrings
- Forming polygons from a collection of noded linestrings

Types of collections

Being inside the Oracle database, we have a limited number of methods or data types that can be used to represent a collection of geometry objects, which can be passed to our Java code for processing. These methods or collection types draw from SQL, PL/SQL and OGC SFA's geometry type hierarchy. They include:

- A REF CURSOR/result set (Java type `Java.sql.ResultSet`):

```
TYPE refcur_t IS REF CURSOR; -- or SYS_REFCURSOR;


- Example PL/SQL function declaration
Function ST_NodeLinestrings(p_resultSet in SPEX.refcur_t,
                            p_precision in number)
  Return SDO_GEOMETRY
         Deterministic;
-- How to create and use ref cursor based functionality:
--
Set serveroutput On Size Unlimited
Declare
  mycur  book.SPEX.refcur_t;
  v_geom SDO_GEOMETRY;
Begin
  Open mycur For
  Select line
    From (
      Select SDO_GEOMETRY('LINESTRING( 0  0, 5 10)') line
        From Dual Union All
      Select SDO_GEOMETRY('LINESTRING(10 10,15  0)') line
        From Dual Union All
      Select SDO_GEOMETRY('LINESTRING( 2  8,13  8)') line
        From Dual Union All
      Select SDO_GEOMETRY('LINESTRING( 0  3, 5 -8)') line
        From Dual Union All
      Select SDO_GEOMETRY('LINESTRING(15  3,10 -8)') line
        From Dual Union All
      Select SDO_GEOMETRY('LINESTRING( 2 -6,13 -6)') line
        From Dual
    ) a;
  v_geom := book.SPEX.ST_NodeLineStrings(mycur,2);
  dbms_output.put_line('ST_NodeLineStrings: ' ||
```

```
                              v_geom.get_Wkt());
     Close myCur;
End;
/
show errors
```

- An array data type (Java type `oracle.sql.ARRAY`):

```
Create Or Replace Type SDO_GEOMETRY_ARRAY
     As VARRAY(10485760) Of SDO_GEOMETRY

-- Example PL/SQL function declaration
--
Function ST_NodeLinestrings(
                p_geomset   in mdsys.SDO_GEOMETRY_ARRAY,
                p_precision in number)
   Return SDO_GEOMETRY
        Deterministic;
-- How to create an SDO_GEOMETRY_ARRAY...
--
With unNoded As (
   Select SDO_GEOMETRY('LINESTRING(0  0, 5 10)') line
     From dual Union All
   Select SDO_GEOMETRY('LINESTRING(10 10,15  0)') line
     From dual Union All
   Select SDO_GEOMETRY('LINESTRING( 2  8,13  8)') line
     From dual Union All
   Select SDO_GEOMETRY('LINESTRING( 0  3, 5 -8)') line
     From dual Union All
   Select SDO_GEOMETRY('LINESTRING(15  3,10 -8)') line
     From dual Union All
   Select SDO_GEOMETRY('LINESTRING( 2 -6,13 -6)') line
     From dual
)
Select book.SPEX.ST_NodeLineStrings(
              CAST(COLLECT(a.line) As
                  mdsys.SDO_GEOMETRY_ARRAY),2)
        as noded_Lines
   From unnoded a;
```

- `GeometryCollection`:

```
GEOMETRYCOLLECTION (POINT(0 0), LINESTRING(10 10,15 0))
-- Which can be created in many ways, here is one.
--
Select SDO_GEOM.Sdo_Union(a.point,a.line,0.05)
              .get_wkt() as gColl
   From (Select SDO_GEOMETRY('POINT(0 0)') as point,
                SDO_GEOMETRY('LINESTRING(10 10,15 0)') line
          From dual) a;
```

The execution of Java spatial processing functions that require a collection of `Geometry` objects first requires us to create the required collection type within the Oracle database (via SQL), and pass it to the Java method, which processes it and passes its result back (as a single `Geometry` which can be a `GeometryCollection`). Examples of each have been supplied earlier.

Intersecting lines – a preliminary discussion

The noding of a collection of linestrings is a common requirement, when processing and using spatial data. By noding we mean finding all possible intersections between the segments that describe a set of linestrings, and adding a vertex to each linestring at each intersection point, where there is none. A picture tells a thousand words; the two linestring in the image on the left are noded returning the linestrings on the right.

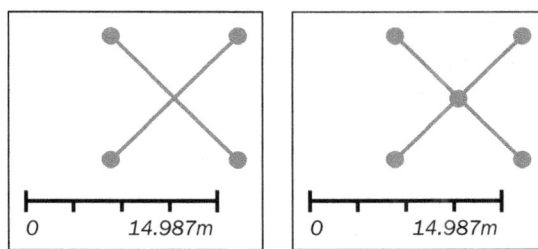

There are pure SQL techniques available that use Oracle Spatial functionality for intersecting two geometries, but these are not straight forward or very efficient. What is needed is an efficient and effective way of dealing with a large number of lines at a time. For example, let's assume we have a collection of lines that define a set of land parcels. From these lines we wish to create polygons that compose the area they define. However, a set of problems arise that must be addressed. These include the fact that the lines:

- Overshoot each other with no node existing between two intersecting segments
- Undershoot each other with no intersection existing between them

The following figure depicts these land parcels and their linestring intersection problems:

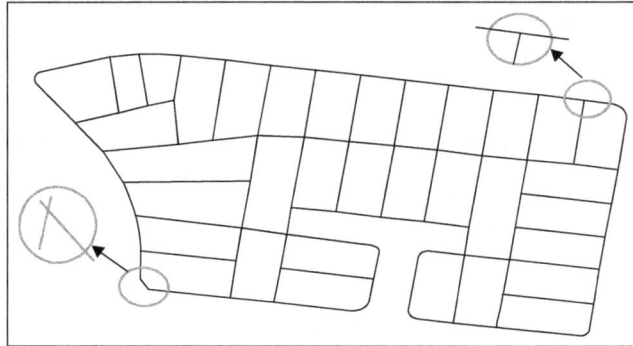

Creating land parcels

At its simplest, the steps to create the land parcel polygons are as follows:

1. Extend all lines such that they all cross each other.
2. Node all the boundary lines creating vertices at all intersection points.
3. Build polygons from resulting lines.

Step 1 – extending the linestrings

JTS does not have linestring extension functionality. However, in *Chapter 7, Editing, Transforming, and Constructing Geometries,* the ST_Extend PL/SQL function was created that is capable of extending a linestring from the start, finish, or both ends. The member function of the T_GEOMETRY method is:

```
Member Function
ST_Extend(p_extend_dist in Number,
          p_start_end   in Varchar2 default 'START',
          p_unit        in Varchar2 default null)
          Return T_GEOMETRY Deterministic
```

Examples of its use are in *Chapter 7, Editing, Transforming, and Constructing Geometries,* and are included in the processing demonstrated in Step 2.

Step 2 – noding linestrings

How can JTS be used to node a collection of linestrings? From the **Frequently Asked Questions (FAQ)** on the JTS website:

> *JTS provides several methods for noding lines. They vary in terms of simplicity, performance, robustness and flexibility. Options are:*
>
> Geometry.union() *(unary union) has the effect of noding and dissolving a set of lines contained in a geometry collection*
>
> GeometryNoder *performs snap-rounding on a collection of geometries*
>
> MCIndexNoder *nodes a set of* SegmentStrings, *without dissolving*

Given that dissolving of the resultant linestring to remove duplicates will be of benefit to the polygon building step, the chosen approach for our function is to use Geometry.union). In order to maximize polygon formation, the ordinates of the noded linestrings will be rounded via GeometryPrecisionReducer.Java, which reduces the precision of a geometry according to the supplied PrecisionModel, ensuring that the result is topologically valid.

PL/SQL implementation

Three PL/SQL functions are added to our SPEX package:

```
/**
 * ST_NodeLinestrings
 * Takes a GeometryCollection of linestring geometries
 * and ensures nodes are created at all topological intersections
**/
Function ST_NodeLinestrings(p_geometry  in SDO_GEOMETRY,
                            p_precision in Number)
  Return SDO_GEOMETRY
        As language Java name          'com.packt.spatial.JTS.ST_
NodeLinestrings(oracle.sql.STRUCT,int)
return oracle.sql.STRUCT';

/** Wrappers….. */
Function ST_NodeLinestrings(p_resultSet in SPEX.refcur_t,
                            p_precision in Number)
  Return SDO_GEOMETRY
        As language Java name
        'com.packt.spatial.JTS.ST_NodeLinestrings(Java.sql.
ResultSet,int) return oracle.sql.STRUCT';
```

```
Function ST_NodeLinestrings(p_geomset in mdsys.SDO_GEOMETRY_ARRAY,
                            p_precision in Number)
   Return SDO_GEOMETRY
        As language Java name         'com.packt.spatial.JTS.ST_
NodeLinestrings(oracle.sql.ARRAY,int)
   return oracle.sql.STRUCT';
```

Implementing _nodeLinestrings

The REF_CURSOR, ARRAY, and SDO_GEOMETRY PL/SQL functions convert their collection objects into a Java collection of linestring geometries, before using a common function (_nodeLinestrings) to node the resultant collection. A single GeometryCollection SDO_GEOMETRY object (x004) is returned.

The full implementation details can be determined by looking at the source code for the methods in the JTS.Java class provided with this book.

The main, common, implementation Java code is as follows:

```
public static STRUCT _nodeLinestrings(Collection     _lines,
                                      GeometryFactory _gf)
throws SQLException
{
   If (_lines == null || _lines.size()==0 || _gf==null) {
     return null;
   }
   try
   {
     // ------- Precision reduce collection ------------
     //
     Collection lineList = new ArrayList();
     Geometry geom = null;
     GeometryPrecisionReducer gpr =
         new GeometryPrecisionReducer(_gf.getPrecisionModel());
     Iterator iter = _lines.iterator();
     while (iter.hasNext() ) {
         geom = (Geometry)iter.next();
         // test geom to ensure only processing linestrings
         geom = gpr.reduce(geom);
         If (geom!=null && geom.isValid()) {
             lineList.add(geom);
         }
     }
     If ( lineList.isEmpty() ) { return null; }
     // -- Node the precision reduced geometry collection ---
     //
```

```
    GeometryCollection nodedLineStrings = new
        GeometryCollection(((Geometry[])
                            lineList.toArray(new Geometry[0])),
                        _gf);
    geom = nodedLineStrings.union();
    If (geom==null || geom.isEmpty()) { return null; }
    OraWriter ow = new OraWriter(Tools.getConnection());
    return ow.write(geom);
  } catch (SQLException sqle) {
    System.err.println(sqle.getMessage());
    throw new SQLException(sqle.getMessage());
  }
}
```

Testing the implementation

The following simple noding example shows how to call one of the Java methods:

```
With unNoded As (
  Select SDO_GEOMETRY('LINESTRING( 0  0, 5 10)') line
    From Dual Union All
  Select SDO_GEOMETRY('LINESTRING(10 10,15  0)') line
    From Dual Union All
  Select SDO_GEOMETRY('LINESTRING( 2  8,13  8)') line
    From Dual Union All
  Select SDO_GEOMETRY('LINESTRING( 0  3, 5 -8)') line
    From Dual Union All
  Select SDO_GEOMETRY('LINESTRING(15  3,10 -8)') line
    From Dual Union All
  Select SDO_GEOMETRY('LINESTRING( 2 -6,13 -6)') line
    From Dual
)
Select SPEX.ST_NodeLineStrings(
            CAST(COLLECT(a.line) As mdsys.SDO_GEOMETRY_ARRAY),
            2)
      As noded_Lines
  From unnoded a;
```

The un-noded and noded linestrings look like this:

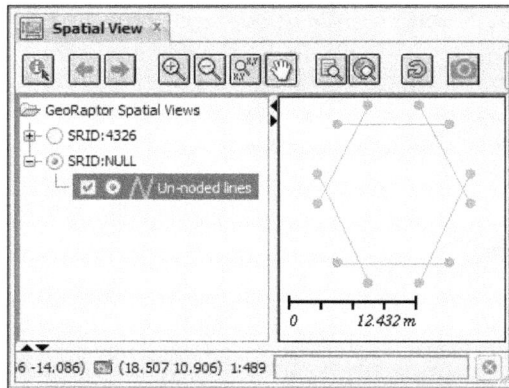

Illustration 10

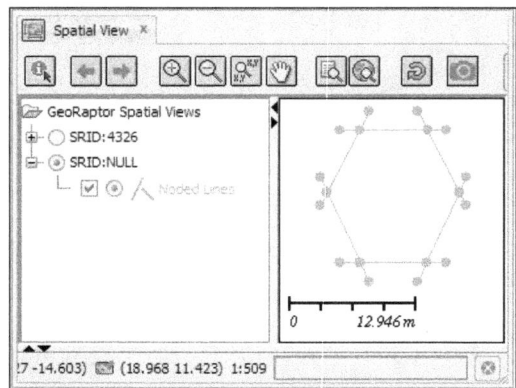

Illustration 11

Step 3 – build polygons

JTS has a simple, but quite powerful, `Polygonizer` class that:

> "*Polygonizers are a set of geometries, which contain linework that represents the edges of a planar graph. All types of Geometry are accepted as input; the constituent linework is extracted as the edges to be polygonized. The processed edges must be correctly noded; that is, they must only meet at their endpoints. The Polygonizer will run on incorrectly noded input but will not form polygons from non-noded edges, and will report them as errors.*"

Even though the Polygonizer will work against all geometry types, this implementation is designed only to work on lines. If one has non-linestring input, appropriate SQL processing should occur to ensure the data supplied conforms to the required input.

Implementing ST_Polygonizer

Three PL/SQL functions are added to our SPEX package that allows polygon building:

```
/**
 * ST_PolygonBuilder
 * Builds polygons from ref cursor containing input linestrings
 * (all else are filtered out).
 * Result, if successful, is Polygon or MultiPolygon.
**/
```

```
Function ST_PolygonBuilder(p_resultSet in SPEX.refcur_t,
                           p_precision in Number)
  Return SDO_GEOMETRY
        Deterministic;


/**
 * ST_PolygonBuilder
 * Builds polygons from array of linestrings
 * (all else are filtered out).
 * Result, if successful, is Polygon or MultiPolygon.
**/
Function ST_PolygonBuilder(p_geomset  in mdsys.SDO_GEOMETRY_ARRAY,
                           p_precision in Number)
  Return SDO_GEOMETRY
        Deterministic;


/**
 * ST_PolygonBuilder
 * Builds polygons from input linestrings in GeometryCollection
 * Result, if successful, is Polygon or MultiPolygon.
**/
Function ST_PolygonBuilder(p_geometry  in SDO_GEOMETRY,
                           p_precision in Number)
  Return SDO_GEOMETRY
        Deterministic;
```

Implementing _polygon=Builder

The REF_CURSOR, ARRAY, and SDO_GEOMETRY PL/SQL functions convert their collection objects into Java collection of lineString geometries, before using a common function (__polygonBuilder) to build polygons from the resultant collection. A single SDO_GEOMETRY object (polygon x003 or multipolygon x007) is returned.

The full implementation details can be determined by looking at the source code for the methods in the JTS.Java class provided with this book.

The main, common, implementation Java code is as follows:

```
private static STRUCT _polygonBuilder(Collection    _lines,
                                      GeometryFactory _gf,
                                      OraWriter       _ow)

throws SQLException
{
```

```
STRUCT resultSDOGeom = null;
// Try and create polygons from linestrings in lines collection
//
Polygonizer polygonizer = new Polygonizer();
polygonizer.add(_lines); // mlines);
Collection polys = polygonizer.getPolygons();
If ( polys != null && polys.size()>0 ) {
  Collection polygons = new ArrayList();
  // Iterate over all formed polygons and create single result
  //
  Iterator it = polys.iterator();
  If ( it != null ) {
    int i = 0;
    Object p = null;
    while ( it.hasNext() )
    {
       p = it.next();
       If ( p instanceof Polygon ) {
         polygons.add(p);
       }
    }
    If ( polygons.size()==0 ) {
      return null;
    } Else If ( polygons.size() == 1 ) {
      Geometry poly = (Geometry)polygons.toArray(
                                        new Geometry[0])[0];
      resultSDOGeom = _ow.write(poly);
    } Else If ( polygons.size() > 1 ) {
      GeometryCollection coll = new
          GeometryCollection(((Geometry[])polygons.toArray(
                                        new Geometry[0])),_gf);
      resultSDOGeom = _ow.write(coll);
    }
  }
} Else {
  Collection remains = polygonizer.getDangles();
  If (remains!=null)
    System.err.println("Dangles "+remains.size());
  remains = polygonizer.getCutEdges();
  If (remains!=null)
    System.err.println("CutEdges "+remains.size());
  remains = polygonizer.getInvalidRingLines();
  If (remains!=null)
```

```
                System.err.println("InvalidRings "+remains.size());
        }
        return resultSDOGeom;
}
```

Putting it all together – forming land parcel polygons

We are now in a position to form our land parcel polygons from the boundary linestrings via the described steps and new Java Stored Procedure based functions. The processing steps can be executed via a single declarative SQL statement as follows:

```
With lines As (
Select a.geom.sdo_srid as srid,
        t_geometry(a.geom,0.05,9,'GEOGRAPHIC')
          .ST_RemoveDuplicateVertices()
          .ST_Extend(0.1,'BOTH','unit=M') as line
   From (Select SDO_GEOMETRY('LINESTRING(
                    144.663009408565 -37.8642206756175,
                    144.663009488508 -37.8642206843752,
                    144.663460702298 -37.8642700441642)',
                    4326) as geom
          From dual Union All
        Select SDO_GEOMETRY('LINESTRING (
                    144.663496479881 -37.8640640556114,
                    144.662952669779 -37.8640728854823,
                    144.662952612966 -37.8640728863805)',
                    4326) as geom
          From dual Union All
         -- .....46 other lines removed
         ) a
)
Select SPEX.ST_PolygonBuilder(CAST(COLLECT(c.vector)
                             As mdsys.SDO_GEOMETRY_ARRAY),
                        max(c.precision)) as polygons
   From (Select b.precision, b.srid,
               v.vector.ST_SdoGeometry(2,b.srid) as vector
          From (Select max(a.line.precision) as precision,
                      max(a.line.ST_SRID()) as srid,
                      SPEX.ST_NodeLineStrings(
                             CAST(COLLECT(a.line.geom)
                               As mdsys.SDO_GEOMETRY_ARRAY),
```

```
                        Max(a.line.precision))
                As vectors
        From lines a
    ) b,
    Table(T_Geometry(b.vectors,0.05,
                        b.precision,'GEOGRAPHIC')
        .ST_Vectorize()) v
    ) c;
```

The steps to creating the land parcel polygons in the preceding SQL include an extra initial step (the vectorization of the noded linestrings into two simple vertex segments or vectors):

- Remove duplicate vertices from linestrings (ST_RemoveDuplicateVertices)

- Extend (0.1m) all lines such that they all cross each other (ST_Extend)

- Node all the boundary lines creating vertices at all intersection points (ST_NodeLineStrings)

- Vectorize the linestrings into two component vertex segments (ST_Vectorize)

- Build polygons from resulting lines (ST_PolygonBuilder)

The result is depicted in the following figure:

A collection of useful functions

The source code that accompanies this chapter includes the following additional Java based functionality:

Function	Code snippet and image

ST_Voronoi
 	
Takes a geometry collection and creates a Voronoi diagram from the geometries' points, returning the resulting triangles as geometries.
	``DelaunayTriangulationBuilder builder =
	new
	  DelaunayTriangulationBuilder();
	builder.setTolerance(tolerance);
	builder.setSites(
	        GeometryCollection);
	QuadEdgeSubdivision subdiv =
	        builder.getSubdivision();
	Geometry gOut =
	    subdiv.getVoronoiDiagram(
	            GeometryFactory);```
	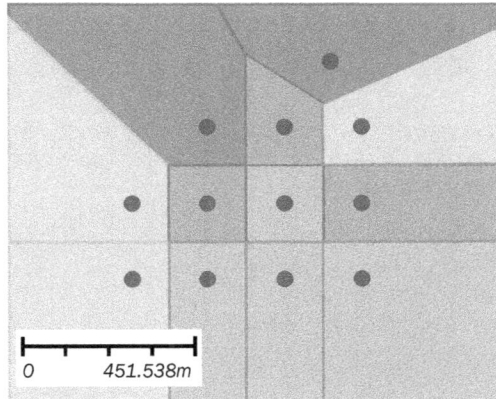

`ST_DelaunayTriangles`

Takes a geometry collection and creates a Delaunay Triangulation from its constituent points returning the resulting triangles as geometries.

```
DelaunayTriangulationBuilder builder =
  new
  DelaunayTriangulationBuilder();
builder.setTolerance(tolerance);
builder.setSites(geo);
Geometry gOut = builder
    .getTriangles(GeometryFactory);
```

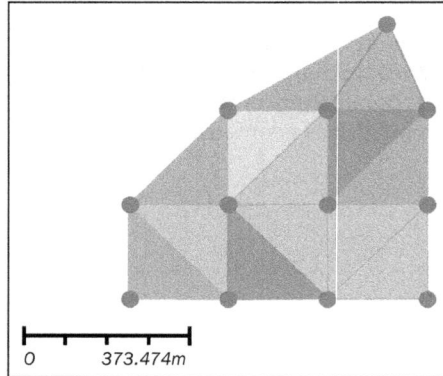

```
0        373.474m
```

`ST_InsertVertex, ST_ UpdateVertex,` and `ST_ DeleteVertex`

Java version of the same functions in *Chapter 7, Editing, Transforming, and Constructing Geometries,* implemented using JASPA code.

```
JTSChangePointFilter filter =
  new JTSChangePointFilter(
              pointIndex,
              pointCoord);
geom.apply(filter);
```

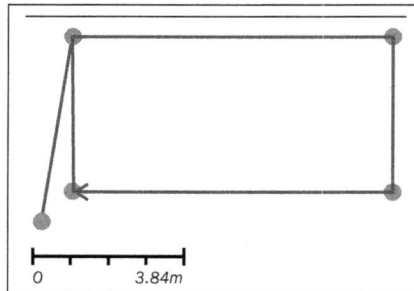

```
0        3.84m
```

`ST_LineMerger`

Takes a collection of linear components and forms maximal-length linestrings.

```
LineMerger lm = new LineMerger();
lm.add(Collection ofLineStrings);
Collection mlines =
lm.getMergedLineStrings();
MultiLineString mergedLineStrings =
    new MultiLineString(
        ((LineString[])
        mlines.toArray(
            new LineString[0])),
        GeometryFactory);
```

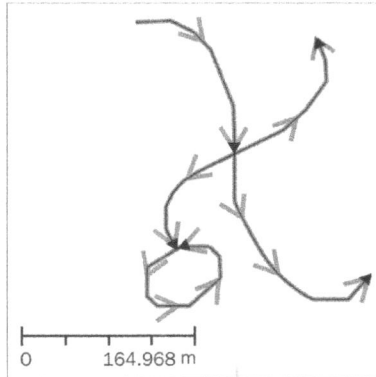

0   164.968 m

`ST_Centroid`

Computes centroid (mathematical or interior) of a geometry. Interior point is different from that computed by Oracle's `SDO_UTIL.INTERIOR_POINT`.

```
Geometry outGeom = null;
if ( Math.abs(_interior) == 0 ) {
   outGeom = inGeom
             .getCentroid();
} Else {
   outGeom = inGeom
             .getInteriorPoint();
}
```

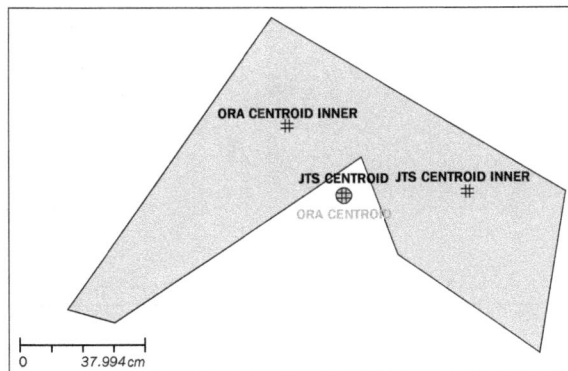

0   37.994 cm

ST_Densify

Densifies a geometry's vertex description using a given distance tolerance, and respecting the input geometry's **PrecisionModel**.

```
Densifier densifier = new
Densifier(geo);
densifier.setDistanceTolerance(
                _distanceTolerance);
Geometry outGeom = densifier.
getResultGeometry();
```

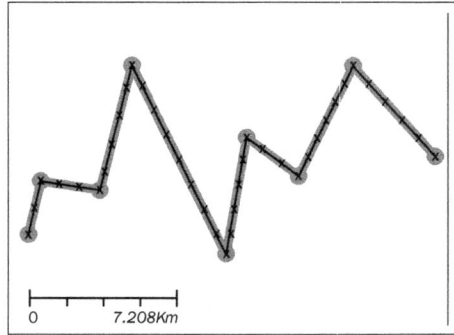

ST_InterpolateZ

Given three points that define a triangular facet (see ST_ DelaunayTriangles, returns the Z value of the provided point that falls within the triangle.

```
double interpolatedZ = Triangle.
interpolateZ(
        point.getCoordinate(),
        geo1.getCoordinate(),
        geo2.getCoordinate(),
        geo3.getCoordinate());
```

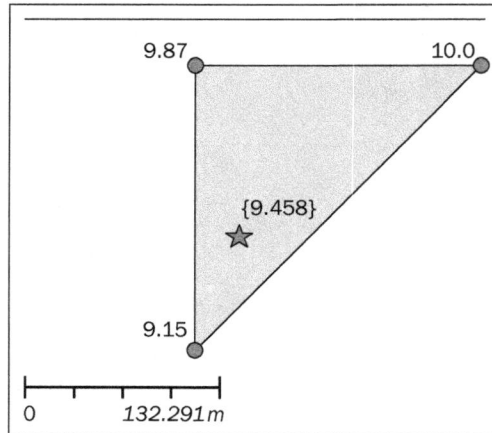

ST_TopologyPreservingSimplify

Simplifies `Geometry`, ensuring that the result is a valid geometry having the same dimension and number of components as the input. The simplification uses a maximum distance difference algorithm similar to the one used in the Douglas-Peucker algorithm.

```
Geometry outGeom = null;
outGeom = TopologyPreservingSimplifier
  .simplify(inGeom,
       _distanceTolerance));
```

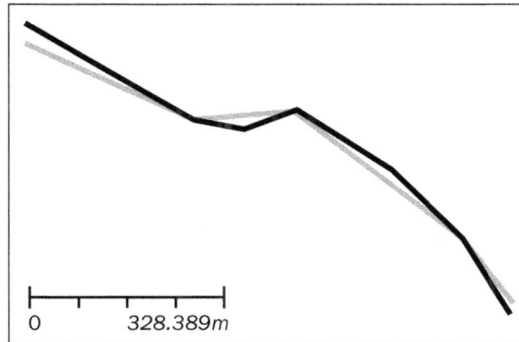

```
0        328.389m
```

ST_MinimumBoundingCircle

"Computes the Minimum Bounding Circle (MBC) for the points in `Geometry`. The MBC is the smallest circle which contains all the input points (this is sometimes known as the Smallest Enclosing Circle). This is equivalent to computing the maximum diameter of the input point set."

```
MinimumBoundingCircle mbc = new
   MinimumBoundingCircle(inGeom);
Geometry circle=mbc.getCircle();
MinimumBoundingCircle mbc = new
     MinimumBoundingCircle(geo);
Geometry circle=mbc.getCircle();
```

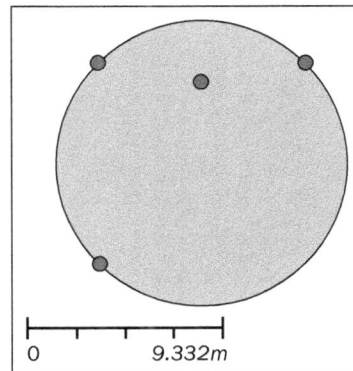

```
0        9.332m
```

| ST_AsEWKT and ST_GeomFromEWKT | ```
Select SPEX.ST_AsEWKT(
    SDO_GEOMETRY(4302,8307,NULL,
      SDO_ELEM_INFO_ARRAY(1,2,1),
      SDO_ORDINATE_ARRAY(0,0,2,3,
            50,50,100,200)))
       as eWKT
  From dual;
-- Result
--
EWKT
-------------------------------
SRID=8307;
LINESTRING(0 0 2 3,50 50 100 200)

Select SPEX.ST_GeomFromEWKT(
'SRID=8307;LINESTRING(0 0 2 3,
        50 50 100 200)')
      as geom
  From dual;
-- Result
--
GEOM
-------------------------------
SDO_GEOMETRY(4002,8307,NULL,
SDO_ELEM_INFO_ARRAY(1,2,1),
SDO_ORDINATE_ARRAY(0,0,2,3,
          50,50,100,200))
``` |
```

Performance of Java-based SQL processing

If a function is not available natively in Oracle, then having it become available via a custom Java function is a reasonable compromise! Even so, to be really useful, such code must operate correctly and efficiently.

Much has been written on the performance of PL/SQL against native SQL functionality, with some also available for Java-based stored procedures. From such work you can discover a set of guidelines for creating fast executing functions.

While little can be done about the cost to call a Java Stored Procedure from SQL (c.f., context swapping), one does have control over the efficiency of the implemented algorithm. JTS is particularly known for this, such that its algorithms are well understood. In addition methods that operate on an array or collection of geometry objects (c.f., ST_PolygonBuilder's processing of x and y collections), will result in faster operation than the methods that process only one geometry at a time, out of many. While ST_Centroid operates on one object, it could be made to operate on multiple.

Compilation to native code

Across all Oracle database versions, when Java source is loaded into the Oracle database JVM, it is compiled to byte code before execution; if you load a jar file of compiled classes this step is not required. However, Oracle provides an additional mechanism called native compilation, which converts Java byte code classes to native code shared libraries. These shared libraries are linked into the kernel, thus resulting in increased performance.

Since Oracle 11*g*, **Just In Time (JIT)** compilation of Java classes to native code has been available via the database's JVM:

Based on dynamically gathered profiling data, this compiler transparently selects Java methods to compile the native machine code and dynamically makes them available to running Java sessions. Additionally, the compiler can take advantage of Oracle JVM's class resolution model to optionally persist compiled Java methods across database calls, sessions, or instances. Such persistence avoids the overhead of unnecessary recompilations across sessions or instances, when it is known that semantically the Java code has not changed.

> *The JIT compiler is controlled by a new Boolean-valued initialization parameter called* JAVA_JIT_ENABLED. *When running heavily used Java methods with* Java_jit_enabled *parameter value as true, the Java methods are automatically compiled to native code by the JIT compiler and made available for use by all sessions in the instance.*

In addition, the loaded Java sources and classes can be forcibly compiled using the DBMS_JAVA package's compile_class method as follows:

```
Set serveroutput On Size Unlimited
Declare
    v_compile Number;
Begin
    For rec In (Select name From user_Java_classes) Loop
    Begin
        v_compile := DBMS_JAVA.compile_class(rec.name);
        dbms_output.put_line('Compiling class ' || rec.name ||
                        ' resulted in ' || v_compile ||
                        ' methods being compiled.');
        Exception
            When OTHERS Then
                dbms_output.put_line('Compile of Java class ' ||
                                rec.name ||
                                ' failed with ' || SQLCODE );
    End;
    End Loop;
End;
/
show errors
```

This code is supplied in the compile_Java_11g.sql file (6365EN\deploy folder) with this book.

Before 11g compilation was possible, but only externally via use of the ncomp utility and is not covered in this book.

Summary

The thrust of this chapter has been to show, in a step-by-step manner, how existing open source Java-based algorithms and solutions can be made available to enhance or compliment Oracle SDO_GEOMETRY based processing. The reasons why recourse to Java-based stored procedures should be considered were covered. The following were identified as the chief criteria for determining a core Java platform on which to build functionality, with JTS Topology Suite being chosen:

- Common geometry type hierarchy
- Conversion methods
- A source of quality existing spatial algorithms

You were then shown how to download, compile, package, and load the required JTS source code into the Oracle database. Source code-based class development was covered as was the creating and loading via a single JAR file JTS. A PL/SQL package called SPatialEXtension (SPEX) was created that provided SQL access to the Java classes in the jar file. A number of JASPA and JTS based functions were documented with examples of how to use them. Finally, a brief discussion on the native compilation of Java classes within the database was presented pointing you to resources for understanding and ensuring high performance Java based solutions. In short, the use of Java geoprocessing within the Oracle database, regardless of version, is possible because of the availability of mature APIs JTS, and solutions built on the common framework that OGC's geometry type hierarchy provides. It is expected that this chapter will help convince you that Java-based processing is a real option for building complementary functions that work with Oracle Locator or Spatial. While not straightforward, this chapter provides a recipe and an outline of how to build and maintain powerful and useable functionality.

The next chapter will investigate in a more detailed manner, the use of the Oracle MDSYS.ST_GEOMETRY object type and API as a framework for implementing cross-database application development.

11
SQL/MM – A Basis for Cross-platform, Inter-operable, and Reusable SQL

While Oracle's SDO_GEOMETRY type is very widely used, even by the open source community, which promotes the virtues of standards conformance and compliance, many in the geospatial industry still criticize Oracle's SDO_GEOMETRY for its lack of perceived standards compliance. Whether this criticism is based on ignorance or maleficence, SDO_GEOMETRY is standards compliant in its storage, geometry description, and with some functions, but its API is not fully compliant.

However, SDO_GEOMETRY is not the whole story. Oracle also provides an ST_GEOMETRY object type, which is an implementation that is based on the **ISO 13249-3, Information Technology – Database Language Spatial – SQL Multimedia and Applications Packages – Part 3: Spatial standard** (hereafter known as SQL/MM). This chapter aims to show that such criticism of standards compliance of Oracle is limited and ill informed, through an exposure of the benefits of ST_GEOMETRY to practitioners.

ST_GEOMETRY is of special importance in situations, where a business IT environment has a heterogeneous database environment (for example, Oracle, SQL Server, and PostgreSQL). It can be a most useful mechanism for implementing cross-platform spatial data processing and developing highly reusable skills. This latter aspect of skills development is important because re-usability and training, which increases and improves an individual's skill-set, is an important ingredient in staff training and development.

This chapter will present two aspects of how the use of the SQL/MM and **OpenGIS® Implementation Specification for Geographic information - Simple Feature Access - Part 2: SQL option** (hereafter known as OGC SFA 1.1 or 1.2 depending on whether a function is only available in one of the two versions) standards aid cross-platform interoperability:

1. A demonstration of how the spatial data types offered by three databases Oracle, SQL Server 2012 (Express Edition); PostgreSQL 9.x with the PostGIS 2.x extension. PostgreSQL can be used to create constrained geometry storage. This does not repeat the material presented in *Chapter 1, Defining a Data Model for Spatial Data Storage*. Rather it presents the material in light of the OGC SFA 1.x and SQL/MM standards, and the benefits for database interoperability.

2. How the standardized methods common to the three databases can be used to develop SQL statements and stored procedures, which are functions that can be ported to other platforms with relative ease.

> There are other database candidates: MySQL 5.6, Ingres (open source), Informix/DB2 Spatial Extenders, Spatialite etc. But this chapter is not meant to present an exhaustive cross-database comparison; rather it is about the basic principles of interoperability and so is limited to three databases that collectively represent the main issues. Additionally, there are other spatial data type implementations for Postgresql other than PostGIS's. PostGIS is, however, the type that is closest to a native type for Postgresql and is the most widely used of those types. Finally, instead of constantly referring to the PostGIS spatial type within PostgreSQL as PostgreSQL/PostGIS, either PostgreSQL or PostGIS on their own will be used depending on context.

Cross-platform representation

There are two aspects to representing spatial data in a standards compliant, cross-platform manner:

- Geometry type definition (what can be represented and stored)
- Internal geometric description (how each object is physically described)

These will be covered in the following sections:

Outlining the SQL/MM ST_GEOMETRY type

The SQL/MM standard provides access for the database designer and developer to a geometry type hierarchy that can be used for both storage and application development. This is shown by the following diagram.

Oracle's MDSYS.ST_GEOMETRY is a database implementation of this SQL/MM geometry hierarchy, whose actual implementation is depicted graphically as in the following diagram (created using SQL Developer's Data Modeling extension).

> The MDSYS schema prefix is actually not needed for the Oracle SQL/MM types due to the presence of global synonyms and as long as no other object has the same name in the current connected schema.

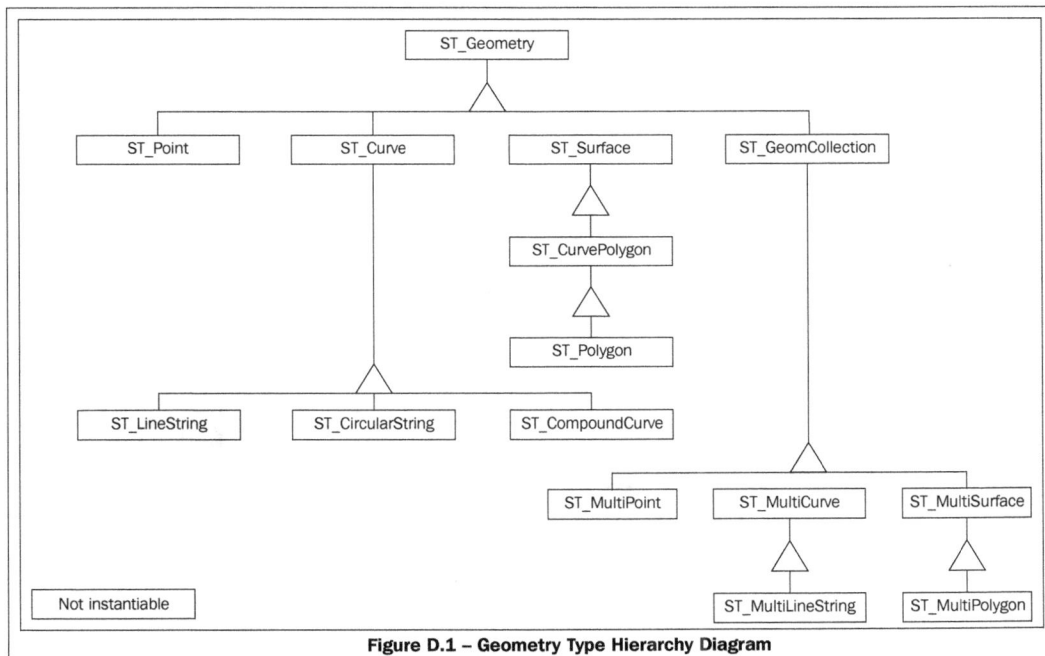

Figure D.1 – Geometry Type Hierarchy Diagram

As can be seen Oracle's ST_GEOMETRY type is multiply inherited in conformance with the standard. ST_GEOMETRY is multiple inherited because it has has many subtypes each of which inherit common properties from it; and each of those may have subtypes, and so on.

Each type and subtype has constructors and a specific set of applicable methods, for example, the ST_Area method only appears in the ST_SURFACE and ST_MULTISURFACE subclasses. Hence, they can only be executed by objects of that type or one of its subclasses, for example, ST_POLYGON. Because ST_LINESTRING is not a subclass of a class containing the ST_Area method, requesting the area of a linestring is invalid, and so cannot be requested (the method does not exist in any of its supertypes). However, ST_LINESTRING's ST_Length method can calculate the length of each linestring object instance as this is a valid method for a linestring.

> In this chapter, object types will be shown in upper case, for example, ST_GEOMETRY; whereas all methods will be shown in mixed case, such as, ST_Area.

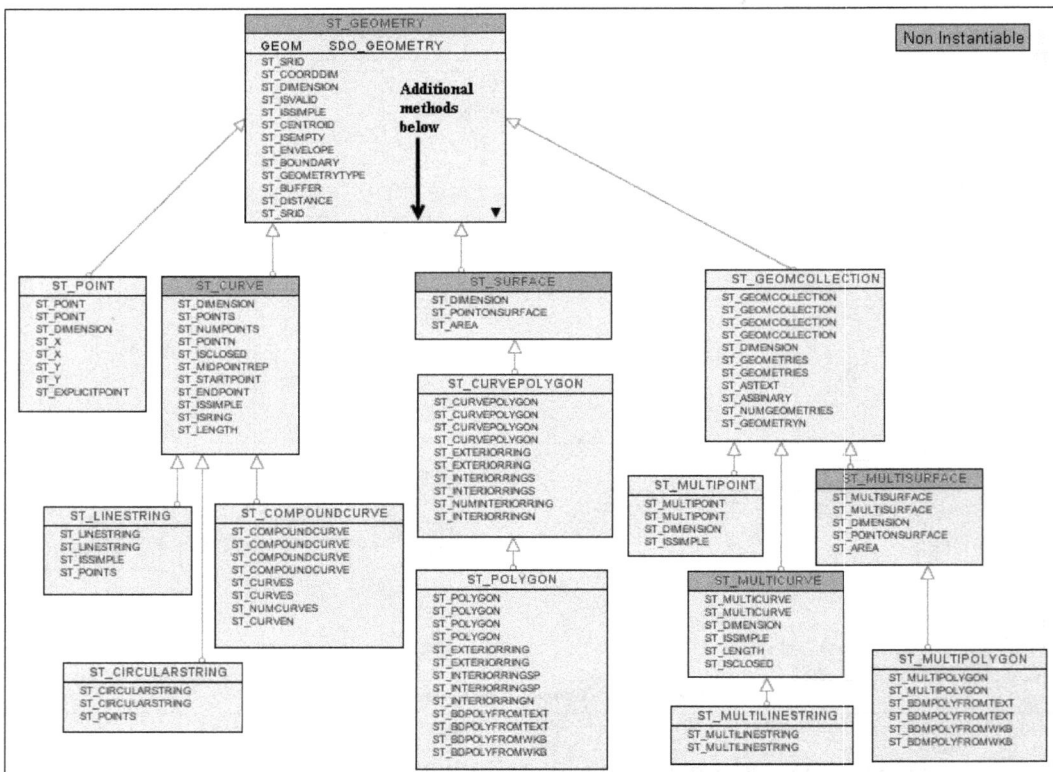

Oracle ISO geometry type hierarchy implementation

The Oracle implementations of ST_GEOMETRY up until 11*g*R2 are 2D only, that is, they do not include methods that relate to Z or M ordinate-based processing. The reason for this is that the Oracle implementations are based on the 2D-only version of the SQL/MM and OGC standards.

For example, `ST_GEOMETRY`'s `ST_AsText` method (implemented as `GET_WKT`) and `ST_POINT`'s `ST_ExplicitPoint` method only support 2D geometries as the following examples show:

```
Select MDSYS.ST_Geometry(
        SDO_GEOMETRY(3001,2872,
                     sdo_point_type(6012578.005,2116495.361,999),
                     null,null)
        ).Get_WKT() as point
  From dual;
```

```
Error report:
SQL Error: ORA-13199: 3D geometries are not supported by geometry WKB/
WKT generation.
```

```
Select p.*
  From Table(book.ST_POINT(
              SDO_GEOMETRY(3001,2872,
                     sdo_point_type(6012578.005,2116495.361,999),
                     null,null)
            ).ST_ExplicitPoint()) p;
```

```
IntValue
------------
 6012578.005
 2116495.361 -- Note that the third Z ordinate is not output.
```

> WKB/WKT stands for Well-Known Binary / Well-Known Text. They are standardized encodings for serializing Geometry objects for import and export.

The `SDO_GEOMETRY`, SQL Server 2012 `GEOMETRY`, and PostgreSQL `GEOMETRY` implementations are in contrast, singly inherited. For Oracle, there is nothing else but `SDO_GEOMETRY`; it has no `SDO_POINT`, or `SDO_CURVE` subtypes. This is the same for SQL Server 2012 and PostgreSQL. Because they are singly inherited all methods are part of the main `GEOMETRY` object and can be executed by any geometry type; thus one can request the area of a linestring. The ramifications of the type of inheritance will be demonstrated and discussed in the section that following internal representation.

> The existence of the GEOMETRY data type for SQL Server 2012 and PostgreSQL (that are provided for geodetic/geographic data storage) are not covered in this Chapter as the issues relating to their GEOMETRY data types also apply to the GEOGRAPHY data type. GEOMETRY and GEOGRAPHY can be thought of as interchangeable.

How are geometry objects internally organized?

The spatial standards not only define a geometry hierarchy and type methods, but also define the way vertices must be organized to correctly define the geometry. Correctness is defined in terms of validity and simplicity. From the SQL/MM standard, the ST_GEOMETRY type defines two methods that implement tests for these characteristics:

- ST_IsValid tests that an ST_GEOMETRY object is well formed. Each ST_GEOMETRY subtype defines the specific conditions that define how well formd it is

- ST_IsSimple tests if an ST_GEOMETRY object value has no anomalous geometric point, such as self-intersection or self-tangency

For example, the specific conditions the SQL/MM standard defines for a ST_CURVEPOLYGON to be well-formed are as follows:

An ST_CURVEPOLYGON value is a planar surface, defined by one exterior boundary and zero or more interior boundaries. Each interior boundary defines a hole in the ST_CURVEPOLYGON value.

ST_CURVEPOLYGON values are topologically closed [start point of each boundary line is equal to its end point]. The boundary of an ST_CURVEPOLYGON consists of an exterior ring and zero or more interior rings. No two rings in the boundary cross. The rings in the boundary of an ST_CURVEPOLYGON value may intersect at a point but only as a tangent. An ST_CURVEPOLYGON shall not have cut lines, spikes or punctures. The interior of every ST_CURVEPOLYGON is a connected point set. The exterior of an ST_CURVEPOLYGON with one or more holes is not connected. Each hole defines a disconnected component of the exterior.

This definition is shared with the OGC SFA 1.2 specification that adds this important element to the definition:

The […] exterior boundary [is a LinearRing defined by three or more vertices organized] in a counter clockwise direction. The interior LinearRings will have the opposite [clockwise] orientation.

Definitions exist that describe valid geometry subtypes. These definitions are implemented by Oracle, SQL Server 2012, and PostgreSQL. Each of these products claims to comply with one or both of the SQL/MM or OGC SFA standards they have implemented. Some standards publish a conformance test suite. Where these are available, vendors may test their implementation against that suite's documented outcomes. Finally, some vendors submit their product to the standards body, and if they pass, they will receive a conformance certificate.

For cross-database development to be possible, any algorithm implementing coordinate processing on one database product, to run correctly on another database, needs those products to at least comply with the relevant standards, even if a conformation certificate does not yet exist. The reader is directed to the documentation of each database product for more detailed discussion and examples of geometry coordinate organization, and to the particular standards bodies to determine levels of conformance.

Storage and function execution using a singly inherited geometry type

We will now look more closely at the use of the three databases' singly inherited geometry types for storage and function execution. Each will be used to store the spatial description of a road in a table called SI_ROAD as follows:

Oracle SDO_GEOMETRY	```Create Table SI_ROAD (
FID   Integer,
STREET_NAME Varchar2(1000),
CLASSCODE   Varchar2(1),
GEOM   SDO_GEOMETRY,
Constraint SI_ROAD_PK Primary Key (FID)
);```
SQL Server	```Create Table SI_ROAD (
FID   Integer,
STREET_NAME Varchar(1000),
CLASSCODE   Varchar(1),
GEOM   GEOMETRY,
Constraint SI_ROAD_PK Primary Key (FID)
);```
PostgreSQL/PostGIS	```Create Table SI_ROAD (
FID   Integer,
STREET_NAME Varchar(1000),
CLASSCODE   Varchar(1),
GEOM   GEOMETRY,
Constraint SI_ROAD_PK Primary Key (FID)
);```

In all three databases the GEOM columns' data types enable generic geometry representation, that is, they can store points, lines, polygons, and so on without restriction. The advantage of this is that, if the business description of a road allowed points, linestrings, and polygons, then this is possible in one column as these `Insert` statements show (WKT constructor is used to maximize cross-database portability):

```
Insert  Into SI_ROAD Values(1,'Main St','C',
SDO_GEOMETRY('LINESTRING(6012759.630 2116512.488, 6012420.596
2116464.910)',2872));
1 rows inserted.
Insert  Into SI_ROAD Values(2,'Main St','E',
SDO_GEOMETRY('POLYGON((6012420.596 2116464.91, 6012458.378
2116435.212, 6012730.642 2116473.282, 6012759.63 2116512.488,
6012720.509 2116541.314, 6012448.247 2116503.244, 6012420.596
2116464.91))',2872));
1 rows inserted.
Insert  Into SI_ROAD Values(3,'Main St','L',
SDO_GEOMETRY('POINT(6012578.005 2116495.361)',2872));
1 rows inserted.
Commit;
Committed.
```

Geometry subtype implementation via constraints

Singly inherited geometry data types are not internally constrained to a particular representation. However, when modeling business objects (for example, a sewer asset), it is normal for that object to be described via a single geometry subtype. For example, a point may be used to represent a sewer manhole, a linestring a sewer main, a polygon, or a sewer catchment area. The only way to enforce this for a single inherited type is by applying to a geometry column, at the model level, a specific database constraint that implements the restriction.

The following table shows constraints named `SI_ROAD_LINESTRING_CK` that restricts the geometry type for the GEOM columns to linestrings only:

SQL Server	`Alter Table SI_ROAD ADD`
	`Constraint SI_ROAD_LINESTRING_CK`
	`Check (GEOM IS NULL`
	`Or (GEOM IS NOT NULL`
	`And GEOM.STGeometrytype() = 'LineString'))`
	`Enable NoCheck;`

| PostgreSQL/<br>PostGIS | ```
Delete From SI_ROAD
   Where ST_Geometrytype(GEOM) = 'ST_LINESTRING';


Alter Table SI_ROAD ADD
  Constraint SI_ROAD_LINESTRING_CK
       CHECK (GEOM IS NULL
           Or (GEOM IS NOT NULL
           And
               ST_Geometrytype(GEOM)='ST_LINESTRING'));
``` |
| --- | --- |

Having added the linestring constraint to the table, executing an insert statement for a point (or polygon) will cause such an insert to fail:

```
Insert  Into SI_ROAD
    Values(4,'Main St','L',
            SDO_GEOMETRY('POINT(6012459.974 2116490.184)',2872));
SQL Error: ORA-02290: check constraint (BOOK.SI_ROAD_LINESTRING_CK)
violated
```

Accessing singly inherited subtype properties

Even having a GEOM column constrained to hold only linestrings does not stop the execution of spatial methods that are inappropriate to that subtype. For example, the ability to inspect the area of geometry is a spatial method that is valid only for the polygon geometry subtypes. Requesting the area of points and linestrings should not be allowed. Yet, in all three databases requesting the area of a linestring is allowed and calculated to be zero (0.0).

| PostgreSQL | ```
Select ST_Area(r.geom)
         as area
   From SI_ROAD r
  Where r.geom
      .ST_Geometrytype()
      = 'ST_LINESTRING';


st_area
-------
0
``` |
| --- | --- |

| Oracle SDO_GEOMETRY | ```
Select Sdo_Geom.
        Sdo_Area(
          r.geom,0.05)
        as area
     From SI_ROAD r
    Where r.geom
         .Get_Gtype() = 2;


AREA
----
0
``` |
|---|---|
| SQL Server 2012 | ```
Select r.geom.STArea()
       as area
    From SI_ROAD r
   Where  r.geom
              .STGeometrytype()
        = 'LineString';



area
----
0
``` |

Examining function or method execution style

The area method execution examples highlight another issue that needs to be addressed when writing cross-database SQL or procedures. That is the way in which methods, for example, area, are executed against geometry objects.

Both PostGIS and Oracle SDO_GEOMETRY implement functionality that is "external" to the geometry type. For example, to return the area of a geometry object in PostGIS one passes a geometry object to the ST_Area function as a parameter. For Oracle an SDO_GEOMETRY object is passed as a parameter to the function SDO_Area, which is a member of the SDO_GEOM package of functions.

The SDO_GEOMETRY object itself, though, also demonstrates the second form of method execution which is where the functions are actual methods of the object via "dot notation". This form is shared with SQL Server 2012 and Oracle ST_GEOMETRY (see later). In the preceding example, SQL Server 2012's ST_Area method is executed by taking a geometry object, and asking it to execute the area method against itself: geom.ST_Area. A method that is common (standardized) between SQL Server 2012 and Oracle SDO_GEOMETRY is ST_IsSimple:

Oracle SDO_GEOMETRY	SQL Server 2012

Select r.geom.ST_IsValid()
    as Valid
From SI_ROAD r
Where r.geom.Get_Gtype()
    = 2;
Valid
----- 	
0
	Select r.geom.STIsValid()
	    as Valid
	From SI_ROAD r
	Where r.geom.STGeometrytype()
	    = 'LineString';
	Valid
	----- 	
	0

As we will see later on, writing SQL or procedures that even use the common methods does require some find/replace processing to rename or convert methods from object "dot notation" to function/parameter. As long as the methods are those common to shared spatial standards, the amount of work is greatly reduced with common processing remaining semantically and functionally correct.

Subtype restriction using PostgreSQL typmod

The preceding discussion on constraints showed how a PostgreSQL geometry column can be constrained to hold only a single geometry type by defining a suitable constraint. However, PostGIS also provides a mechanism for constraining a geometry column to a particular type at the point of its declaration through the use of PostgreSQL's typmod system. This alternate method for constraining a geometry column to hold a linestring can be seen in this alternate method for creating the SI_ROAD table:

```
Create Table SI_ROAD (
    FID          Integer,
    STREET_NAME  Varchar(1000),
    CLASSCODE    Varchar(1),
    GEOM         GEOMETRY(LINESTRING,4326),
    Constraint ROAD_PK Primary Key (FID)
);
```

The basic type GEOMETRY is modified (typmod) by the addition of two parameters:

- The type of geometry (LINESTRING)
- The SRID (4326)

(Note that this is similar to NUMERIC (precision, scale), whose type modifiers are precision and scale.) However, it is important to note that the base type being modified is still GEOMETRY, thus even with a typmod-ed geometry, one can still do this:

```
Select ST_Area(geom) as area
  From SI_ROAD;
-- Results
--
st_area
double precision
----------------
0
```

This shows that typmod is a form of constraint on a geometry type that is still singly inherited. The underlying geometry type is still not multiply-inherited therefore, it cannot constrain method execution to the appropriate subtype.

Storage and function execution using a multiply inherited geometry type

Because PostgreSQL and SQL Server 2012 do not provide multiply inherited data types, discussion in this chapter will be limited to Oracle's ST_GEOMETRY implementation. This implementation resolves two issues described earlier:

* Use of subclasses removes the need for user constraints
* Subclass types enable the execution of allowable methods only

Geometry type restriction using a subtype

Using subclassed geometry types, the SI_ROAD table's geom column can be declared to hold only linestrings as follows:

```
Create Table ST_ROAD (
  FID         Integer,
  STREET_NAME Varchar2(1000),
  CLASSCODE   Varchar2(1),
  GEOM        MDSYS.ST_LINESTRING,
  Constraint ST_ROAD_PK Primary Key (FID)
);
```

No constraint, ROAD_LINESTRING_CK, is required to stop non-linestring geometries from being stored as ST_LINESTRING is a subtype of ST_GEOMETRY and so can only store and process linestrings.

In the following, the first insert of a ST_LINESTRING succeeds, but the insert of an ST_POINT fails. (Because the ST_LINESTRING subtype calls the from_WKT method, which it inherits from its parent ST_GEOMETRY, the returned geometry type is ST_GEOMETRY and not ST_LINESTRING even though its contents the LINESTRING WKT string is of that subtype. The TREAT operator (see *Appendix, Use of TREAT and IS OF Type with ST_GEOMETRY* for a full discussion) has to be used to ensure that the converted geometry is correctly interpreted as a ST_LINESTRING.)

```
Insert  Into ST_ROAD Values(1,'Main St','C',
TREAT (MDSYS.ST_LINESTRING.from_WKT ('LINESTRING(6012759.63041794
2116512.48842026, 6012420.59599103 2116464.90977527)',2872) As MDSYS.
ST_LINESTRING));
1 rows Inserted.
Insert  Into road Values(3,'Main St','L',
TREAT(MDSYS.ST_Geometry.From_WKT('POINT(6012578.005
2116495.361)',2872) As MDSYS.ST_Point))
SQL Error: ORA-00932: inconsistent datatypes: expected MDSYS.ST_
LINESTRING got MDSYS.ST_POINT
commit;
committed.
```

> For some reason, Oracle chose not to implement ST_AsText or ST_GeomFromText as per the standard, instead exposing the SDO_GEOMETRY object methods: GET_WKT and From_WKT. This is a minor problem for cross-database programming perspective, which will be dealt with later in the chapter.

Like the PostGIS typmod, the use of ST_LINESTRING implements a powerful declarative method for describing geometry columns and their contents. This declarative power allows one to succinctly create a ST_ROAD table that has multiple geometry columns with suitable names (not the generic) and data types. In the following **Data Definition Language (DDL)**, we declare that the ST_ROAD table has the following features:

- A label point
- A center-line describing the middle of the road surface
- A polygon describing the easement or casement within which the road pavement and footpath lie

```
Create Table ST_ROAD (
  FID         Integer,
  STREET_NAME Varchar2(1000),
  CLASSCODE   Varchar2(1),
  LABEL       MDSYS.ST_POINT,
```

```
    CENTERLINE  MDSYS.ST_LINESTRING,
    CASEMENT    MDSYS.ST_POLYGON,
    Constraint ST_ROAD_PK Primary Key (FID)
);
```

Into which we can write our geometries:

```
Insert Into ST_ROAD Values(1,'Main St','C',
            TREAT (MDSYS.ST_Point.From_WKT('POINT(
                6012578.005 2116495.361)',
                2872) As MDSYS.ST_Point),
            TREAT(MDSYS.ST_LINESTRING.From_WKT('LINESTRING(
                6012759.63041794 2116512.48842026,
                6012420.59599103 2116464.90977527)',
                2872) As MDSYS.ST_LINESTRING),
            TREAT(MDSYS.ST_Polygon.From_WKT('POLYGON((
                6012420.596 2116464.91, 6012458.378 2116435.212,
                6012730.642 2116473.282, 6012759.63 2116512.488,
                6012720.509 2116541.314, 6012448.247 2116503.244,
                6012420.596 2116464.91))',
                2872) as MDSYS.ST_Polygon));
1 rows inserted.
Commit;
Committed.
```

Unlike PostGIS however, each subtype knows what methods are associated with it as the following demonstrates:

```
Select a.label.ST_Length() as lLen
  From ROAD a;
SQL Error: ORA-00904: "A"."LABEL"."ST_Length": inValid identifier

Select a.centerline.ST_Length() as Clen
  From ROAD a;
    CLEN
----------
104.3505131
Select a.casement.ST_Length() as SLen
  From ROAD a;
SQL Error: ORA-00904: "A"."CASEMENT"."ST_Length": inValid identifier
```

This clearly demonstrates that each subtype knows what methods it inherits and those it exposes.

Subtype inheritance issues

The SQL/MM standard's geometry type hierarchy defines an inheritance that includes some restrictions. For example, while an ST_LINESTRING subtype inherits methods from ST_CURVE, it does not inherit from ST_MULTILINESTRING. Thus one cannot store an ST_LINESTRING in an ST_MULTLINESTRING column, unless it is converted to being an ST_MULTILINESTRING containing one ST_LINESTRING! The following example shows how the hierarchy does not allow a user to execute the ST_GeometryN and ST_NumGeometries methods against a ST_LINESTRING in Oracle, yet SQL Server 2012 and PostgreSQL allow it, because their type system is singly inherited from the base geometry type object:

SQL Server 2012	```Select a.geom.STEquals(
        a.geom.STGeometryN(b.IntValue)) as line
    From (Select geometry::STLineFromText(
                'LINESTRING (0 0, 10 0, 0 10)',0)
            as geom) a
    Cross Apply
    Generate_Series(1,
                a.geom.STNumGeometries(),
            1) b;

line

1
``` |
| PostgreSQL | ```Select ST_Equals(b.geom,b.geomN) as line
    From (Select a.geom,ST_GeometryN(
                    a.geom,
                    Generate_Series(1,
                    ST_NumGeometries(a.geom),
                    1))
                as geomN
        From (Select ST_LineFromText(
                'LINESTRING (0 0, 10 0, 0 10)',
                0) as geom) a
) as b;

line
boolean
----
t
``` |

| Oracle ST | ```
Select a.geom.ST_Equals(
            a.geom.ST_GeometryN(b.IntValue))
        as line
    From (Select book.ST_LINESTRING.ST_LineFromText(
            'LINESTRING (0 0, 10 0, 0 10)',
            0) as geom
        From dual
    ) a,
        Table(Generate_Series(1,
            a.geom.ST_NumGeometries(),1)) b


ORA-00904: "A"."GEOM"."ST_NumGeometries": inValid
identifier
``` |

One also cannot do the following:

```
Select a.geom.ST_Equals(
            a.geom.ST_GeometryN(b.IntValue))
        as line
    From (Select book.ST_MultiLineString(
            'LINESTRING (0 0, 10 0, 0 10)',
            0) as geom
        From dual) a,
        Table(Generate_Series(1,
            a.geom.ST_NumGeometries(),1)) b


ORA-20120: Result is not ST_MultiLineString
ORA-06512: at "BOOK.ST_MULTILINESTRING", line 20
```

However, if the SQL/MM standard type hierarchy had defined ST_LINESTRING as a subtype of ST_MULTILINESTRING as in the following diagram, then such storage would be possible.
```

As it stands, the storage of a `ST_LINESTRING` subtype in a `ST_MULTILINESTRING` column involves additional conversion processing. Although it is unlikely that the relevant standards will be modified, it is possible to define an alternate hierarchy that would implement such a representation as shown in the earlier diagram by *Dr. Knut Stolze*. This geometry type hierarchy could be used for processing geometry data stored using a standards compliant geometry type.

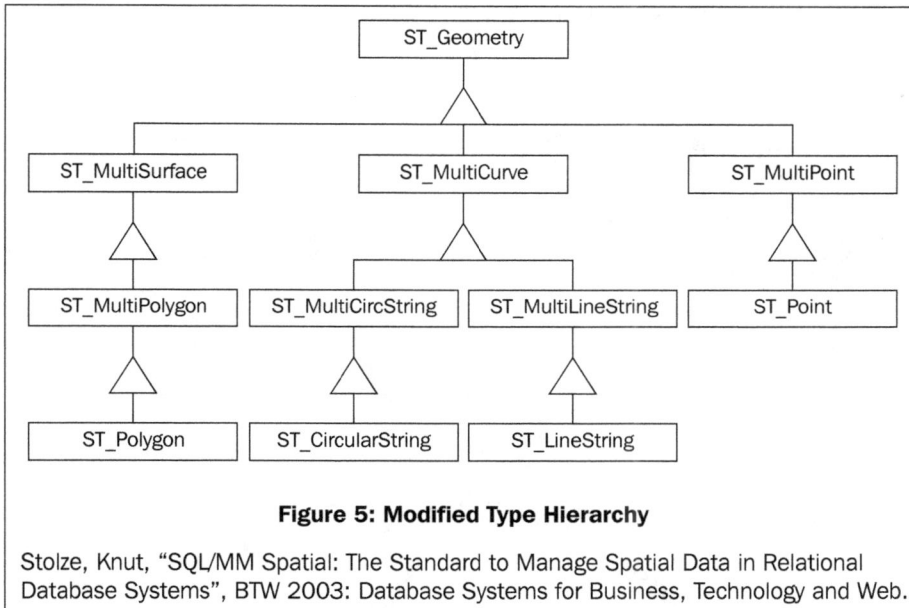

Figure 5: Modified Type Hierarchy

Stolze, Knut, "SQL/MM Spatial: The Standard to Manage Spatial Data in Relational Database Systems", BTW 2003: Database Systems for Business, Technology and Web.

Geometry type implementation matrix

From what we have learned so far, a table/column implementation matrix can be produced that will allow us to implement all possible variations of the SQL/MM Spatial standard. In the following diagram, the matrix is implemented only for the ST_POINT subtype. However, included with this book are a set of SQL files that provide an implementation of the standard's ST_* geometry types. No Primary Key attribute and constraint text is included for brevity's sake in any of the implementations shown in the following diagram:

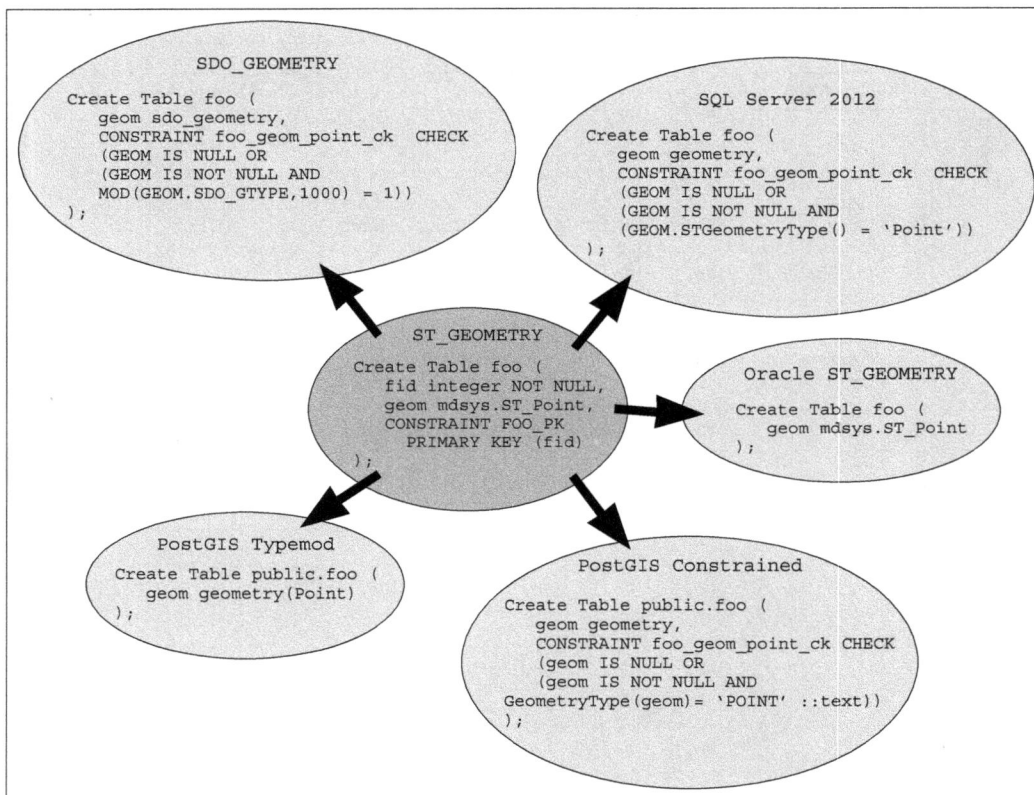

```
SDO_GEOMETRY
Create Table foo (
  geom sdo_geometry,
  CONSTRAINT foo_geom_point_ck  CHECK
  (GEOM IS NULL OR
  (GEOM IS NOT NULL AND
  MOD(GEOM.SDO_GTYPE,1000) = 1))
);
```

```
SQL Server 2012
Create Table foo (
  geom geometry,
  CONSTRAINT foo_geom_point_ck  CHECK
  (GEOM IS NULL OR
  (GEOM IS NOT NULL AND
  (GEOM.STGeometryType() = 'Point'))
);
```

```
ST_GEOMETRY
Create Table foo (
  fid integer NOT NULL,
  geom mdsys.ST_Point,
  CONSTRAINT FOO_PK
  PRIMARY KEY (fid)
);
```

```
Oracle ST_GEOMETRY
Create Table foo (
  geom mdsys.ST_Point
);
```

```
PostGIS Typemod
Create Table public.foo (
  geom geometry(Point)
);
```

```
PostGIS Constrained
Create Table public.foo (
  geom geometry,
  CONSTRAINT foo_geom_point_ck CHECK
  (geom IS NULL OR
  (geom IS NOT NULL AND
  GeometryType(geom)= 'POINT' ::text))
);
```

Because some Oracle geometry subtypes (CircularString) can only be identified by functions that cannot be used by CHECK constraints (such as Get_WKT), a trigger must be constructed to implement the constraint:

```
Create Or Replace Trigger st_road_circularstring_biu
Before Insert or update on ST_ROAD
Referencing old As old new As new
```

```
For Each Row
Begin
   If (:new.GEOM IS NULL) Then
      Return;
   End If;
   If (:new.GEOM IS NOT NULL And
       :new.GEOM.GET_WKT() like 'CIRCULARSTRING%') Then
      Return;
   Else
      raise_application_error(-20000,'GEOM is not a
CIRCULARSTRING',false);
   End If;
End;
```

Alternate geometry type implementation

Oracle's ST_GEOMETRY implementation has a number of problems. Some of these are trivial and relate to simple naming, others relate to missing constructors and methods. The ST_GEOMETRY's Get_WKT method is one example of a misnamed method, which to align with the SQL/MM standard (and SQL Server 2012 and PostgreSQL), should be renamed to ST_GeomFromText or ST_TextToSQL.

All the geometry subtypes are missing WKT/WKB constructors, for example:

```
ST_LINESTRING(AWKT Varchar2, ASRID Integer).
```

In addition, the ST_GEOMETRY subtype constructors do not throw exceptions when an incorrect geometry is provided to a constructor. Thus, it is possible to construct an ST_LINESTRING using an ST_POINT WKT as follows:

```
Select MDSYS.ST_LINESTRING.From_WKT('POINT(0 0)').geom as line
   From DUAL;

LINE
-----------------------------------------------------------------
MDSYS.SDO_GEOMETRY(2001,NULL,MDSYS.SDO_POINT_type(0,0,NULL),NULL,NULL)
```

TREATing the object is still inadequate as no error is thrown:

```
Select TREAT(MDSYS.ST_LINESTRING.From_WKT('POINT(0 0)')
          As MDSYS.ST_LINESTRING).geom as line
   From DUAL;

LINE
---------
(nothing)
```

The subtypes are also missing functions, such as ST_LineFromText (awkt Varchar2, asrid integer) that return an ST_LINESTRING object (similar to constructor). Most of these missing or misnamed functions can be lived with.

However, ST_GEOMCOLLECTION is missing two vital methods: ST_NumGeometries and ST_GeometryN (n integer). As we will see later, these two functions are vital when constructing cross-platform SQL or functions, that extract geometry collection elements for processing.

There are some methods, constructors and functions missing from the Oracle ST_GEOMETRY type (most of these are also not available in the SQL Server and PostgreSQL implementations). In addition, Oracle's ST_GEOMETRY implements an earlier 2D version of the SQL/MM standard, as such the WKB/WKT conversion functions only work for 2D unmeasured data. For the same reason, the ST_M and ST_Z methods in a later version of the standard do not exist.

For the purpose of this book, an enhanced version of the ST_GEOMETRY type has been created that allows for better alignment with SQL Server 2012 and PostgreSQL (or any other geometry type system that includes the common or missing methods), and implements some missing methods and constructors. This type is not for storage, but should be used in company with SDO_GEOMETRY or ST_GEOMETRY storage, when creating cross-database compliant SQL or functionality.

Constructing cross-database SQL

Constructing SQL that can run across multiple database vendor databases is a design goal for a lot of applications in the hope that their development investment can create better return on investment through streamlined porting and reducing maintenance. The goal of standards (whether SQLMM, OGC SFA or SQL92) is to facilitate the achievement of these design goals. The presentation of the SQL/MM geometry type hierarchy and function set in this chapter is to introduce the only foundation that exists to support cross-platform spatial SQL today. But, as we shall see, nothing is perfect.

Is database independence possible?

It is common to meet developers and architects who believe that the only sound approach to using a database is to aim for database independence. That is, that all application database logic must be implemented such that it can run on any database vendor's product.

As Thomas Kyte observed in his book, *Effective Oracle by Design*.

> *Many times, people approach the database as if it were a black box – a commodity as interchangeable as a battery in a radio. According to this approach, you use the database, but you avoid, at all costs, doing anything that would make you database-dependent, as if that were a bad thing. In the name of database independence, you will refuse to use the features of the database, to exploit its capabilities [....] your customers [or employers] have paid for. [Thomas Kyte, Page 11, Chapter 1: The Right Approach to Building Applications, "Effective Oracle By Design", Oracle Press2003].*

Kyte reasonably argues that it is not possible (or defensible) to avoid using database-specific functionality regardless of the vendor. However, that is not to say that there isn't some merit to implementing database independence, where it is reasonable and possible. This is the domain that SQL standards inhabit. Where standards compliance is high across databases, then use of the features of the relevant standard should be encouraged.

Some spatial issues, such as the lack of self-constraining spatial subtypes in a singly inherited geometry type, can be "hidden" from the application or end user though suitable use of the SQL data definition language constraint system.

Oracle's `SDO_GEOMETRY` does not provide function type signatures that on first inspection align with the SQL/MM or OGC SFA standards; yet some are still capable of being used in a manner that supports the other two databases. SQL Server and PostGIS have a much closer alignment, but they introduce other issues relating to their names and how to call them.

Yet when one considers this issue from other than the purely spatial, one notices that all the database vendors provide a greater range of non-spatial functionality that is different from each other, and creates a greater barrier to achieving easy cross-platform development as the following limited set of examples show:

- Oracle's transactional consistency and concurrency controls differ sufficiently with other databases, so that the implementation of exactly the same SQL on one database could return a different result on another.

- Oracle requires use of the DUAL table for non-table based queries, but both SQL Server and PostgreSQL do not. Some like this, some hate it, but it is so minor as to not lose sleep over!

- Oracle and SQL Server do not implement the LIMIT Select operator found in PostgreSQL. SQL Server provides the TOP operator and Oracle provides the ROWNUM pseudo column. All however, support the Row_Number SQL99 analytic.

- Oracle's CONNECT BY SQL extension implements some high performance advantages over those databases that do not have such a construct.

- Some data types don't exist on other databases for historical reasons, or as a result of specific platform issues. Oracle's NUMBER data type, for example, doesn't exist on PostgreSQL; rather one has to use the NUMERIC data type. Similarly, PostgreSQL's BOOLEAN data type is not supported in SQL Server (BIT is commonly used instead) or Oracle. BOOLEAN is not a standardized SQL data type.

- PostgreSQL requires a greater use of casts (for example, 123::int, The rain in...::text) in SQL (using nonstandard naming convention "::") that are not required in Oracle or SQL Server.

- Oracle's partitioning is built very differently to the approach taken by PostgreSQL (table inheritance), neither being based on any common SQL standard.

- The only standardized SQL language which was designed for cross-database use is SQL-92/PSM (ISO/IEC 9075-4:1996). However, it is not supported by SQL Server or Oracle. There is a PSM plugin for PostgreSQL, but it appears to be rarely used. SQL Server's main database development language, T-SQL, is incredibly different from Oracle's PL/SQL or PostgreSQL's pl/pgSQL (which are very similar to each other making cross-database function migration relatively straightforward). But PostgreSQL offers a large range of stored procedure languages, such as PHP which the others do not provide. So, if cross-database stored procedure development is part of your goal for standardization and vendor neutrality, the choice of language becomes a major issue. In this chapter, the more common T-SQL, PL/SQL and pl/pgSQL languages are used.

- • The handling of set returning functions is nonstandard and different in each database. Oracle allows its `SQL Select` statement to "send" attributes from one table in the `from` clause as arguments to a set returning function within its `Table` function, which enables access to the rows returned by that set returning function. SQL Server implements access via a special `CROSS APPLY` construct, while PostgreSQL's approach is to simply select the values directly from the function itself as a "column" that represents not one value but a set of values. The function can itself be used as the source for a single valued parameter of another function. (See later examples and discussion around the use of the `Generate_Series` set returning function.)

Programming for cross-database deployment

The main mechanism for implementing object-relational querying is, of course, SQL. As such it is the principle element through which we should aim to use for maximal cross-database querying. In this section, we will first look at querying the basic geometry types, and then finish by putting all we have learned together to implement something of greater use in day-to-day spatial processing.

Querying ST_POINT geometries

The querying of points is quite simple:

```
-  SQL Server 2012
Select a.geom.STX as x, a.geom.STY as y
   From (Select geometry::STPointFromText('POINT(301301.0
5201201.0)',28355)
                  as geom) a;


-- PostgreSQL
Select ST_X(a.geom) as x, ST_Y(a.geom) as y
   From (Select ST_PointFromText('POINT(301301.0 5201201.0)',28355)
                  as geom) a;


-- Oracle ST
Select a.geom.ST_X() as x, a.geom.ST_Y() as y
   From (Select book.ST_Point('POINT(301301.0 5201201.0)',28355)
                  as geom
         From dual) a;
-- Or
Select a.geom.ST_X() as x, a.geom.ST_Y() as y
   From (Select book.ST_Point.ST_PointFromText(
                  'POINT(301301.0 5201201.0)',28355) as geom
         From dual) a;
```

As you can see all three forms are quite similar. Oracle and SQL Server 2012 are probably closest given the "dot notation" method of access. However, SQL Server 2012 and PostgreSQL both do not require the Oracle proprietary "From DUAL" SQL construct.

Querying ST_LINESTRING geometries

Querying linestring geometries increases the complexity of our querying and introduces the need for additional functions to aid in cross-database portability. The following queries implement the following querying:

- Converts a WKT string to an ST_LINESTRING

- Iterates over the points that define the linestring via use of the SQLMM/ OGC standard functions ST_PointN and ST_NumPoints and the nonstandard Generate_Series function.

- For each point its X and Y ordinates are extracted, and each point is compared to the original geometry's first and last point using the ST_Equals and ST_StartPoint and ST_EndPoint methods.

- SQL Server 2012

```
Select c.geom.STStartPoint().STEquals(c.point) as isStartPt,
       c.geom.STPointN(2).STEquals(c.point)    as ismidPt,
       c.geom.STEndPoint().STEquals(c.point)    as isendPt,
       c.point.STX as x,
       c.point.STY as y
   From (Select a.geom,
                a.geom.STPointN(b.IntValue) as point
           From (Select geometry::STLineFromText(
                        'LINESTRING(0 0,1 1,2 2)',0)
                        as geom) a
                Cross Apply
                Generate_Series(1,a.geom.STNumPoints(),1) b
        ) c;
```

```
-- PostgreSQL
Select ST_Equals(ST_StartPoint(b.geom),b.point)  as isStartPt,
       ST_Equals(   ST_PointN(b.geom,2),b.point) as ismidPt,
       ST_Equals(  ST_EndPoint(b.geom),b.point)   as isendPt,
       ST_X(b.point) as x,
       ST_Y(b.point) as y
   From (Select a.geom,
                ST_PointN(a.geom,
```

```
                    Generate_Series(1,ST_NumPoints(a.geom),
                            1))
                as point
        From (Select ST_LineFromText(
                'LINESTRING(0 0,1 1,2 2)',0) as geom) a
    ) b;


-- Oracle ST
Select c.geom.ST_StartPoint().ST_Equals(c.point) as isStartPt,
       c.geom.ST_PointN(2).ST_Equals(c.point)    as ismidPt,
       c.geom.ST_EndPoint().ST_Equals(c.point)   as isendPt,
       c.point.ST_X() as x,
       c.point.ST_Y() as y
  From (Select a.geom,
            Treat(a.geom.ST_PointN(b.IntValue)
                as book.ST_Point)
            as point
        From (Select book.ST_LINESTRING(
                'LINESTRING(0 0,1 1,2 2)',0) as geom
            From dual) a,
            Table(book.Generate_Series(1,
                        a.geom.ST_NumPoints(),1)) b
    ) c;
```

The basic structure of the queries is essentially the same. The lack of "dot notation" in PostgreSQL is the main difference in the geometry type method invocations. In essentials, the SQLMM/OGC SFA method/function names are clearly translatable from one database to the other.

The `Generate_Series` function (standard with PostgreSQL) is chosen to provide a portable method for generating the required point IDs as the SQL standards do not standardize a method for the generation of series of integers. Some databases provide their own proprietary methods (for example, PostgreSQL's `Generate_ Series`, Oracle's `CONNECT BY LEVEL`, the latter cannot be used in all situations). The need to create such a function is of little concern; implementations for SQL Server 2012 and Oracle are provided with this book.

However, when working with Oracle and SQL Server 2012 the provided `Generate_Series` function for Oracle has been written to use a dedicated type `T_IntValue`, which allows the returned value to be referenced by the SQL Server 2012 name `IntValue` rather than via Oracle's default `Column_Value` name.

```
Create Type T_IntValue  As Object  ( IntValue Integer );
/
show errors
Create Type T_IntValues Is Table Of BOOK.T_IntValue;
/
show errors
Create Or Replace
Function Generate_Series(p_start in pls_integer,
                         p_end   in pls_integer,
                         p_step  in pls_integer default 1)
Return book.T_IntValues Pipelined
As
    v_i PLS_Integer := p_start;
Begin
    While ( v_i <= p_end) Loop
        Pipe Row ( BOOK.T_IntValue(v_i) );
        v_i := v_i + p_step;
    End Loop;
    Return;
End Generate_Series;
/
show errors
-- Testing
--
Select a.IntValue
  From Table(Generate_Series(1,2,1)) a;
-- Result
--
INTVALUE
--------
       1
       2
```

From a cross-database perspective the main hurdle is not spatial, it is the method via which the `Generate_Series` set returning function is called for each geometry being processed. The issue is that for each geometry object, we wish to pass it to a table-valued (or set returning) function that, as the name implies, returns a table of computed objects, in this case points from a single object (think of it as a one to many join where the many is provided by a function that returns a set or array of objects).

In Oracle, the `Table` function implements the required access to the rows of a table-valued function. In SQL Server the access is provided by CROSS APPLY. With PostgreSQL a set returning function can only be used in a `Select` statement's list of "attributes"; the attribute is special in that it returns more than one row, rather than a value of a column in a table being to a function that acts like a table. The first example shows the `Generate_Series` function action on the values of an in-line table (correct), while the second is what Oracle and SQL Server user might expect.

```
Select Generate_Series(1,            Select b.IntValue
                  a.count,1)         From (Select 2 as count) as a
   From (Select 2 as count) as       Cross Apply /*or TABLE()*/
a;                                   Generate_Series(1,
                                                 a.count,1) as b;
```

As one can see, the difference in handling of table-valued functions is a greater obstacle to cross-database spatial than the spatial functionality itself.

Querying ST_CIRCULARSTRING geometries

The creation of queries for the processing of ST_CIRCULARSTRINGS is fairly straight forward even given the lack of support in OGC SFA 1.2 for ST_CIRCULARSTRING and ST_COMPOUNDCURVE (see *Annex B, Comparison of Simple feature access/SQL and SQL/ MM – Spatial, OGC 06-104r3, 2006*. A copy is included as an appendix to this book). SQL Server 2012 implements OGC SFS 1.1. This standard does not have a ST_CircularStringFromText method, and STLineFromText does not recognize ST_CIRCULARSTRINGS WKT values. However, the creation of ST_CIRCULARSTRING is possible via SQL Server 2012's STGeomFromText constructor.

```
-- SQL Server 2012
Select Case When c.geom.STStartPoint().STEquals(c.point)=1
            Then 'StartPoint'
            when c.geom.STEndPoint().STEquals(c.point)=1
            Then 'EndPoint'
            when (c.pid % 2)=0
                And
                c.geom.STPointN(c.pid).STEquals(c.point)=1
            Then 'MidPoint'
            Else 'StartPoint'
        End As PointType,
     c.point.STAsText() as point
  From (Select a.geom, b.IntValue as pid,
           a.geom.STPointN(b.IntValue) as point
        From (Select geometry::STGeomFromText(
```

```
                          'CIRCULARSTRING(0 0,10 0,10 10,5.5 9,3 1)',0)
                                as geom) a
                    Cross Apply
                    Generate_Series(1,a.geom.STNumPoints(),1) b
            ) c;
```

PostgreSQL also does not have a `ST_CIRCULARSTRINGFromText` method, but `ST_GeomFromText` can convert `CIRCULARSTRING` WKT values. In addition, `PostgreSQL` requires an additional in-line `Select Generate_Series` statement in order to expose the point identifier.

```
-- PostgreSQL
Select Case When ST_Equals(ST_StartPoint(c.geom),c.point)
            then 'StartPoint'
            when ST_Equals(ST_EndPoint(c.geom),c.point)
            then 'EndPoint'
            when (c.pid % 2)=0
                And
                ST_Equals(ST_PointN(c.geom,c.pid),c.point)
            then 'MidPoint'
            Else 'StartPoint'
        end as PointType,
        ST_AsText(c.point) as point
    From (Select b.geom, b.pid, ST_PointN(b.geom,b.pid) as point
        From (Select a.geom,
                    Generate_Series(1,ST_NPoints(a.geom),1)
                        as pid
                From (Select ST_GeomFromText(
                    'CIRCULARSTRING(0 0,10 0,10 10,5.5 9,3 1)',0)
                            as geom) as a
            ) as b
        ) c;
```

The Oracle implementation is very similar to the SQL Server 2012 version:

```
-- Oracle ST
Select Case When c.geom.ST_StartPoint().ST_Equals(c.point)=1
            then 'StartPoint'
            when c.geom.ST_EndPoint().ST_Equals(c.point)=1
            then 'EndPoint'
            when mod(c.pid,2)=0
             And c.geom.ST_PointN(c.pid).ST_Equals(c.point)=1
            then 'MidPoint'
            Else 'StartPoint'
        end as PointType,
```

```
              c.point.ST_AsText() as point
     From (Select a.geom, b.IntValue as pid,
                   Treat(a.geom.ST_PointN(b.IntValue)
                         as book.ST_Point)
                      as point
              From (Select book.ST_CIRCULARSTRING(
                         'CIRCULARSTRING(0 0,10 0,10 10,5.5 9,3 1)',0)
                         as geom
                      From dual) a,
                  Table(book.Generate_Series(1,
                            a.geom.ST_NumPoints(),1)) b
          ) c;

POINTType    POINT
----------   ------------------
StartPoint   POINT (0.0 0.0)
MidPoint     POINT (10.0 0.0)
StartPoint   POINT (10.0 10.0)
MidPoint     POINT (5.5 9.0)
EndPoint     POINT (3.0 1.0)
```

The SQL/MM standard has an additional method, ST_MidPointRep, for the ST_CIRCULARSTRING subclass, which returns, for the first circular arc segment of the curve, the starting vertex, a computed mid-vertex exactly half way between the first and third (last) vertex, and the last vertex. For each subsequent segments the method returns the mid end points (as each start vertex is the end vertex of the last circular arc).

```
-- Compute CircularString with computed mid points
Select p.geom as point
  From (Select book.ST_CIRCULARSTRING(
             'CIRCULARSTRING(0 0,10 0,10 10,5.5 9,3 1)',0)
             as geom
         From dual) a,
     Table(a.geom.ST_MidPointRep()) p;
POINT
-----------------------------------------------------------------
SDO_GEOMETRY(2001,NULL,SDO_POINT_Type(0,0,NULL),NULL,NULL)
SDO_GEOMETRY(2001,NULL,SDO_POINT_Type(10,0,NULL),NULL,NULL)
SDO_GEOMETRY(2001,NULL,SDO_POINT_Type(10,10,NULL),NULL,NULL)
SDO_GEOMETRY(2001,NULL,SDO_POINT_Type(3.9,7.53,NULL),NULL,NULL)
SDO_GEOMETRY(2001,NULL,SDO_POINT_Type(3,1,NULL),NULL,NULL)
```

Querying ST_COMPOUNDCURVE geometries

Processing ST_COMPOUNDCURVE geometries requires a few small issues to be resolved. Most result from the lack of support in OGC SFA 1.2 for ST_CIRCULARSTRING and ST_COMPOUNDCURVE geometry types.

SQL Server 2012's STGeomFromText function does not support ST_COMPOUNDCURVES, but its generic and nonstandard Parse function does. Fortunately, SQL Server 2012 provides the STNumCurves and STCurveN methods for extracting the individual parts of ST_COMPOUNDCURVE. Note though that SQL Server 2012 allows these methods to execute against all three linestring types, while the SQL/MM standard limits them to ST_COMPOUNDCURVE geometry subtypes only.

Notice also that the STGeometryType method returns a mixed case string without the "ST_" prefix which Oracle and PostgreSQL include:

```
-- SQL Server 2012
Select Upper(c.line.STGeometryType())    as geometryType,
       c.line.STStartPoint().STAsText() as StartPoint,
       Case When c.line.STGeometryType() = 'CircularString'
            then c.line.STPointN(2).STAsText()
            Else NULL
        end                               as midPoint,
       c.line.STEndPoint().STAsText()    as endPoint
  From (Select a.geom.STCurveN(b.IntValue) as line
          From (Select geometry::Parse(
'COMPOUNDCURVE ((10.0 45.0, 20.0 45.0),
CIRCULARSTRING (20.0 45.0, 23.0 48.0, 20.0 51.0),
(20.0 51.0, 10.0 51.0))') as geom) a
                Cross Apply
                Generate_Series.(1,a.geom.STNumCurves(),1) b
        ) c;
```

```
geometryType   StartPoint    midpoint           endPoint
LINESTRING     POINT (10 45) NULL               POINT (20 45)
CIRCULARSTRING POINT (20 45) POINT (22.4 49.8)  POINT (23 48)
CIRCULARSTRING POINT (23 48) POINT (21.9 45.7)  POINT (20 51)
LINESTRING     POINT (20 51) NULL               POINT (10 51)
```

PostgreSQL's implementation for ST_COMPOUNDCURVEs has missing functionality that limits what can be done with the standard functions. There are no ST_NumCurves and ST_CurveN functions. However, ST_GeomFromText will construct a ST_COMPOUNDCURVE geometry from suitable WKT (there are no ST_CompoundFromTxt/WKB methods).

The following SQL illustrates these issues with the current functionality:

```
-- PostgreSQL
Select Upper(ST_GeometryType(B.line))       as geometryType,
       ST_AsText(ST_StartPoint(b.line))  as StartPoint,
       Case When ST_GeometryType(B.line) = 'ST_CIRCULARSTRING'
            then ST_AsText(ST_PointN(b.line,2))
            Else NULL
        end                               as midPoint,
       ST_AsText(ST_EndPoint(b.line))     as endPoint
  From (Select (ST_Dump(a.geom)).geom as line
          From (Select ST_GeomFromText('COMPOUNDCURVE ((10.0 45.0,
20.0 45.0), CIRCULARSTRING (20.0 45.0, 22.4 49.8, 23.0 48.0, 21.9
45.7, 20.0 51.0), (20.0 51.0, 10.0 51.0))',0) as geom) a
       ) b;
GeometryType       StartPoint    midpoint          endPoint
ST_LINESTRING      POINT(10 45)  NULL              POINT(20 45)
ST_CIRCULARSTRING  POINT(20 45)  POINT(22.4 49.8)  POINT(20 51)
ST_LINESTRING      POINT(20 51)  NULL              POINT(10 51)
```

The `ST_CIRCULARSTRING` and its mid-point returned are only from the first circular arc in the `ST_CIRCULARSTRING`. On investigation, one discovers that the whole (5 vertex) `ST_CIRCULARSTRING` is returned by `ST_Dump` instead of its individual curves.

```
CIRCULARSTRING(20 45,22.4 49.8,23 48,21.9 45.7,20 51)
```

Now, from the point of view of the whole objects within the `ST_COMPOUNDCURVE` (`ST_LINESTRING`, `ST_CIRCULARCURVE` and `ST_LINESTRING`) this is correct. `ST_Dump` deals with whole geometry objects, whereas SQL Server 2012's `STNumCurves` and `STCurveN` extract the individual curves in the `ST_CIRCULARSTRING`.

However, this can be corrected by writing the missing curve functions as follows:

```
Create Or Replace Function ST_NumCurves(p_geom in geometry)
Returns Integer
As $$
Select sum(g.elems)::Integer
  From (Select Case When ST_GeometryType(f.geom) =
                    'ST_CIRCULARSTRING'
                then ((ST_NPoints(f.geom)-1)/2)
                Else 1
           end as elems
    From (Select (ST_Dump($1)).*
          Where ST_GeometryType($1) = 'ST_COMPOUNDCURVE'
          Union All
```

```
            Select ARRAY[1], $1
              Where ST_GeometryType($1) <> 'ST_COMPOUNDCURVE') as f
       ) as g
$$
Language SQL
Volatile
Returns Null On Null Input;


-- Testing we get....
--
Select ST_NumCurves(ST_GeomFromText('COMPOUNDCURVE ((10.0 45.0, 20.0
45.0), CIRCULARSTRING (20.0 45.0, 22.4 49.8, 23.0 48.0, 21.9 45.7,
20.0 51.0), (20.0 51.0, 10.0 51.0))',0)) as curves;


curves
integer
-------
4
```

PostgreSQL does not have any function that allows access to the individual circular curves in an ST_CIRCULARSTRING, therefore, one has to be constructed.

```
Create Or Replace Function ST_curves(p_geom in geometry)
Returns SETOF geometry
As
$BODY$
Declare
   v_numCurves Integer;
   v_wkt       Text := '';
Begin
   If ( ST_GeometryType(p_geom) = 'ST_CIRCULARSTRING' ) Then
       v_numCurves := ((ST_NPoints(p_geom)-1)/2)::Integer;
       For v_i IN 1..ST_NPoints(p_geom) Loop
          v_wkt := v_wkt || ST_X(ST_PointN(p_geom,v_i)) || ' ' ||
                            ST_Y(ST_PointN(p_geom,v_i));
          If ( v_i=3 Or (v_i-3>0 And ((v_i-3) % 2)=0) ) then
             Return next ST_GeomFromText(
                             'CIRCULARSTRING(' || v_wkt || ')',
                             ST_SRID(p_geom));
             v_wkt := ST_X(ST_PointN(p_geom,v_i)) || ' ' ||
                      ST_Y(ST_PointN(p_geom,v_i)) || ',';
          Else
             v_wkt := v_wkt || ',';
          End If;
       End Loop;
```

```
   End If;
   Return;
 End
$BODY$
Language plpgsql Volatile
Cost 100;

-- Testing we get....
--
Select ST_AsText(ST_Curves(ST_GeomFromText('CIRCULARSTRING (20.0 45.0,
   22.4 49.8, 23.0 48.0, 21.9 45.7, 20.0 51.0)',0)));

st_astext
text
---------
CIRCULARSTRING(20 45,22.4 49.8,23 48)
CIRCULARSTRING(23 48,21.9 45.7,20 51)
```

Now that we have these two functions, we can write ST_CurveN, which will allow us to extract a specific ST_CIRCULARSTRING from within an ST_CIRCULARSTRING geometry that is composed of more than one curve.

```
Create Or Replace Function ST_CurveN(p_geom      in geometry,
                                     p_curve_num in Integer)
Returns geometry
As $$
Select h.geom
  From (Select row_Number() over (order by 1) as gid,
               g.geom
          From (Select Case When ST_GeometryType(f.geom) =
                               'ST_CIRCULARSTRING'
                            then ST_CURVEs(f.geom)
                            Else f.geom
                       end as geom
                  From (Select (ST_Dump($1)).*
                          Where ST_GeometryType($1) =
                               'ST_COMPOUNDCURVE'
                        Union All
                        Select ARRAY[1], $1
                          Where ST_GeometryType($1) <>
                               'ST_COMPOUNDCURVE'
                       ) as f
               ) as g
       ) as h
 Where h.gid = $2
```

```
$$
Language SQL
Volatile
Returns Null On Null Input;
```

Putting this all together, we can now construct a version of our original SQL that aligns more closely with the previous SQL Server 2012 version.

```
Select Upper(ST_GeometryType(B.line))      as geometryType,
       ST_AsText(ST_StartPoint(b.line))  as StartPoint,
       Case When ST_GeometryType(B.line) = 'ST_CIRCULARSTRING'
            then ST_AsText(ST_PointN(b.line,2))
            Else NULL
        end                                 as midPoint,
       ST_AsText(ST_EndPoint(b.line))      as endPoint
  From (Select ST_CurveN(a.geom,
                     Generate_Series(1,
                         ST_NumCurves(a.geom),1)) as line
        From (Select ST_GeomFromText('COMPOUNDCURVE ((10.0 45.0,
20.0 45.0), CIRCULARSTRING (20.0 45.0, 22.4 49.8, 23.0 48.0, 21.9
45.7, 20.0 51.0), (20.0 51.0, 10.0 51.0))',0) as geom) a
        ) b;
```

GeometryType	Startpoint	Midpoint	Endpoint
text	text	text	text
ST_LINESTRING	POINT(10 45)	NULL	POINT(20 45)
ST_CIRCULARSTRING	POINT(20 45)	POINT(22.4 49.8)	POINT(23 48)
ST_CIRCULARSTRING	POINT(23 48)	POINT(21.9 45.7)	POINT(20 51)
ST_LINESTRING	POINT(20 51)	NULL	POINT(10 51)

Oracle's MDSYS.ST_GEOMETRY provides ST_CurveN, ST_NumCurves and ST_Curves methods, but these methods do not return the same results as its PostgreSQL and SQL Server 2012 equivalents. As such, a re-coding of these methods was carried out in the enhanced version of the geometry type library, which is supplied with this book. Here are the re-coded methods (the body of ST_Curves is not shown):

```
Create Or Replace
Type BODY ST_COMPOUNDCURVE
As
  [...]
  Member Function ST_Curves
  Return MDSYS.ST_CURVE_ARRAY
  As
    [...]
  Begin
    [...]
```

```
End;

Member Function ST_NumCurves
        Return Integer
As
Begin
  -- sdo_elem_info triplet
  v_offset         pls_integer := 0;
  v_eType          pls_integer := 0;
  v_interpretation pls_integer := 0;
  -- working variables
  v_elements       pls_integer := 0;
  v_sub_elem_count pls_integer := 0;
  v_last_ord       pls_integer := 0;
  v_nCoords        pls_integer := 0;
Begin
  v_elements := ( ( SELF.geom.sdo_elem_info.COUNT / 3 ) - 1 );
  <<element_extraction>>
  For v_i IN 0 .. v_elements Loop
      v_offset         := SELF.geom.sdo_elem_info(v_i * 3 + 1);
      v_eType          := SELF.geom.sdo_elem_info(v_i * 3 + 2);
      v_interpretation := SELF.geom.sdo_elem_info(v_i * 3 + 3);
      If ( v_eType in (4,1005,2005) ) Then
         -- Compound elements with sub-elements follow
         v_sub_elem_count := v_sub_elem_count +
                             v_interpretation;
      ElsIf ( v_interpretation = 2 ) Then
         -- Always count the arcs whether compound or not
         If ( v_i = v_elements /* last triplet */ ) then
           v_last_ord := SELF.geom.sdo_ordinates.count;
         else
           v_last_ord := SELF.geom.sdo_elem_info((v_i+1)*3 + 1);
         end If;
         v_nCoords := ((v_last_ord - v_offset) /
                      SELF.geom.get_dims())+1;
         /* Next coord is part of circular arc */
         v_sub_elem_count :=v_sub_elem_count+((v_nCoords-1)/2)-1;
         /*One circular arc counted in v_interpretation above*/
      End If;
  End Loop element_extraction;
  Return v_sub_elem_count;
End;

Member Function ST_CurveN(aposition Integer)
```

```
            Return BOOK.ST_CURVE
    As
      v_geom mdsys.SDO_GEOMETRY;
    Begin
      Select v.geom
        Into v_geom
        From Table(SELF.ST_CURVEs()) v
       Where v.gid = aposition;
      Return BOOK.ST_CURVE(v_geom);
      Exception
        When NO_DATA_FOUND Then
            Return NULL;
    [...]
End;


-- Oracle ST_Geometry
Select c.line.ST_GeometryType() as geometryType,
       c.line.ST_StartPoint().ST_AsText() as StartPoint,
       Case When c.line.ST_GeometryType() = 'ST_CIRCULARSTRING'
            then c.line.ST_PointN(2).ST_AsText()
            Else NULL
        end                           as midPoint,
       c.line.ST_EndPoint().ST_AsText()    as endPoint
  From (Select a.geom.ST_CURVEN(b.IntValue) as line
          From (Select BOOK.ST_COMPOUNDCURVE('COMPOUNDCURVE ((10.0
45.0, 20.0 45.0), CIRCULARSTRING (20.0 45.0, 22.4 49.8, 23.0 48.0,
21.9 45.7, 20.0 51.0), (20.0 51.0, 10.0 51.0))') as geom
                From dual) a,
              Table(book.Generate_Series(1,
                         a.geom.ST_NumCurves(),1)) b
        ) c;


GEOMETRYType       STARTPOINT    MIDPOINT           ENDPOINT
----------------   ------------- ----------------   -------------
ST_LINESTRING      POINT (10 45) NULL               POINT (20 45)
ST_CIRCULARSTRING  POINT (20 45) POINT (22.4 49.8)  POINT (23 48)
ST_CIRCULARSTRING  POINT (23 48) POINT (21.9 45.7)  POINT (20 51)
ST_LINESTRING      POINT (20 51) NULL               POINT (10 51)
```

In summary, even though functions were missing or incorrectly coded in PostgreSQL and Oracle, a bit of "can do" attitude, coupled with the coding of some very simple functions, ensures that cross-database support for ST_COMPOUNDCURVE access is possible.

Querying ST_CURVEPOLYGON geometries

The ST_CURVEPOLYGON object that will be processed is as follows. This object will be "exploded" into its individual ST_CURVE elements in a manner that allows for cross-database processing.

```
-- SQL Server 2102
With poly As (
Select geometry::STGeomFromText(
'CURVEPOLYGON (
  COMPOUNDCURVE (
    CIRCULARSTRING(1.0 -0.7,1.95 -0.1,2 1,2 3,4 3),
    (4 3, 4 5, 1 4, 0 0),
    CIRCULARSTRING(0 0,0.4 -0.5,1.0 -0.7)
  ),
  (0.4 0.1,0.9 0.9,1.5 0.2,0.4 0.1),
  CIRCULARSTRING(1.5 0.4,1.2 0.8,1.7 1,1.6 0.5,
                1.6 0.4,1.6 0.4,1.5 0.4)
 )',0) as geom
)
Select a.ringN, a.lineN, a.line.STAsText() as line
  From ( Select g.ringN, p.IntValue as lineN,
                g.geom.STCurveN(p.IntValue) as line
          From (Select 1 as ringN,
                       p.geom.STExteriorRing() as geom
                  From poly p
                Union All
                Select r.[IntValue] + 1 as ringN,
                       p.geom.STInteriorRingN(r.[IntValue])
                          as geom
                  From poly p
                Cross Apply
                Generate_Series(1,
                        p.geom.STNumInteriorRing(),1) r
                Where r.[IntValue] <= p.geom.STNumInteriorRing()
               ) g
                Cross Apply
                Generate_Series(1,g.geom.STNumCurves(),1) p
        ) a
  Order By a.ringN, a.lineN;

ringN lineN line
1     1     CIRCULARSTRING (1 -0.7, 1.95 -0.1, 2 1)
1     2     CIRCULARSTRING (2 1, 2 3, 4 3)
```

```
1       3       LINESTRING (4 3, 4 5)
1       4       LINESTRING (4 5, 1 4)
1       5       LINESTRING (1 4, 0 0)
1       6       CIRCULARSTRING (0 0, 0.4 -0.5, 1 -0.7)
2       1       LINESTRING (0.4 0.1, 0.9 0.9)
2       2       LINESTRING (0.9 0.9, 1.5 0.2)
2       3       LINESTRING (1.5 0.2, 0.4 0.1)
3       1       CIRCULARSTRING (1.5 0.4, 1.2 0.8, 1.7 1)
3       2       CIRCULARSTRING (1.7 1, 1.6 0.5, 1.6 0.4)
3       3       CIRCULARSTRING (1.6 0.4, 1.6 0.4, 1.5 0.4)
```

```sql
-- PostgreSQL
WITH poly As (
Select ST_GeomFromText(
'CURVEPOLYGON (
  COMPOUNDCURVE (
    CIRCULARSTRING (1.0 -0.7, 1.95 -0.1, 2 1, 2 3, 4 3),
    (4 3, 4 5, 1 4, 0 0),
    CIRCULARSTRING (0 0, 0.4 -0.5, 1.0 -0.7)
  ),
  (0.4 0.1, 0.9 0.9, 1.5 0.2, 0.4 0.1),
  CIRCULARSTRING(1.5 0.4,1.2 0.8,1.7 1,1.6 0.5,
                 1.6 0.4,1.6 0.4,1.5 0.4)
 )',0)  As GEOM
)
Select a.ringN, a.lineN,
       ST_AsText(ST_CurveN(a.geom,a.lineN)) as line
  From (Select g.ringN,
               Generate_Series(1,ST_NumCurves(g.geom),1) as lineN,
               g.geom
          From (Select 1 as ringN,
                       ST_ExteriorRing(p.geom) as geom
                  From poly p
                Union All
                Select r.ringN + 1 as ringN,
                       ST_InteriorRingN(p.geom,r.ringN) as geom
                  From poly p,
                      (Select Generate_Series(1,
                              ST_NumInteriorRing(p.geom),1)
                                  as ringN
                         From poly p) as r
               ) g
        ) a
  Order By a.ringN, a.lineN;
```

The resulting table is the same as SQL Server 2012 so is not repeated for brevity's sake.

With Oracle we have a lot more work to do, because at the time of writing this chapter, the test WKT will not convert correctly on 11gR2. As such, we have to generate it from its components as shown in the following code:

```
set serveroutput on size unlimited
With poly As (
Select book.ST_CURVEPOLYGON(
        SDO_GEOM.Sdo_Difference(
          SDO_GEOM.Sdo_Difference(
            book.ST_CURVEPOLYGON('CURVEPOLYGON (COMPOUNDCURVE
(CIRCULARSTRING (1.0 -0.7, 1.95 -0.1, 2 1, 2 3, 4 3), (4 3, 4 5, 1 4,
0 0), CIRCULARSTRING (0 0, 0.4 -0.5, 1.0 -0.7)) )',null).geom,
            book.ST_POLYGON('POLYGON((0.4 0.1, 0.9 0.9, 1.5 0.2, 0.4
0.1))',NULL).geom,
            0.005),
          book.ST_CURVEPOLYGON('CURVEPOLYGON(CIRCULARSTRING (1.5 0.4,
1.2 0.8, 1.7 1, 1.5 0.7, 1.5 0.4))',null).geom,
            0.005)) as geom
  From dual
)
Select a.ringN,a.lineN, a.line.ST_AsText() as line
  From (Select g.ringN, p.IntValue as lineN,
        case when g.geom.ST_GeometryType() = 'ST_COMPOUNDCURVE'
          then book.ST_COMPOUNDCURVE(g.geom.geom)
                .ST_CurveN(p.IntValue)
          Else g.geom
        end as line
        From (Select b.ringN, b.geom
              From (Select 1 as ringN,
                           p.geom.ST_ExteriorRing() as geom
                    From poly p
                    Union All
                    Select r.IntValue + 1 as ringN,
                           p.geom.ST_InteriorRingN(r.IntValue)
                             as geom
                    From poly p,
                         Table(book.Generate_Series(1,
                               p.geom.ST_NumInteriorRing(),
                               1)) r
                    ) b
              ) g,
              Table(book.Generate_Series(1,
```

[497]Chapter 11

```
              case when g.geom.ST_GeometryType() =
                             'ST_COMPOUNDCURVE'
                  then book.ST_COMPOUNDCURVE(g.geom.geom)
                               .ST_NumCurves()
                  Else 1
              end,1)) p
         ) a
  Order By a.ringN, a.lineN;


RINGN LINEN LINE
----- ----- ---------------------------------------------------
  1     1   CIRCULARSTRING (0.0 0.0,0.4 -0.5,1.0 -0.7)
  1     2   CIRCULARSTRING (1.0 -0.7,1.94 -0.11,2.0 1.0)
  1     3   CIRCULARSTRING (2.0 1.0,2.0 3.0,4.0 3.0)
  1     4   LINESTRING (4.0 3.0,4.0 5.0,1.0 4.0,0.0 0.0,0.4 0.1)
  1     5   LINESTRING (0.4 0.1,0.9 0.9,1.5 0.2,0.4 0.1,1.5 0.4)
  1     6   CIRCULARSTRING (1.5 0.4,1.21 0.83,1.7 1.0)
  2     1   LINESTRING (0.4 0.1,1.5 0.2,0.9 0.9,0.4 0.1)
  3     1   CIRCULARSTRING (1.5 0.4,1.51 0.73,1.7 1.0,
                            1.21 0.83,1.5 0.4)
```

Other than the WKT import problem, additional difficulties were encountered.

In Postgesql/SQL Server 2012's case, their ST_NumCurves/STNumCurves methods handle all three types of ST_CURVE (ST_CIRCULARCURVE, ST_LINESTRING and ST_COMPOUNDCURVE) even though only curve and linestring are a part of the OGC SFA 1.x standards. Oracle does not support all three because in the ISO/IEC 13249:2003 SQL/MM standard, the methods are only supported on the ST_COMPOUNDCURVE geometry subtype as can be seen in the following diagram:

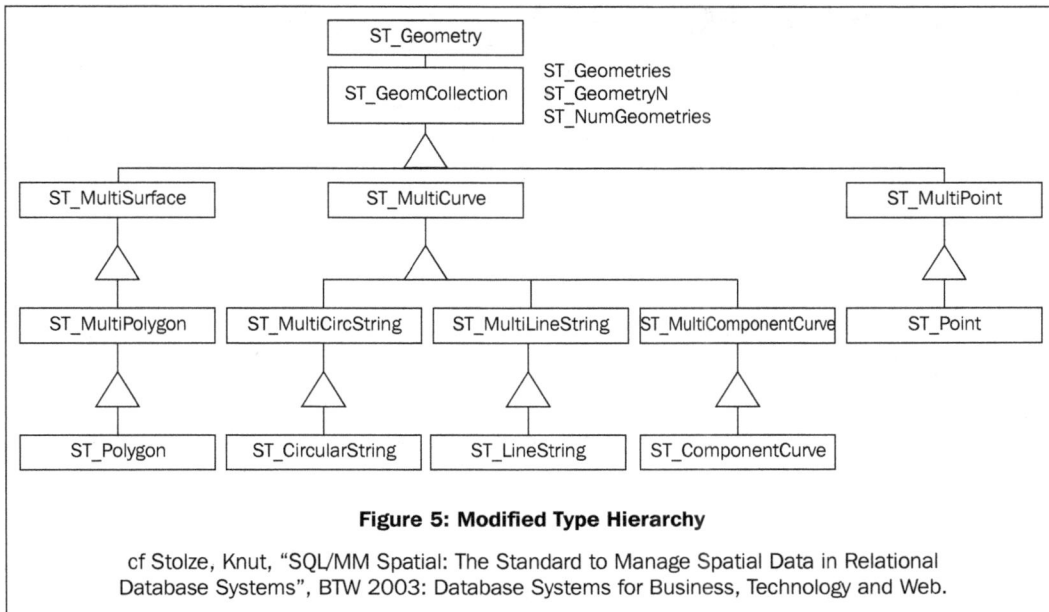

Figure 5: Modified Type Hierarchy

cf Stolze, Knut, "SQL/MM Spatial: The Standard to Manage Spatial Data in Relational Database Systems", BTW 2003: Database Systems for Business, Technology and Web.

In Oracle's case, examination of the result of the ring extraction is needed, so that ST_NumCurves is only applied to ST_COMPOUNDCURVE, for example:

```
Case When g.geom.ST_GeometryType() = 'ST_COMPOUNDCURVE'
     then book.ST_COMPOUNDCURVE(g.geom.geom).ST_NumCurves()
     Else 1
end
```

Querying ST_POLYGON geometries

The querying of ST_POLYGON objects is simpler than ST_CURVEPOLYGON in that we do not need to process the individual ST_LINESTRING and ST_CIRCULARSTRING objects. The SQL Server 2012 query is neat and straightforward though it requires the use of the CROSS APPLY join to drive the Generate_Series set returning function.

```
-- SQL Server 2102
use [GISDB]
go
With poly As  (
  Select geometry::STPolyFromText(
              'POLYGON((0 0,100 0,100 100,0 100,0 0),
                  (10 10,10 20,20 20,20 10,10 10),
                  (12 73, 12 86, 27 86, 27 73, 12 73))',0)
          as geom
)
Select a.ringN,
       a.vectorN,
       a.sp.STX as sx, a.sp.STY as sy,
       a.ep.STX as ex, a.ep.STY as ey
   From (Select g.ringN,
              p.IntValue as vectorN,
              g.geom.STPointN(p.[IntValue]) as sp,
              g.geom.STPointN(p.[IntValue]+1) as ep
         From (Select 1 as ringN, p.geom.STExteriorRing() as geom
               From poly p
               Union All
               Select r.[IntValue] + 1 as ringN,
                      p.geom.STInteriorRingN(r.[IntValue])
                        as geom
                From poly p
                Cross Apply
                Generate_Series(1,
                          p.geom.STNumInteriorRing(),1) r
              Where r.[IntValue] <= p.geom.STNumInteriorRing()
              ) g
              Cross Apply
              Generate_Series(1,g.geom.STNumPoints(),1) p
        Where p.IntValue < g.geom.STNumPoints()
       ) a
 Order By a.ringN, a.vectorN;
-- Results
--
ringN vectorN sx  sy  ex  ey
```

```
1     1          0    0 100    0
1     2        100    0 100  100
1     3        100  100    0  100
1     4          0  100    0    0
2     1         10   10   20   10
2     2         10   20   20   20
2     3         20   20   20   10
2     4         20   10   10   10
3     1         12   73   12   86
3     2         12   86   27   86
3     3         27   86   27   73
3     4         27   73   12   73
```

The PostgreSQL query is straightforward given the usual limits of its function call notation and the different methods for executing the `Generate_Series` set returning function:

```
-- PostgreSQL
With poly As (
  Select ST_PolyFromText('POLYGON((0 0,100 0,100 100,0 100,0 0),(10
10,10 20,20 20,20 10,10 10),(12 73, 12 86, 27 86, 27 73, 12 73))',0)
as geom
)
Select a.ringN,
       a.vectorN,
       ST_X(a.sp) as sx,ST_Y(a.sp) as sy,
       ST_X(a.ep) as ex,ST_Y(a.ep) as ey
  From (Select g.ringN,
               Generate_Series(1,ST_NumPoints(g.geom)-1,1)
                 as vectorN,
               ST_PointN(g.geom,
                       Generate_Series(1,
                           ST_NumPoints(g.geom)-1,1)) as sp,
               ST_PointN(g.geom,
                       Generate_Series(2,
                           ST_NumPoints(g.geom)  ,1)) as ep
     From (Select 1 as ringN, ST_ExteriorRing(p.geom) as geom
           From poly p
          Union All
           Select r.ringN+1 as ringN,
                  ST_InteriorRingN(p.geom,r.ringN) as geom
           From poly p,
              ( Select Generate_Series(1,
                          ST_NumInteriorRing(p.geom),1)
                        as ringN
```

```
                          From poly p) as r
            ) g
         ) a
  Order By a.ringN, a.vectorN;
```

The resulting table is exactly the same as the preceding SQL Server 2012.

For Oracle, the existing MDSYS.ST_CURVEPOLYGON's ST_NumInteriorRings method returns the wrong answer as shown:

```
Select mdsys.ST_Polygon(SDO_GEOMETRY(
               'POLYGON((0 0,100 0,100 100,0 100,0 0),
                        (10 10,10 20,20 20,20 10,10 10),
                        (12 73, 12 86, 27 86, 27 73, 12 73))'
               )).ST_NumInteriorRing() As iRingCount
   From dual;

-- Results
--
IRINGCOUNT
----------
         3
```

This is easy to correct and is available in the parallel ST_GEOMETRY type implementation shipped with this book. The solution is to replace the faulty functions in the ST_CURVEPOLYGON geometry subtype.

```
Create Or Replace Type Body ST_CURVEPOLYGON
As
  [...]
  Member Function ST_NumInteriorRing
  Return Integer
  As
    v_elements   pls_integer := 0;
    v_ring_count pls_integer := 0;
  Begin
     If (SELF.GEOM IS NULL) Then
        Return NULL;
     End If;
     v_elements := ( ( SELF.geom.sdo_elem_info.COUNT / 3 ) - 1 );
     <<element_extraction>>
     For v_i IN 0 .. v_elements Loop
        If ( SELF.geom.sdo_elem_info(v_i * 3 + 2)
             in (2003,2005) ) Then
           v_ring_count := v_ring_count + 1;
        End If;
```

```
          End Loop element_extraction;
          Return v_ring_count;
      End;
      [...]
End;

-- Testing we get
--
Select book.ST_Polygon(SDO_GEOMETRY(
                'POLYGON((0 0,100 0,100 100,0 100,0 0),
                        (10 10,10 20,20 20,20 10,10 10),
                        (12 73, 12 86, 27 86, 27 73, 12 73))'
                )).ST_NumInteriorRing() As iRingCount
      From dual;

-- Results
--
IRINGCOUNT
----------
          2
```

Now that we have fixed the wrong result we can execute our query as follows:

```
Select a.ringN,
       a.vectorN,
       a.sp.ST_X() as sx, a.sp.ST_Y() as sy,
       a.ep.ST_X() as ex, a.ep.ST_Y() as ey
  From (Select g.ringN,
               p.IntValue as vectorN,
               g.geom.ST_PointN(p.IntValue  ) as sp,
               g.geom.ST_PointN(p.IntValue+1) as ep
         From (Select 1 as ringN,
                      p.geom.ST_ExteriorRing() as geom
                 From poly p
               Union All
               Select p.IntValue+1 as ringN,
                      p.geom.ST_InteriorRingN(p.IntValue) as geom
                 From poly p,
                      Table(book.Generate_Series(1,
                              p.geom.ST_NumInteriorRing(),1)) p
                Where p.IntValue <= p.geom.ST_NumInteriorRing()
               ) g,
               Table(booK.Generate_Series(1,
                              g.geom.ST_NumPoints(),1)) p
         Where p.IntValue < g.geom.ST_NumPoints()
       ) a
Order By a.ringN, a.vectorN;
```

The resulting table is exactly the same as the preceding SQL Server 2012.

Querying ST_MULTIPOINT geometries

ST_MULTIPOINT geometry access is quite simple, being effectively repeated access to individual ST_POINT objects. All the necessary OGC/SQL/MM functions are available across all three databases.

```
-- SQL Server 2012
USE [GISDB]
GO
Select c.point.STX as x, c.point.STY as y
  From (Select a.geom.STGeometryN(b.IntValue) as point
        From (Select geometry::STMPointFromText(
                        'MULTIPOINT((0 0),(1 1),(2 2))',0) as geom) a
            Cross Apply
            Generate_Series(1,a.geom.STNumGeometries(),1) b
       ) c;

x y
0 0
1 1
2 2


-- PostgreSQL
Select ST_X(b.geom) as x, ST_Y(b.geom) as y
  From (Select ST_GeometryN(a.geom,
                            Generate_Series(1,
                                ST_NumGeometries(a.geom),1))
                as geom
        From (Select ST_GeomFromText(
                        'MULTIPOINT((0 0),(1 1),(2 2))',0) as geom
             ) a
       ) b;
```

The result of this query is identical to the earlier SQL Server 2012 table.

```
-- Oracle ST
Select c.point.ST_X() as x, c.point.ST_Y() as y
  From (Select Treat(a.geom.ST_GeometryN(b.IntValue)
                    as book.ST_Point)
               as point
        From (Select book.ST_MULTIPOINT(
                        'MULTIPOINT((0 0),(1 1),(2 2))',0)
```

```
                as geom
          From dual) a,
        Table(book.Generate_Series(1,
                        a.geom.ST_NumGeometries(),1)) b
    ) c;
```

The result of this query is identical to the earlier SQL Server 2012 table.

Querying ST_MULTILINESTRING geometries

The example chosen to demonstrate ST_MULTILINESTRING access is to firstly extract all disjointed linestrings, then for each linestring extract and display its point in the WKT format.

```
-- SQL Server 2012
Select c.lineN, d.IntValue as pointN,
       c.line.STPointN(d.IntValue).STAsText() as point
  From (Select b.IntValue as lineN,
               a.geom.STGeometryN(b.IntValue) as line
          From (Select geometry::STMLineFromText(
                      'MULTILINESTRING((0 0,1 1),(2 1,3 1,3 0))',0)
                     as geom
                ) a
          Cross Apply
          Generate_Series(1,a.geom.STNumGeometries(),1) b
       ) c
       Cross Apply
       Generate_Series(1,c.line.STNumPoints(),1) d;
lineN pointN point
1     1      POINT (0 0)
1     2      POINT (1 1)
2     1      POINT (2 1)
2     2      POINT (3 1)
2     3      POINT (3 0)
```

The PostgreSQL equivalent requires additional in-line SQL to generate the line and point numbers from the Generate_Series set returning function:

```
-- PostgreSQL
Select d.lineN, d.pointN,
       ST_AsText(ST_PointN(d.line,d.pointn)) as point
  From (Select c.linen,
               Generate_Series(1,ST_NumPoints(c.line),1) as pointN,
               c.line
```

```
From (Select b.lineN,
                ST_GeometryN(b.geom,b.lineN) as line
        From (Select a.geom,
                    Generate_Series(1,
                        ST_NumGeometries(a.geom),1)
                    as linen
                From (Select ST_MLineFromText(
                'MULTILINESTRING((0 0,1 1),(2 1,3 1,3 0))',
                0) as geom) as a
                ) as b
        ) as c
) as d;
```

The result of this query is identical to the SQL Server 2012, so the resulting
table is not shown. Oracle's implementation is very similar to SQL Server 2012's
implementation. Tasks required are the addition of the From DUAL, changing CROSS
APPLY to its Table equivalent, adding an underscore after the ST function prefix and
adding in the appropriate constructors and TREATs.

```
-- Oracle ST
Select c.lineN, d.IntValue as pointN,
        c.line.ST_PointN(d.IntValue).ST_AsText() as point
    From (Select b.IntValue as linen,
                book.ST_LINESTRING(a.geom.
                    ST_GeometryN(b.IntValue).geom)
                as line
            From (Select book.ST_MULTILINESTRING
                    .ST_MLineFromText(
                    'MULTILINESTRING((0 0,1 1),(2 1,3 1,3 0))',
                    0) as geom
                From dual
                ) a,
                Table(book.Generate_Series(1,
                        a.geom.ST_NumGeometries(),1)) b
            ) c,
        Table(book.Generate_Series(1,
                Treat(c.line as book.st_linestring)
                .ST_NumPoints(),1)) d;
```

The result of this query is identical to the SQL Server 2012, so the resulting table is
not shown.

Querying ST_MULTIPOLYGON geometries

The example chosen to demonstrate ST_MULTIPOLYGON access is to firstly extract all disjointed polygons, then for each polygon, extract its exterior and interior rings. Each ring's points are extracted, and an ST_LINESTRING formed from each pair (a two point segment or vector); finally, each ST_LINESTRING has its length measured via the ST_Length method:

```
--SQL Server
USE [gisdb]
GO
With poly As (
  Select a.polyN, a.geom
    From (Select t.IntValue as polyN,
             o.geom.STGeometryN(t.IntValue) as geom
          From (Select geometry::STMPolyFromText(
          'MULTIPOLYGON(((10 10, 15 10, 15 15, 10 15, 10 10),
                    (12 12, 12 14, 14 14, 14 12, 12 12)),
              ((100 100, 150 100, 150 150, 100 150, 100 100)))',0)
                   as geom
               ) o
               Cross Apply
               Generate_Series(1,o.geom.STNumGeometries(),1) t
       ) a
)
Select a.polyN, a.ringN, a.vectorN,
       a.sp.STX as sx, a.sp.STY as sy,
       a.ep.STX as ex, a.ep.STY as ey,
       geometry::STLineFromText('LINESTRING(' +
               str(a.sp.STX,5,3)+ ' ' + str(a.sp.STY,5,3) + ',' +
               str(a.ep.STX,5,3)+ ' ' + str(a.ep.STY,5,3) +
               ')',0).STLength() as vLen
   From (Select g.polyN, g.ringN, p.IntValue as vectorN,
               g.geom.STPointN(p.IntValue  ) as sp,
               g.geom.STPointN(p.IntValue+1) as ep
         From (Select polyN, 1 as ringN,
                    p.geom.STExteriorRing() as geom
               From poly p
               Union All
               Select polyN, t.IntValue+1 as ringN,
                    p.geom.STInteriorRingN(t.IntValue) as geom
                 From poly p
```

```
                              Cross Apply
                              Generate_Series(1,
                                      p.geom.STNumInteriorRing(),1) t
                          ) g
                          Cross Apply
                          Generate_Series(1,g.geom.STNumPoints(),1) p
                  Where p.IntValue < g.geom.STNumPoints()
              ) a
      Order By a.polyN, a.ringN, a.vectorN;
```

```
polyN ringN vectorN sx   sy   ex   ey   vLen
1     1     1       10   10   15   10   5
1     1     2       15   10   15   15   5
1     1     3       15   15   10   15   5
1     1     4       10   15   10   10   5
1     2     1       12   12   12   14   2
1     2     2       12   14   14   14   2
1     2     3       14   14   14   12   2
1     2     4       14   12   12   12   2
2     1     1       100  100  150  100  50
2     1     2       150  100  150  150  50
2     1     3       150  150  100  150  50
2     1     4       100  150  100  100  50
```

The result is correct; the SQL is well-formed and readable. From this, a PostgreSQL version can be created, but it requires considerable reformatting from the SQL Server 2012 "dot notation" approach, and to accommodate PostgreSQL's handling of the `Generate_Series` set returning function.

```
-- PostgreSQL
With poly As (
  Select a.polyN, a.geom
    From (Select t.polyN, ST_GeometryN(t.geom,t.polyN) as geom
        From (Select Generate_Series(1,
                          ST_NumGeometries(o.geom),1)
                      as polyN,
                    o.geom
              From (Select ST_MultiPolygonFromText(
          'MULTIPOLYGON(((10 10, 15 10, 15 15, 10 15, 10 10),
                      (12 12, 12 14, 14 14, 14 12, 12 12)),
            ((100 100, 150 100, 150 150, 100 150, 100 100)))',0)
                          as geom
                    ) o
              ) t
        ) a
```

```
)
Select i.polyN, i.ringN, i.vectorN,
       ST_X(i.sp) as sx,ST_Y(i.sp) as sy,
       ST_X(i.ep) as ex,ST_Y(i.ep) as ey,
       ST_Length(ST_MakeLine(ARRAY[i.sp,i.ep])) as vLen
   From (Select h.polyN, h.ringN, h.IntValue as vectorN,
               ST_PointN(h.geom,h.IntValue) as sp,
               ST_PointN(h.geom,h.IntValue+1) as ep
         From (Select Generate_Series(1,
                         ST_NumPoints(g.geom),1) as IntValue,
                     g.polyN, g.ringN, g.geom
               From (Select p.polyN, 1 as ringN,
                           ST_ExteriorRing(p.geom) as geom
                     From poly p
                     Union All
                     Select p.polyN, r.ringN + 1,
                           ST_InteriorRingN(p.geom,r.ringN)
                               as geom
                     From poly p,
                           (Select Generate_Series(1,
                                   ST_NumInteriorRing(p.geom),1)
                                       as ringN
                            From poly p) as r
                     ) g
               ) h
         Where h.IntValue < ST_NumPoints(h.geom)
         ) i
   Order By i.polyN, i.ringN, i.vectorN;
```

The resulting PostgreSQL table is the same as that for SQL Server 2012, so it is not shown.

Oracle SQL translates reasonably easily from the SQL Server 2012 version as follows:

```
-- Oracle ST
With poly As  (
  Select a.polyN, a.geom
    From (Select t.IntValue as polyN,
                book.ST_POLYGON(o.geom.ST_GeometryN(t.IntValue)
                    .geom) as geom
          From (Select book.ST_MULTIPOLYGON(
            'MULTIPOLYGON(((10 10, 15 10, 15 15, 10 15, 10 10),
                          (12 12, 12 14, 14 14, 14 12, 12 12)),
                ((100 100, 150 100, 150 150, 100 150, 100 100)))')
```

```
                              as geom
                    From dual
                  ) o,
                  Table(book.Generate_Series(1,
                            o.geom.ST_NumGeometries(),1)) t
          ) a
      )
    Select a.polyN, a.ringN, a.vectorN,
          a.sp.ST_X() as sx, a.sp.ST_Y() as sy,
          a.ep.ST_X() as ex, a.ep.ST_Y() as ey,
          book.ST_LINESTRING(
              MDSYS.ST_POINT_ARRAY(
                  MDSYS.ST_Point(a.sp.ST_X(),a.sp.ST_Y()),
                  MDSYS.ST_Point(a.ep.ST_X(),a.ep.ST_Y())))
              .ST_Length() as vLen
        From (Select g.polyN, g.ringN, p.IntValue as vectorN,
                  g.geom.ST_PointN(p.IntValue  ) as sp,
                  g.geom.ST_PointN(p.IntValue+1) as ep
            From (Select p.polyN, 1 as ringN,
                      p.geom.ST_ExteriorRing() as geom
                From poly p
                Union All
                Select p.polyN, p.IntValue+1 as ringN,
                      p.geom.ST_InteriorRingN(p.IntValue) as geom
                From poly p,
                    Table(book.Generate_Series(1,
                          p.geom.ST_NumInteriorRing(),1)) p
              ) g,
              Table(book.Generate_Series(1,
                      g.geom.ST_NumPoints(),1)) p
          Where p.IntValue < g.geom.ST_NumPoints()
          ) a
    Order By a.polyN, a.ringN, a.vectorN;
```

The resulting table is the same as that for PostgreSQL and SQL Server 2012, so it is not shown.

Oracle's ST_POLYGON does not inherit from ST_MULTIPOLYGON, and thus ST_GEOMCOLLECTION. The result of extracting each polygon from the multi-polygon via ST_GeometryN therefore results in an ST_GEOMETRY object (not an ST_POLYGON object), which does not support the ST_ExteriorRing and ST_InteriorRing methods. To access these methods, each ST_GEOMETRY object has to be cast as a ST_Polygon. This can be done in many ways; the one chosen is by passing the o.geom.ST_GeometryN (t.IntValue) value as an SDO_GEOMETRY to the book.ST_POLYGON (geom) constructor book.ST_POLYGON (aWKT) could also have been used as follows:

```
book.ST_POLYGON(o.geom.ST_GeometryN(t.IntValue).geom) as geom
```

Again, the implementation of access to set returning functions (CROSS APPLY/ Table and Select) provides the greatest point of difference between the three implementations.

Finally, the requirement to compute the length of each vector is handled differently for each database.

SQL Server 2012	```geometry::STLineFromText('LINESTRING(' + str(a.sp.STX,5,3) + ' ' + str(a.sp.STY,5,3) + ',' + str(a.ep.STX,5,3) + ' ' + str(a.ep.STY,5,3) + ')',0).STLength()```
PostGIS	```ST_Length(ST_MakeLine(ARRAY[i.sp,i.ep])) as vLen```
Oracle	```book.ST_LINESTRING(MDSYS.ST_POINT_ARRAY(MDSYS.ST_Point(a.sp.ST_X(),a.sp.ST_Y()), MDSYS.ST_Point(a.ep.ST_X(),a.ep.ST_Y()))).ST_Length()```

The SQL Server 2012 version is based on constructing a valid LINESTRING WKT string, and then converting it; PostGIS uses a specialist function ST_MakeLine and Oracle uses the SQL/MM ST_LINESTRING (appoint array ST_POINT_ARRAY) constructor. PostGIS and Oracle could be modified to implement SQL Server 2012's WKT based approach. This is left to the reader to implement.

Putting it all together – gridding a vector object

The examples presented show how one can use a standardized geometry type and methods to query geometry objects. Those examples include SQL Select statements, and some functions or methods that can be constructed from them.

The main purpose of all these types and methods is to provide a consistent basis for solving more complicated business or algorithmic problems. The following example shows how to generate grid cells that cover a polygon object (with interior rings). This problem requires the ability to calculate the bounding box or minimum bounding rectangle (MBR) of a geometry object. The SFA 1.1 and 1.2 standards make an envelope function available for both PostgreSQL and SQL Server 2012 users, which can return the envelope of an existing geometry. Oracle's SDO_GEOMETRY uses proprietary methods (SDO_MBR) to return the envelope of a function, but its SQL/MM compliant type library makes the standardized SQL/MM ST_Envelope method available.

SQL Server 2012	`Select geometry::STGeomFromText(`
	`    'LINESTRING(0 0,2 2,0 2,2 0)',0)`
	`        .STEnvelope().STAsText();`

PostgreSQL	`Select ST_AsText(`
	`    ST_Envelope(`
	`    ST_GeomFromText(`
	`    'LINESTRING(0 0,2 2,0 2,2 0)',0)`
	`    )`
	`) as mbr;`

Oracle SDO_GEOMETRY	`Select sdo_aggr_mbr(`
	`    SDO_GEOMETRY(`
	`    'LINESTRING(0 0,2 2,0 2,2 0)',null)`
	`    ).get_wkt() as mbr`
	`From dual;`

Oracle ST_Geometry	`Select book.ST_LINESTRING(`
	`    'LINESTRING(0 0,2 2,0 2,2 0)')`
	`    .ST_Envelope().ST_AsText() as mbr`
	`From dual;`

While the extraction of the envelope of existing geometries is important, the creation of new geometries that are identical to its envelope are missing in the OGC and SQL/MM type libraries. The only method offered is to create a new polygon object from the vertices that make up the square/rectangular grid cells.

This approach is wordy and convoluted. For Oracle, the replacement ST_GEOMETRY type provides an additional constructor:

```
Create Body ST_POLYGON
As
    [...]
Constructor Function ST_POLYGON(xmin Number, ymin Number,
                                xmax Number, ymax Number,
                                asrid Number default null)
Return SELF As Result
As
```

```
    Begin
        SELF.GEOM := SDO_GEOMETRY(2003,ASRID,NULL,
                        SDO_ELEM_INFO_ARRAY(1,1003,3),
                        SDO_ORDINATE_ARRAY(xmin,ymin,xmax,ymax));
        Return;
    End;
      […]
End;
```

For SQL Server 2012, a new function called `ST_Envelope` was created that returns polygon geometry from its lower left and upper right ordinates:

```
Create Function [dbo].[ST_Envelope] (
  @p_minx  float,
  @p_miny  float,
  @p_maxx  float,
  @p_maxy  float,
  @p_srid  Int )
  Returns geometry
As
Begin
  Return geometry::STGeomFromText('POLYGON((' +
                Convert(Varchar(50),
                    CAST(@p_minx as DECIMAL(24,12))) + ' ' +
                Convert(Varchar(50),
                    CAST(@p_miny as DECIMAL(24,12))) + ',' +
                Convert(Varchar(50),
                    CAST(@p_maxx as DECIMAL(24,12))) + ' ' +
                Convert(Varchar(50),
                    CAST(@p_miny as DECIMAL(24,12))) + ',' +
                Convert(Varchar(50),
                    CAST(@p_maxx as DECIMAL(24,12))) + ' ' +
                Convert(Varchar(50),
                    CAST(@p_maxy as DECIMAL(24,12))) + ',' +
                Convert(Varchar(50),
                    CAST(@p_minx as DECIMAL(24,12))) + ' ' +
                Convert(Varchar(50),
                    CAST(@p_maxy as DECIMAL(24,12))) + ',' +
                Convert(Varchar(50),
                    CAST(@p_minx as DECIMAL(24,12))) + ' ' +
                Convert(Varchar(50),
                    CAST(@p_miny as DECIMAL(24,12))) + '))',
                @p_srid);
End;
Go
```

PostgreSQL already offers a ST_MakeEnvelope method that constructs a polygon from the bottom left and upper right corners of a grid cell. A "wrapper" function was created for this called ST_Envelope which can be seen in the following code:

```
Create Or Replace Function st_envelope(double precision,
                                       double precision,
                                       double precision,
                                       double precision,
                                       integer DEFAULT 0)
   ReturnS geometry
As $$
Begin
   Return ST_MakeEnvelope($1,$2,$3,$4,$5);
End;
$$ Language 'plpgsql';
```

In creating these functions it is important to observe another impediment to creating cross-database solutions; the lack of a single native (stored) procedure language implementation across all three databases. However, careful design of the functions ensures that the specific language issues will be "encapsulated" or "hidden" from view.

To create all the grid cells (square polygons) that cover geometry, we have to create a set returning function for each of the databases. This function will take geometry of any type, and a tile size (expressed in x and y geometry units); from these inputs it will create all possible grid cells that cover the envelope of the geometry object.

The functions and the SQL that executes them for a specific "donut polygon" geometry is described in the following sections.

SQL Server 2012

The created function is as follows:

```
Create Function [dbo].[ST_GEOM2GRID](@p_geom  geometry,
                                     @p_TileX float,
                                     @p_TileY float )
Returns @table Table (col Int,row Int, geom geometry)
As
Begin
   Declare
     @v_srid   Int = case when @p_geom is null
                          then 0
                          Else @p_geom.STSrid
                     end,
```

```
    @v_loCol   int,
    @v_hiCol   int,
    @v_loRow   int,
    @v_hiRow   int,
    @v_col     int,
    @v_row     int,
  @v_envelope geometry;
Begin
  If ( @p_geom is null ) Begin
    Return;
  End;
  If ( UPPER(@p_geom.STGeometryType()) = 'POINT' ) Begin
   Insert  Into @table
        Values(1,1,[dbo].[ST_Envelope](
                        @p_geom.STX - (@p_TileX/2.0),
                        @p_geom.STY - (@p_TileY/2.0),
                        @p_geom.STX + (@p_TileX/2.0),
                        @p_geom.STY + (@p_TileY/2.0),
                        @v_srid));
  End Else Begin
      Set @v_envelope = @p_geom.STEnvelope().STExteriorRing()
      Set @v_loCol = FLOOR(  @v_envelope.STPointN(1).STX
                   / @p_TileX);
      Set @v_hiCol = CEILING(@v_envelope.STPointN(3).STX
                   / @p_TileX) - 1;
      Set @v_loRow = FLOOR(  @v_envelope.STPointN(1).STY
                   / @p_TileY);
      Set @v_hiRow = CEILING(@v_envelope.STPointN(3).STY
                   / @p_TileY) - 1;
      Set @v_col = @v_loCol;
      While ( @v_col <= @v_hiCol ) Begin
        SET @v_row = @v_loRow;
        While ( @v_row <= @v_hiRow ) Begin
          Insert  Into @table
                Values(@v_col,@v_row,
                      [dbo].[ST_Envelope](
                          @v_col * @p_TileX,
                          @v_row * @p_TileY,
                          (@v_col * @p_TileX) + @p_TileX,
                          (@v_row * @p_TileY) + @p_TileY,
                          @v_srid ) );
          Set @v_row = @v_row + 1;
        End;
        Set @v_col = @v_col + 1;
```

```
                End;
           End;
           Return;
        End;
End
Go
```

We can now execute it against a generated polygon as follows:

```
With geomQuery As (
Select a.geom, 0.050 as gridX, 0.050 as gridY
  From (Select a.geom.STBuffer(1.0)
                   .STSymDifference(a.geom.STBuffer(0.50))
              as geom
        From (Select geometry::STGeomFromText(
                   'MULTIPOINT((09.25 10.00),(10.75 10.00),
                             (10.00 10.75),(10.00 9.25))',0)
                   as geom ) as a
       ) a
)
Select f.col, f.row, f.geom
  From (Select c.col, c.row,
             case when Upper(a.geom.STGeometryType())
                      In ('POLYGON','MULTIPOLYGON')
                  then a.geom.STIntersection(c.geom)
                  Else a.geom
              end as geom
        From geomQuery a
             Cross Apply
             [dbo].ST_GEOM2GRID(a.geom,a.gridX,a.gridY) c
         Where a.geom.STIntersects(c.geom) = 1
       ) f
 Where Upper(f.geom.STGeometryType())
        In ('POLYGON','MULTIPOLYGON','CURVEPOLYGON')
       /* Don't want point or line tiles */;
```

The result looks like this:

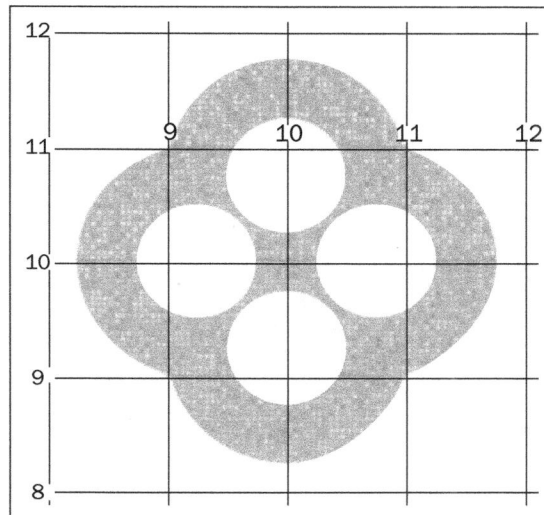

PostgreSQL

The PostgreSQL ST_Geom2Grid function is displayed in the following code. The difference between pgPlSQL and T-SQL is quite stark:

```
Create Or Replace Function
ST_Geom2Grid(p_geom   geometry,
             p_TileX numeric,
             p_TileY numeric)
Returns SETOF ST_GridCell Immutable
As $$
Declare
    v_loCol    int4;
    v_hiCol    int4;
    v_loRow    int4;
    v_hiRow    int4;
    v_envelope geometry;
    v_geom     geometry;
    v_grid     st_gridCell;
Begin
    If ( p_geom is null ) Then
       Return;
    End If;
```

```
      If ( UPPER(ST_GeometryType(p_geom)) = 'ST_POINT' ) Then
        v_geom := ST_Envelope(ST_X(p_geom) - (p_TileX / 2.0),
                              ST_Y(p_geom) - (p_TileY / 2.0),
                              ST_X(p_geom) + (p_TileX / 2.0),
                              ST_Y(p_geom) + (p_TileY / 2.0),
                              ST_Srid(p_geom));
        Select 1::int4,1::int4,v_geom::geometry Into v_grid;
        Return Next v_grid;
      Else
        v_envelope := ST_ExteriorRing(ST_ENVELOPE(p_geom));
        v_loCol := trunc( (ST_X(ST_PointN(v_envelope,1))
                      / p_TileX)::numeric );
        v_hiCol := ceil(  (ST_X(ST_PointN(v_envelope,3))
                      / p_TileX)::numeric ) - 1;
        v_loRow := trunc( (ST_Y(ST_PointN(v_envelope,1))
                      / p_TileY)::numeric );
        v_hiRow := ceil(  (ST_Y(ST_PointN(v_envelope,3))
                      / p_TileY)::numeric ) - 1;
        For v_col in v_loCol..v_hiCol Loop
          For v_row in v_loRow..v_hiRow Loop
              v_geom := ST_Envelope( (v_col * p_TileX),
                                     (v_row * p_TileY),
                                    ((v_col * p_TileX)+p_TileX),
                                    ((v_row * p_TileY)+p_TileY),
                                     ST_Srid(p_geom));
              Select v_col::int4,v_row::int4,v_geom::geometry
                Into v_grid;
              Return Next v_grid;
          End Loop;
        End Loop;
      End If;
  End;
  $$ Language 'plpgsql';
```

The gridding SQL is as follows:

```
With geomQuery As (
Select a.geom, 0.050 as gridX, 0.050 as gridY
  From (Select ST_SymDifference(ST_Buffer(a.geom,1.0),
                               ST_Buffer(a.geom,0.50)) as geom
        From (Select ST_GeomFromText(
```

```
                    'MULTIPOINT((09.25 10.00),(10.75 10.00),
                                (10.00 10.75),(10.00 9.25))',0)
                    as geom ) as a
        ) a
)
Select f.gcol,f.grow,f.geom
  From (Select b.gcol, b.grow,
              Case When ST_GeometryType(b.geom)
                       In ('ST_Polygon',
                           'ST_MultiPolygon',
                           'ST_CURVEPOLYGON')
                   then ST_Intersection(b.ageom,b.geom)
                   Else b.geom
               end as geom
         From (Select a.geom as ageom,
                  (ST_Geom2Grid(a.geom, a.gridX,a.gridY)).*
               From geomQuery as a
              ) as b
        Where ST_Intersects(b.ageom,b.geom)
       ) as f
 Where Upper(ST_GeometryType(f.geom))
       In ('ST_POLYGON','ST_MULTIPOLYGON','ST_CURVEPOLYGON');
```

The result looks like this:

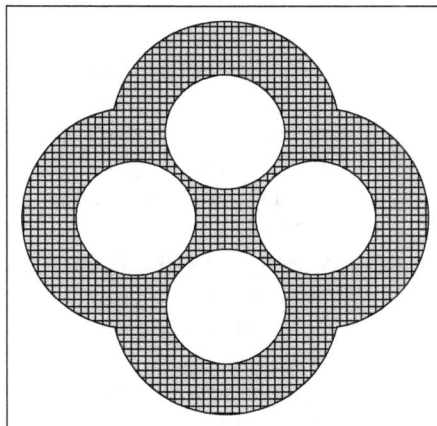

Oracle ST

The Oracle PL/SQL method ST_GEOM2GRID is very similar to PostgreSQL's as the pgPlSql and PL/SQL languages are drawn from the same common SQL standard:

```
Member Function
ST_Geom2Grid(p_TileX In Number,
             p_TileY In Number)
Return book.ST_GridCell_Array Pipelined
Is
  v_loCol     PLS_Integer;
  v_hiCol     PLS_Integer;
  v_loRow     PLS_Integer;
  v_hiRow     PLS_Integer;
  v_point     book.ST_Point;
  v_envelope  book.ST_LINESTRING;
Begin
  If ( SELF.geom is null ) Then
    Return;
  End If;
  If ( SELF.ST_GeometryType() = 'ST_POINT' ) Then
      v_point := TREAT(SELF as book.ST_Point);
      Pipe Row (book.ST_GridCell(1,1,
              book.ST_POLYGON(v_point.ST_X() - (p_TileX / 2.0),
                             v_point.ST_Y() - (p_TileY / 2.0),
                             v_point.ST_X() + (p_TileX / 2.0),
                             v_point.ST_Y() + (p_TileY / 2.0),
                             SELF.ST_SRID()).geom));
  Else
      v_envelope := book.ST_Polygon(SELF.ST_Envelope().ST_AsText())
                      .ST_ExteriorRing();
      v_loCol    := TRUNC(v_envelope.ST_PointN(1).ST_X()
                      / p_TileX);
      v_hiCol    := CEIL( v_envelope.ST_PointN(3).ST_X()
                      / p_TileX) - 1;
      v_loRow    := TRUNC(v_envelope.ST_PointN(1).ST_Y()
                      / p_TileY);
      v_hiRow    := CEIL( v_envelope.ST_PointN(3).ST_Y()
                      / p_TileY) - 1;
      <<column_interator>>
      For v_col in v_loCol..v_hiCol Loop
        <<row_iterator>>
        For v_row in v_loRow..v_hiRow Loop
           Pipe Row (book.ST_GridCell(v_col, v_row,
                   book.ST_POLYGON((v_col * p_TileX),
```

```
                                (v_row * p_TileY),
                                ((v_col * p_TileX) + p_TileX),
                                ((v_row * p_TileY) + p_TileY),
                                SELF.ST_SRID()).geom));


            End Loop row_iterator;
          End Loop col_iterator;
      End If;
      Return;
  End ST_Geom2Grid;
```

The SQL required for the construction and gridding of the polygon is as follows:

```
With geomQuery As (
Select b.geom, 0.050 as gridX, 0.050 as gridY
   From (Select a.geom.ST_Buffer(1.0)
                  .ST_SymDifference(a.geom.ST_Buffer(0.50))
                      as geom
          From (Select book.ST_MULTIPOINT(
                   'MULTIPOINT((09.25 10.00),(10.75 10.00),
                              (10.00 10.75),(10.00 9.25))',NULL)
                             as geom
                  From DUAL
                ) a
        ) b
)
Select f.gcol,f.grow,f.geom
   From (Select b.gcol, b.grow,
                book.ST_Geometry(
                case when a.geom.ST_GeometryType()
                        In ('ST_POLYGON',
                            'ST_MULTIPOLYGON',
                            'ST_CURVEPOLYGON')
                     then a.geom
                        .ST_Intersection(
                                    book.ST_Geometry(b.geom))
                        .geom
                     Else b.geom
                 end) as geom
          From geomQuery a,
               Table(a.geom.ST_Geom2Grid(a.gridX,a.gridY)) b
        Where a.geom.ST_Disjoint(book.ST_Geometry(b.geom)) = 0
        ) f
  Where f.geom.ST_GeometryType()
        In ('ST_POLYGON','ST_MULTIPOLYGON','ST_CURVEPOLYGON')
        /* Don't want point or line tiles */;
```

The result looks like this.

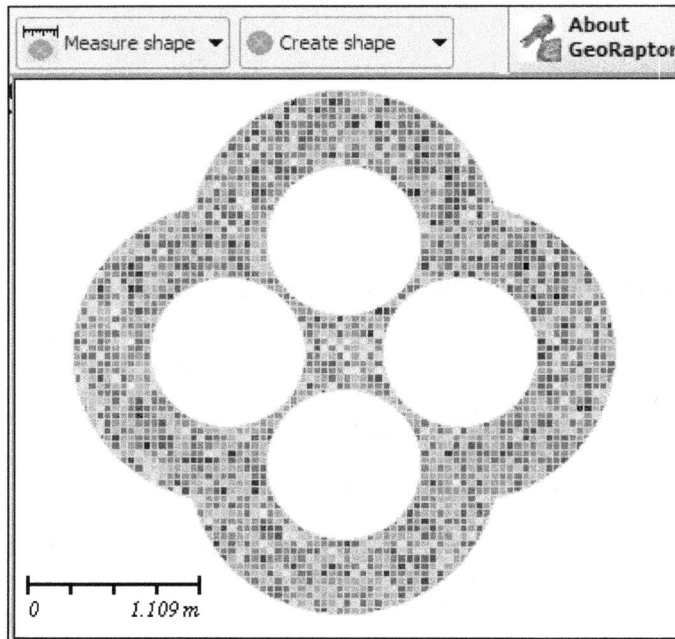

Summary

The chapter first showed how a standardized geometry type and methods can aid in the creation of solutions to spatial problems that can not only run on just one database, but also with editing, any other database supporting that common geometry data type, internal description, and methods that derive from the SQL/MM and OGC SFA 1.x standards.

The issues for data storage that arise from having singly inherited and multiply-inherited geometry data types were examined with a common approach for using all three databases and their types presented.

Oracle's ST_GEOMETRY hierarchy imposes a highly standardized approach to object declaration and use, whereas PostgreSQL's and SQL Server 2012's singly inherited types provide partial implementations of the same standards. This can be seen in the execution of methods like ST_Area against non-polygon geometry objects, while Oracle does not allow inappropriate method execution. In addition, Oracle does not allow ST_GeometryN and ST_NumGeometries to be used against an ST_LINESTRING geometry object, whereas PostgreSQL and SQL Server 2012 do.

When looking at SQL integration, Oracle and SQL Server 2012, through their common, and visually pleasing, "dot notation" method invocation, are the most integrated. In addition, their approach to accessing set returning functions though `Table` and `CROSS APPLY` constructs are also quite similar. PostgreSQL's differences in both these areas provide greater obstacles to integration.

Programming stored procedures in SQL Server and Oracle revealed large differences in that T-SQL and PL/SQL are very different languages. However, Oracle's PL/SQL and PostgreSQL's PL/pgSQL are as similar as to providing little obstacles to cross-database stored procedure development.

In conclusion, while this chapter shows that Oracle's `ST_GEOMETRY` type provides a better and more complete implementation of the SQL/MM standard, the obstacles to cross-database spatial storage and processing between all three databases are very small, because all provide functions or methods that adhere to the common OGC's SFA 1.x and SQL/MM standards to sufficient levels to enable integration. All three describe and store vector geometry data in the same way c.f., and all three describe the rings of a multi-polygon object in the same way. The implementation of inherited methods of a geometry object provided by each database is important, but not a significant obstacle. The main issues for cross-database support are those identified in the book *Effective Oracle by Design*, by *Thomas Kyte*, which when coupled to the various ways that vendors have implemented important non-spatial technologies (for example, set returning functions and analytics) that are far greater obstacles to cross-database independence. In truth, spatial isn't that special, and should never be the stated reason for favoring one database against another.

A

Table Comparing Simple Feature Access/SQL and SQL/MM–Spatial

This appendix provides a comparison of SFA-SQL 1.2 (`http://portal.opengeospatial.org/files/?artifact_id=25354`) and SQL/MM-Spatial (ISO 13249-3, Information technology - Database languages - SQL Multimedia and Application Packages - Part 3: Spatial) by reproducing Table B 1 from Annex B of SFA-SQL 1.2 Part 2:

	SQL with geometry type	ISO/IEC 13249-3:2003 (SQL/MM-Spatial)	Description
Geometry types	Point	`ST_Point`	The type `ST_PolyhedralSurface` is currently not in SQL/ MM, but will be proposed as a result of this document
	Curve	`ST_Curve`	
	Linestring	`ST_Linestring` `ST_Circularstring` `ST_CompoundCurve`	
	Surface	`ST_Surface` `ST_CurvePolygon`	
	Polygon	`ST_Polygon`	
	PolyhedralSurface	`ST_PolyhedralSurface`	
	GeomCollection	`ST_Collection`	
	Multipoint	`ST_Multipoint`	

	SQL with geometry type	ISO/IEC 13249-3:2003 (SQL/MM-Spatial)	Description
	Multicurve	ST_MultiCurve	
	Multilinestring	ST_Multilinestring	
	Multisurface	ST_Multisurface	
	Multipolygon	ST_Multipolygon	
Storage	Binary Type, Text Type, Object Type	Object Type	
Operations	Equals	ST_Equals	
	Disjoint	ST_Disjoint	
	Touches	ST_Touches	
	Within	ST_Within	
	Overlaps	ST_Overlaps	
	Crosses	ST_Crosses	
	Intersects	ST_Intersects	
	Contains	ST_Contains	
	Relate	ST_Relate	
Functions	—	—	—
Point	—	ST_Point()	It returns the point
	X()	ST_X()	It returns the X-coordinate of the point
	Y()	ST_Y()	It returns the Y-coordinate of the point
	Z()	ST_Z()	It return the Z-coordinate of the point
	M()	ST_M()	It returns the M-coordinate of the point
	—	ST_ExplicitPoint()	—
Curve	Length()	ST_Length()	It returns the length of the curve
	StartPoint()	ST_StartPoint()	It returns the first point of the curve
	EndPoint()	ST_EndPoint()	It returns the last point of the curve
	IsClosed()	ST_IsClosed()	It checks whether the curve is closed

	SQL with geometry type	ISO/IEC 13249-3:2003 (SQL/MM-Spatial)	Description
	IsRing()	ST_ISRing()	It check if the curve is closed and simple
	—	ST_CurveToLine	It transforms a curve to LineString
LineString	—	ST_LineString	It returns LineString
	—	ST_Points	It returns a collection of points
	NumPoints()	ST_NumPoints	It returns the number of points
	PointN()	ST_PointN	It returns the nth point of LineString

B

Use of TREAT and IS OF TYPE with ST_GEOMETRY

Understanding the TREAT operator

Chapter 11, SQL/MM – A Basis for Cross-platform, Inter-operable, and Reusable SQL introduced the Oracle MDSYS ST_GEOMETRY type hierarchy. In that chapter, the Oracle TREAT operator was required to ensure that a subtype object such as a point, when created by the ST_GEOMETRY type's GET_WKT method, was correctly understood to be an instance of that subtype (that is, ST_POINT), so that the methods particular to its subtype (for example, ST_X) can be called. This appendix examines the need for TREAT in more detail.

In the ST_GEOMETRY hierarchy, a POINT object can be created in the following two ways:

```
MDSYS.ST_GEOMETRY.FROM_WKT('POINT(6012578.005 2116495.361)',2872)
MDSYS.ST_POINT.FROM_WKT('POINT(6012578.005 2116495.361)',2872)
```

The result in both cases is not an ST_POINT, rather it is an ST_GEOMETRY object. Why is this? It happens in the first because the FROM_WKT method of the ST_GEOMETRY super type is called directly; and in the second, the ST_POINT subtype of ST_GEOMETRY does not have a FROM_WKT method, but its ST_GEOMETRY super type does and so its FROM_WKT method is called instead. Thus in neither case an ST_POINT is returned. The returned objects are not ST_POINT types and can be demonstrated by trying to execute the ST_POINT type's ST_X member function:

```
Select MDSYS.ST_Geometry.FROM_WKT(
            'POINT(6012578.005 2116495.361)',2872)
        .ST_X() as point
```

```
    From dual;
SQL Error: ORA-00904: "MDSYS"."ST_GEOMETRY"."ST_X": invalid identifier

Select MDSYS.ST_Point.FROM_WKT('POINT(6012578.005 2116495.361)',2872)
        .ST_X()
        as point
    From dual;
SQL Error: ORA-00904: "MDSYS"."ST_GEOMETRY"."ST_X": invalid identifier
```

To correct this, the TREAT function must be used as follows:

```
Select TREAT(MDSYS.ST_Geometry.FROM_WKT('POINT(6012578.005
        2116495.361)',2872)
        As MDSYS.ST_Point).ST_X() as x
    From dual;
        X
----------
6012578.005

Select TREAT(MDSYS.ST_Point.FROM_WKT('POINT(6012578.005
        2116495.361)',2872)
        As MDSYS.ST_Point).ST_X() as x
    From dual

        X
----------
6012578.005
```

Why is TREAT needed? The Oracle documentation (Object-Relational Developer's Guide http://docs.oracle.com/cd/E16655_01/appdev.121/e16801/adobjbas.htm#i479093) defines the TREAT operator. **The Puget Sound Oracle Users Group (PSOUG)** website is an excellent additional source for documentation and help http://psoug.org/definition/TREAT.htm as follows:

> *[...] allow[ing] you to change the declared type of the expression used in TREAT. This function comes in handy when you have a subtype that is more specific to your data and you want to convert the parent type to the more specific one.*

Whenever a subtype, such as ST_LINESTRING, calls methods, for example, FROM_WKT, that are inherited from the ST_GEOMETRY parent type, the returned geometry type will always be the generic ST_GEOMETRY even though its contents (LINESTRING WKT) are from a particular subtype. Thus the TREAT operator has to be used to ensure that the converted geometry is correctly interpreted as a ST_LINESTRING.

In the final example, drawn from *Chapter 11, SQL/MM – A Basis for Cross-platform, Inter-operable, and Reusable SQL* the insertion of an ST_LINESTRING object into a table whose geometry column is defined as ST_LineString (not its generic super type ST_GEOMETRY) requires use of the TREAT function:

```
Create table ST_ROAD (
  FID          Integer,
  STREET_NAME  Varchar2(1000),
  CLASSCODE    Varchar2(1),
  GEOM         MDSYS.ST_LINESTRING,
  Constraint ST_ROAD_PK Primary Key (FID)
);
Insert into ST_ROAD Values (1,'Main St','C',
                    TREAT(MDSYS.ST_LINESTRING.FROM_WKT (
                        'LINESTRING(6012759.63041794
                                2116512.48842026,
                                6012420.59599103
                                2116464.90977527)',2872)
                    As MDSYS.ST_LINESTRING));
```

> For some reason, Oracle chose not to implement ST_AsText() or ST_GeomFromText() as per the standard, instead exposing the SDO_GEOMETRY object methods GET_WKT(), FROM_WKT(). This is a minor problem for cross-database programming perspective which will be dealt with in *Chapter 11, SQL/MM – A Basis for Cross-platform, Inter-operable, and Reusable SQL.*

A way to correct these issues is to override the MDSYS.ST_GEOMETRY type's FROM_WKT method in every subtype that needs it (for example, MDSYS.ST_LINESTRING). Because the standard MDSYS.ST_GEOMETRY hierarchy does not do this, TREAT must be used to obtain the correct subtype identification. Only then can subtype methods such as ST_Length (ST_CURVE) be called. However, to avoid the continual use of TREAT, *Chapter 11, SQL/MM – A Basis for Cross-platform, Inter-operable, and Reusable SQL,* introduced its own ST_GEOMETRY type. This type implements subtype constructors and WKT additional conversion methods such as ST_LineFromText that avoid the need to use TREAT. The following snippet from the source code of the ST_GEOMETRY type shows the constructors and the ST_LineFromText method for the ST_LineString subtype as follows:

```
Create or Replace Type ST_CURVE
Under ST_GEOMETRY (
  Overriding Member Function ST_DIMENSION Return Integer
Deterministic,
  Member Function ST_Points Return mdsys.ST_Point_Array Deterministic,
```

```
Member Function ST_NumPoints Return Integer Deterministic,
  Member Function ST_PointN(aposition integer) Return Book.ST_Point
    Deterministic,
[...]
  Member Function ST_Length Return Number Deterministic
) NOT FINAL;

Create or Replace Type ST_LINESTRING
Under ST_CURVE (
[...]
  Constructor Function ST_LINESTRING(AWKT varchar2, ASRID integer
    DEFAULT NULL)
              Return SELF As Result,
[...]
  Static Function ST_LineFromText(AWKT varchar2, ASRID integer
    DEFAULT NULL)
          Return Book.ST_LINESTRING Deterministic,
[...]
);
```

Understanding the IS OF TYPE comparison operator

The Oracle IS OF TYPE comparison operator (IS NULL) allows for the type of an object to be examined. The following example demonstrates its use:

```
Select Case When Then End MDSYS.ST_Point.FROM_WKT('POINT(6012578.005
                            2116495.361)',
                                2872) is of type (MDSYS.ST_Point)
          then 'POINT'
          Else 'GEOMETRY'
      end as ofTypePoint,
      case when MDSYS.ST_Point.FROM_WKT('POINT(6012578.005
                            2116495.361)',
                              2872) is of type (MDSYS.ST_
                                  Geometry)
          then 'GEOM'
          Else 'Something Else'
      end as ofTypeGeom,
      case when MDSYS.ST_Point.FROM_WKT('POINT(6012578.005
                            2116495.361)',
                              2872) is of type (ONLY MDSYS.ST_
                                  Geometry)
```

```
        then 'GEOM'
        Else 'Something Else'
     end as ofTypeOnlyGeom
  From dual;

-- Results
--
OFTYPEPOINT OFTYPEGEOM     OFTYPEONLYGEOM
----------- --------------  --------------
POINT       GEOM            Something Else
```

The first two results are perfectly correct as an ST_POINT type is both an ST_POINT and an ST_GEOMETRY. The use of the ONLY prefix reports that ST_POINT is not just an ST_GEOMETRY type if IS OF TYPE reports that the result of ST_POINT.FROM_WKT is an ST_POINT.

Index

E

EDS
  enabling, for geometry tables 134, 135
entity-relationship (E-R) diagram 16
ETL processing
  with GeoKettle 47, 48
ETL tools
  about 46, 47, 255
  GeoKettle 47, 48
  Map Builder 49
  Shapefile 49
Excel
  using 52-54
Extended Data Type Support. *See* **EDS**
Extended Well Known Text (EWKT) 402
external tables, CSV files 55, 56
Extract, Transform, and Load tools. *See* **ETL tools**

F

flashback database 94
flashback drop 94, 96
flashback queries
  about 94
  flashback database 94
  flashback drop 94, 96
  flashback table 94-96
  flashback transaction 94
  flashback transaction query 94
  flashback versions query 94-100
flashback query 94-98
flashback table 94-96
flashback transaction 94
flashback transaction query 94
flashback version query 94, 99, 100
formats
  exporting 76
Frequently Asked Questions (FAQ), JTS website 441
functions
  calling 236, 238
  custom object type, creating 243, 245
  packaging 236

  PL/SQL package, implementing 238-241
  user object type, implementing 241, 242
functions, linear referencing
  ST_Add_Measure 317
  ST_End_Measure 317
  ST_Find_Measure 317
  ST_Find_Offset 318
  ST_Is_Measure_Decreasing 318
  ST_Is_Measure_Increasing 318
  ST_Locate_Measure 318
  ST_Locate_Measures (no offset) 319
  ST_Locate_Measures (with offset) 319
  ST_Locate_Point 318
  ST_Measure_Range 317
  ST_Measure_To_Percentage 318
  ST_Percentage_To_Measure 318
  ST_Project_Point 318
  ST_Scale_Measures 317
  ST_Snap 316
  ST_Split 316
  ST_Start_Measure 317
functions, packaging
  methods 252
  summary 253

G

GDAL
  about 47, 366
  URL 47
  used, for loading raster data 367-369
gdalinfo command 367
generalized representation
  using 29, 30
geographic information system. *See* **GIS**
GeoJSON format
  generating 78-80
GeoKettle
  about 47, 48
  URL, for downloading 47
geometries
  extracting, after intersection 209-211
  sorting 248
  vectorizing, with linestrings 226, 227
geometries, extracting
  ST_Extract function, implementing 211, 212

V

vector geometry
  tiling 297, 298
vector object
  gridding 511-514
version-enabling tables 110, 111
visualization applications,
    GeoRaster 372, 373

W

Web Mapping Service (WMS) 373
Well Known Text (WKT) 19, 76, 402
Workload capture 104-106
Workload processing 104-107
Workload Replay 104, 108
Workspace locking 114
Workspace Manager
  about 109
  conflict resolution 113, 114
  DDL operations, on version-enabled
      tables 115
  valid-time support 115, 116
  version-enabling tables 110, 111
  workspace, creating 111, 112
  Workspace locking 114
  workspace, using 111, 112
workspaces
  creating 111, 112
  using 111, 112
work tablespace 75

X

XMLAGG function 78
XMLFOREST function 78

Z

Z ordinate problems
  fixing 311

[PACKT] enterprise ⊞
PUBLISHING
professional expertise distilled

Thank you for buying
Applying and Extending Oracle Spatial

About Packt Publishing

Packt, pronounced 'packed', published its first book "Mastering phpMyAdmin for Effective MySQL Management" in April 2004 and subsequently continued to specialize in publishing highly focused books on specific technologies and solutions.

Our books and publications share the experiences of your fellow IT professionals in adapting and customizing today's systems, applications, and frameworks. Our solution based books give you the knowledge and power to customize the software and technologies you're using to get the job done. Packt books are more specific and less general than the IT books you have seen in the past. Our unique business model allows us to bring you more focused information, giving you more of what you need to know, and less of what you don't.

Packt is a modern, yet unique publishing company, which focuses on producing quality, cutting-edge books for communities of developers, administrators, and newbies alike. For more information, please visit our website: www.packtpub.com.

About Packt Enterprise

In 2010, Packt launched two new brands, Packt Enterprise and Packt Open Source, in order to continue its focus on specialization. This book is part of the Packt Enterprise brand, home to books published on enterprise software – software created by major vendors, including (but not limited to) IBM, Microsoft and Oracle, often for use in other corporations. Its titles will offer information relevant to a range of users of this software, including administrators, developers, architects, and end users.

Writing for Packt

We welcome all inquiries from people who are interested in authoring. Book proposals should be sent to author@packtpub.com. If your book idea is still at an early stage and you would like to discuss it first before writing a formal book proposal, contact us; one of our commissioning editors will get in touch with you.

We're not just looking for published authors; if you have strong technical skills but no writing experience, our experienced editors can help you develop a writing career, or simply get some additional reward for your expertise.

Oracle Business Intelligence Enterprise Edition 11*g*: A Hands-On Tutorial

ISBN: 978-1-84968-566-5   Paperback: 620 pages

Leverage the latest Fusion Middleware Business Intelligence offering with this action-packed implementation guide

1. Get to grips with the OBIEE 11*g* suite for analyzing and reporting on your business data

2. Immerse yourself in BI upgrading techniques, using Agents and the Action Framework and much more in this book and e-book

3. A practical, from the coalface tutorial, bursting with step by step instructions and real world case studies to help you implement the suite's powerful analytic capabilities

Oracle Business Intelligence Enterprise Edition 11*g*: A Hands-On Tutorial

Leverage the latest Fusion Middleware Business Intelligence offering with this action-packed implementation guide

Haroun Khan   Christian Screen
Adrian Ward   [PACKT] enterprise
PUBLISHING

Oracle Service Bus 11*g* Development Cookbook

ISBN: 978-1-84968-444-6   Paperback: 522 pages

Over 80 practical recipes to develop service and message-oriented solutions on the Oracle Service Bus

1. Develop service and message-oriented solutions on the Oracle Service Bus following best practices using this book and ebook

2. Extend your practical knowledge of building solutions on the Oracle Service Bus

3. Packed with hands-on cookbook recipes, with the complete and finished solution as an OSB and SOA Suite project, made available electronically for download

Oracle Service Bus 11*g* Development Cookbook

Over 80 practical recipes to develop service and message-oriented solutions on the Oracle Service Bus

Guido Schmutz   Edwin Biemond Jan van Zoggel
Mischa Kölliker   Eric Elzinga   [PACKT] enterprise
PUBLISHING

Please check **www.PacktPub.com** for information on our titles

www.ingramcontent.com/pod-product-compliance
Lightning Source LLC
Chambersburg PA
CBHW060947210326
41598CB00031B/4748